Theory of High Temperature Superconductivity

Fundamental Theories of Physics

An International Book Series on The Fundamental Theories of Physics: Their Clarification, Development and Application

Volume 121

Theory of High Temperature Superconductivity

by

Shigeji Fujita
University at Buffalo, SUNY, Buffalo, N.Y., U.S.A.

and

Salvador Godoy
Universidad Nacional Autónoma de México, México, D.F., México

SPRINGER-SCIENCE+BUSINESS MEDIA, B.V.

A C.I.P. Catalogue record for this book is available from the Library of Congress.

ISBN 978-90-481-5883-6 ISBN 978-0-306-48216-8 (eBook)
DOI 10.1007/978-0-306-48216-8

Printed on acid-free paper

Preface

Flux quantization experiments indicate that the carriers, Cooper pairs (pairons), in the supercurrent have charge magnitude $2e$, and that they move independently. Josephson interference in a Superconducting Quantum Interference Device (SQUID) shows that the centers of masses (CM) of pairons move as *bosons* with a *linear* dispersion relation. Based on this evidence we develop a theory of superconductivity in conventional and high-T_c materials from a unified point of view. Following Bardeen, Cooper and Schrieffer (BCS) we regard the phonon exchange attraction as the cause of superconductivity. For cuprate superconductors, however, we take account of both optical- and acoustic-phonon exchange. BCS started with a Hamiltonian containing "electron" and "hole" kinetic energies and a pairing interaction with the phonon variables eliminated. These "electrons" and "holes" were introduced formally in terms of a free-electron model, which we consider unsatisfactory. We define "electrons" and "holes" in terms of the curvatures of the Fermi surface. "Electrons" (1) and "holes" (2) are different and so they are assigned with different effective masses: $m_1 \neq m_2$. Blatt, Schafroth and Butler proposed to explain superconductivity in terms of a Bose-Einstein Condensation (BEC) of electron pairs, each having mass M and a size. The system of free massive bosons, having a quadratic dispersion relation: $\epsilon = p^2/(2M)$, and moving in three dimensions (3D) undergoes a BEC transition at $T_c = 3.31\hbar^2(Mk_B)^{-1}n_0^{2/3}$, where n_0 is the pair density. This model met difficulties because the interpair separation estimated from this formula is much smaller than the experimental pair size ($\sim 10^3$ Å for Pb), and hence these pairs cannot be regarded as free moving. We show from first principles that the CM of pairons move as bosons with a linear dispersion relation: $w_p = w_0 + cp$,where w_0 is a pairon ground-state energy, and $c \equiv \alpha v_F$, $\alpha = 1/2$ (3D), $2/\pi$ (2D), $v_F \equiv (2\varepsilon_F/m^*)^{1/2}$ = Fermi velocity. The systems of free pairons, moving with the linear dispersion relation, undergo a BEC in 3D and 2D. The critical temperature T_c is given by $T_c = 1.01\,\hbar v_F k_B^{-1} n_0^{1/3}\ (3D)$, $\quad 1.24\,\hbar v_F \mathrm{k}_B n_0^{1/2}$ (2D). The interpairon distance $r_0 \equiv n_0^{-1/3}$ (3D), $\quad n_0^{-1/2}$ (2D)] is several times greater than the BCS pairon size $\xi_0 = 0.181\,\hbar v_F(\mathrm{k}_B T_c)^{-1}$. Hence the BEC occurs without the pairon overlap, which justifies the notion of free pairon motion. The superconducting transition will be regarded as a BEC transition. In the currently predominant BCS theory the superconducting temperature T_c is identified as the temperature at which the stationary (zero-momentum) pairons break up and disappear in the system. But the electronic heat capacity in $YBa_2Cu_3O_{7-\delta}$ has a maximum at T_c with a shoulder above T_c, which can only be explained naturally in terms of a model in which many pairons participate in the phase transition with no dissociation. (No feature above T_c is predicted by the BCS theory.) The dissociation take place one by one just as hydrogen molecules H_2 break up in a gas mixture of (H_2, H). In the B-E condensation a great number of pairons cooperatively participate and exhibit a macroscopic change of phase. Hence in our view the pairons do not break up at T_c. Above T_c pairons move independently in all allowed directions and they contribute to the resistive conduction. Below T_c condensed pairons move without resistance, following the quantum laws described in terms of the Ginzburg-Landau wavefunction. Non-condensed pairons, unpaired electrons and quantum vortices contribute to the normal resistive conduction.

In 1986 Bednorz and Müller reported a discovery of high-temperature (high-T_c) cuprate superconductors (BaLaCuO, $T_c \sim 30\,\mathrm{K}$). Since then many investigations have been carried out on high-T_c cuprates. These cuprates possess all basic superconducting properties: zero resistance, Meissner effect, flux quantization, Josephson interference, and gaps in elementary excitation energy spectra, meaning that the same superconducting state exists in high-T_c as in conventional superconductors. In addition these cuprate superconductors exhibit 2D conduction, short zero-temperature coherence length ξ_0 ($\sim 10\,\text{Å}$), high critical temperature T_c ($\sim 100\,\mathrm{K}$), two energy gaps, d-wave pairon, unusual transport and magnetization behaviors above T_c, and doping dependence of T_c. In the present book we present a theory of high-T_c superconductivity, starting with a generalized BCS Hamiltonian, taking account of energy band structures, and calculating all properties by standard quantum statistical methods, that is, using a grand canonical ensemble for equilibrium properties and using a quantum Liouville equation or Heisenberg equation of motion for non-equilibrium problems.

Because the supercondensate can be described in terms of free moving pairons, all of the properties of a superconductor including the ground state energy, critical temperature, quasiparticle energy spectra containing gaps, condensed pairon density, specific heat, and supercurrent density can be computed without mathematical complexities. This simplicity is in great contrast to the far more complicated treatment required for the phase transition in a ferromagnet or for the familiar vapor-liquid transition.

The authors believe that everything essential about superconductivity can be presented to second-year graduate students. Students are assumed to be familiar with basic differential, integral and vector calculuses, and partial differentiation at the sophomore-junior level. Knowledge of mechanics, electromagnetism, solid state, and statistical physics at the junior-senior level and quantum theory at the first-year graduate level are prerequisite.

A substantial part of the difficulty that students face in learning the theory of superconductivity lies in the fact that they should have not only a good background in many branches of physics but also be familiar with a number of advanced physical concepts such as bosons, fermions, Fermi surface, "electrons", "holes", phonons, density of states, phase transitions. To make all of the needed concepts clear, we have included three preparatory chapters 2-4 in the text, chapters 5 through chapter 15 deal with the general theory of superconductivity. All basic thermodynamic properties are described and discussed; all important formulas are derived without omitting steps. The ground state is treated closely following the original BCS theory. To treat quasi-particles including Bloch electrons, quasi-electrons and pairons, Heisenberg's equation-of-motion method is used, which reduces a quantum many-body problem to a one-body problem when the system-Hamiltonian is a sum of single-particle Hamiltonians. No Green's function techniques are used, which makes the text elementary and readable. Type II compound superconductors are discussed in chapter 16. High T_c superconductors are treated in Chapters 17 through 26. A brief summary and overview are given in the first and last chapters. Second quantization formalism (quantum field theory) may or may not be covered in the first-year quantum theory course. But this theory is indispensable in the microscopic theory of superconductivity, it is reviewed in Appendix for completeness.

The book is written in a self-contained manner. Thus, non-physics majors who want to learn the microscopic theory of superconductivity step by step with no particular hurry may find it useful as a self-study reference. Many fresh, and some provocative, views are presented. Experimental and theoretical researchers in the field are also invited to examine the test. Problems at the end of a section are usually of the straightforward exercise type, directly connected with the material presented in that section. By doing these problems one by one, the reader may be able to grasp the meanings of the newly introduced subjects more firmly.

The authors thank the following individuals for valuable criticisms, discussions and readings: Professor M. de Llano, Universidad Nacional Autónoma de México; Professor T. George, University of Wisconsin at Stevens Point;

Professor T. Obata, Gunma National College of Technology, Maebashi, Japan. They thank Sachiko, Amelia, Michio, Isao, Yoshiko, Eriko, George Redden and Brent Miller for their encouragement and reading the drafts.

Shigeji Fujita and *Salvador Godoy*
Buffalo, NY, March 2001.

Table of Contents

Constants, Signs, and Symbols

Useful Physical Constants

Quantity	Symbol	Value
Absolute zero on Celsius scale		- 273.16°C
Avogadro's number	N_0	6.02×10^{23} mol^{-1}
Boltzmann constant	k_B	1.38×10^{-16} erg K^{-1}
Bohr magneton	μ_B	9.22×10^{-21} erg gauss^{-1}
Bohr radius	a_0	5.29×10^{-9} cm
Electron mass	m	9.11×10^{-28} g
Electron charge (magnitude)	e	4.80×10^{-10} esu
Gas constant	R	8.314 J mol^{-1} K^{-1}
Molar volume (gas at STP)		2.24×10^{4} cm^3 = 22.4 liter
Mechanical equivalent of heat		4.186 J cal^{-1}
Permeability constant	μ_0	1.26×10^{-6} H m^{-1}
Permittivity constant	ε_0	8.854×10^{-12} F m^{-1}
Planck's constant	h	6.63×10^{-27} erg sec
Planck's constant/2π	$\hbar$	1.05×10^{-27} erg sec
Proton mass	m_p	1.67×10^{-24} g
Speed of light	c	3.00×10^{10} cm sec^{-1}

Mathematical Signs and Symbols

$=$	equal to
$\simeq$	equal to approximately
$\neq$	not equal to
$\equiv$	identical to, defined as
$>$	greater than
$\gg$	much greater than
$<$	less than
$\ll$	much less than
$\geq$	greater than or equal to
$\leq$	less than or equal to
$\propto$	proportional to
$\sim$	represented by, of the order
$\langle x \rangle$, $\overline{x}$	average value of x

ln	natural logarithm
Δx	increment in x
dx	infinitesimal increment in x
z^*	complex conjugate of a number z
$\alpha^\dagger$	Hermitean conjugate of operator (matrix) α
α^T	transpose of matrix α
P^{-1}	inverse of P
$\delta_{ab} = \begin{cases} 1 & \text{if } a = b \\ 0 & \text{if } a \neq b \end{cases}$	Kronecker's delta
$\delta(x)$	Dirac's delta function
$\boldsymbol{\nabla}$	nabla (or del) operator
$\dot{x} \equiv dx/dt$	time derivative
$\operatorname{grad}\phi \equiv \nabla\phi$	gradient of ϕ
$\operatorname{div}\mathbf{A} \equiv \boldsymbol{\nabla} \cdot \mathbf{A}$	divergence of $\mathbf{A}$
$\operatorname{curl}\mathbf{A} \equiv \boldsymbol{\nabla} \times \mathbf{A}$	curl of $\mathbf{A}$
∇^2	Laplacian operator

List of Symbols

Å	Ångstrom ($= 10^{-8}$ cm $= 10^{-10}$ m)
$\mathbf{A}$	vector potential
$\mathbf{B}$	magnetic field (magnetic flux density)
C	heat capacity
c	velocity of light
c	specific heat
$\mathcal{D}(p)$	density of states in momentum space
$\mathcal{D}(\omega)$	density of states in angular frequency
E	total energy
E	internal energy
$\mathbf{E}$	electric field
e	base of natural logarithm
e	electron charge (absolute value)
F	Helmholtz free energy
f	one-body distribution function
f_B	Bose distribution function
f_F	Fermi distribution function
f_0	Planck distribution function
G	Gibbs free energy

H	Hamiltonian
H_c	critical magnetic field
$\mathbf{H}_a$	applied magnetic field
$\mathcal{H}$	Hamiltonian density
h	Planck's constant
h	single-particle Hamiltonian
$\hbar$	Dirac's h
$i \equiv \sqrt{-1}$	imaginary unit
$\mathbf{i}$, $\mathbf{j}$, $\mathbf{k}$	Cartesian unit vectors
J	Jacobian of transformation
$\mathbf{J}$	total current
$\mathbf{j}$	single-particle current
$\mathbf{j}$	current density
$\mathbf{k}$	angular wave vector $\equiv$ k-vector
k_B	Boltzmann constant
L	Lagrangian function
L	normalization length
ln	natural logarithm
$\mathcal{L}$	Lagrangian density
l	mean free path
M	molecular mass
m	electron mass
m^*	effective mass
N	Number of particles
$\hat{N}$	number operator
$\mathcal{N}(\varepsilon)$	Density of states in energy
n	particle-number density
P	pressure
$\mathbf{P}$	total momentum
$\mathbf{p}$	momentum vector
p	momentum (magnitude)
Q	quantity of heat
R	resistance
$\mathbf{R}$	position of the center of mass
r	radial coordinate
$\mathbf{r}$	position vector
S	entropy
T	kinetic energy

T	absolute temperature
T_c	critical (condensation) temperature
T_F	Fermi temperature
t	time
TR	sum of N particle traces $\equiv$ grand ensemble trace
Tr	many-particle trace
tr	one-particle trace
V	potential energy
V	volume
$\mathbf{v}$	velocity (field)
W	work
Z	partition function
$e^{\alpha} \equiv z$	fugacity
$\beta \equiv (k_B T)^{-1}$	reciprocal temperature
Δx	small variation in x
$\delta(x)$	Dirac delta-function
$\delta_P = \begin{cases} +1 & \text{if } P \text{ is even} \\ -1 & \text{if } P \text{ is odd} \end{cases}$	parity sign of permutation P
ε	energy
ε_F	Fermi energy
η	viscosity coefficient
Θ_D	Debye temperature
Θ_E	Einstein temperature
θ	polar angle
λ	wavelength
λ	penetration depth
κ	curvature
μ	linear mass density of a string
μ	chemical potential
μ_B	Bohr magneton
ν	frequency
Ξ	grand partition function
ξ	dynamical variable
ξ	coherence length
ρ	mass density
ρ	(system) density operator
ρ	many-particle distribution function

σ	total cross section
σ	electrical conductivity
$\sigma_x, \sigma_y, \sigma_z$	Pauli spin matrices
τ	tension
τ_d	duration of collision
τ_c	average time between collisions
ϕ	azimuthal angle
ϕ	scalar potential
Ψ	quasi wave function for condensed bosons
ψ	wave function for a quantum particle
$d\Omega = \sin\theta d\theta d\phi$	element of solid angle
$\omega \equiv 2\pi\nu$	angular frequency
ω_c	rate of collision
ω_D	Debye frequency
$[,] \equiv [,]_-$	commutator brackets
$\{,\} \equiv [,]_+$	anticommutator brackets
$\{,\}$	Poisson brackets
$[A]$	dimension of A

Chapter 1

Introduction

In this chapter we describe basic properties of a superconductor, occurrence of superconductors, theoretical background, and kinetic theory of conduction electrons.

1.1 Basic Properties of a Superconductor

Superconductivity is characterized by the following five basic properties: zero resistance, Meissner effect, magnetic flux quantization, Josephson effects, and gaps in elementary excitation energy spectra. We shall briefly describe these properties.

1.1.1 Zero Resistance

The phenomenon of superconductivity was discovered in 1911 by Kamerlingh Onnes [1] who measured extremely small electric resistance in mercury below a certain critical temperature $T_c (\approx 4.2\,\mathrm{K})$. His data are reproduced in Fig. 1.1. This *zero resistance* property can be confirmed by a never-decaying supercurrent ring experiment described in Section 1.1.3.

1.1.2 Meissner Effect

Substances that become superconducting at finite temperatures will be called *superconductors* in the present text. If a superconductor below T_c is placed under a weak magnetic field, it repels the magnetic flux (field) $\mathbf{B}$ completely

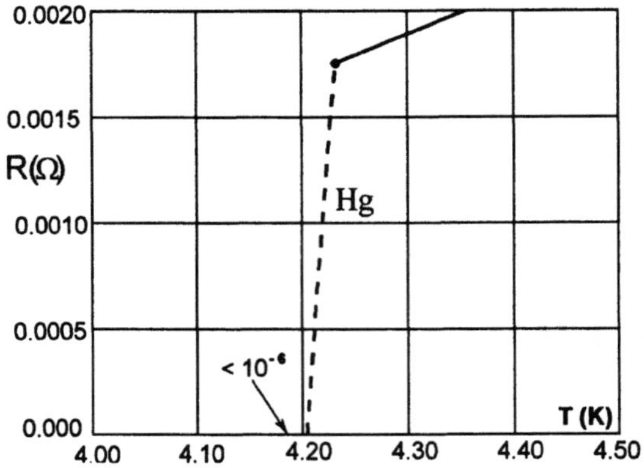

Figure 1.1: Resistance versus temperature, after Kamerling Onnes [1].

from its interior as shown in Fig. 1.2. This is called the *Meissner effect*, and it was discovered by Meissner and Ochsenfeld [2] in 1933.

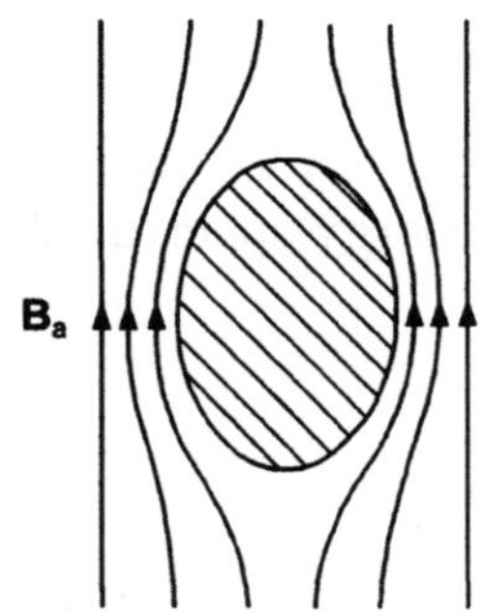

Figure 1.2: A superconductor expels a weak magnetic field.

The Meissner effect can be demonstrated dramatically by a floating magnet as shown in Fig. 1.3. A small bar magnet above T_c simply rests on a superconductor dish. If temperature is lowered below T_c, the magnet will float as indicated. The gravitational force exerted on the magnet is balanced by the magnetic pressure (part of electromagnetic stress tensor) due to the inhomogeneous magnetic field (B-field) surrounding the magnet, which is represented by the magnetic flux lines.

The later more refined experiments reveal that the field $\mathbf{B}$ penetrates into the superconductor within a very thin surface layer. Consider the boundary

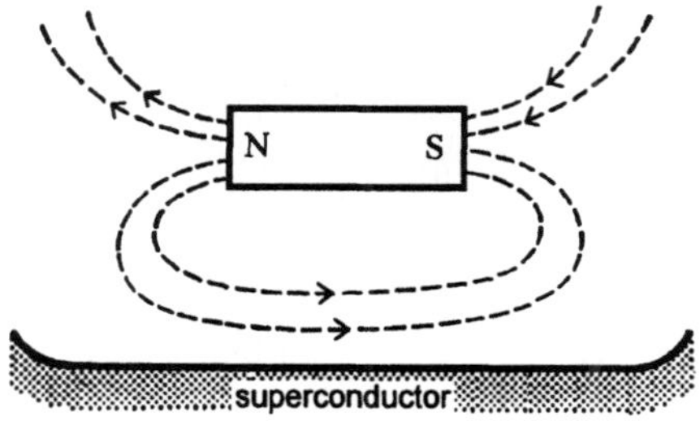

Figure 1.3: A floating magnet.

of a semi-infinite slab. When the external field is applied parallel to the boundary, the B-field falls off exponentially:

$$B(x) = B(0)e^{-x/\lambda}, \tag{1.1}$$

as indicated in Fig. 1.4. Here, λ is called a *penetration depth*, and it is on the order of 500 Å in most superconductors at lowest temperatures. Its small value on a macroscopic scale allows us to speak of the superconductor as being perfectly diamagnetic. The penetration depth λ plays a very important role in the description of the magnetic properties.

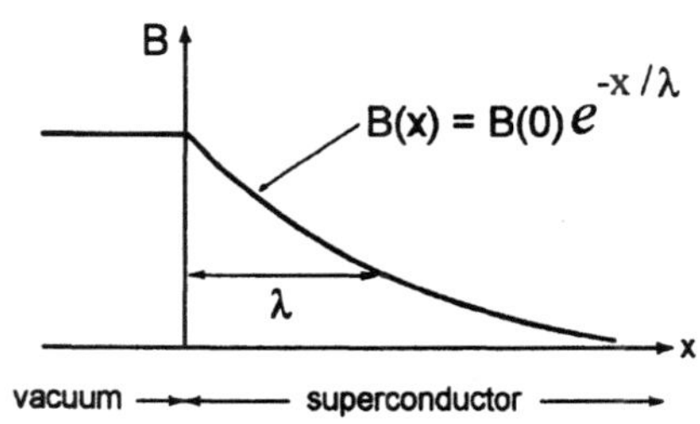

Figure 1.4: Penetration of the magnetic field into a superconductor slab.

1.1.3 Ring Supercurrent and Flux Quantization

Let us take a ring-shaped cylindrical superconductor. If a weak magnetic field **B** is applied along the ring axis and temperature is lowered below T_c, the field is expelled from the ring due to the Meissner effect. If the field is then slowly reduced to zero, part of the magnetic flux lines may be trapped

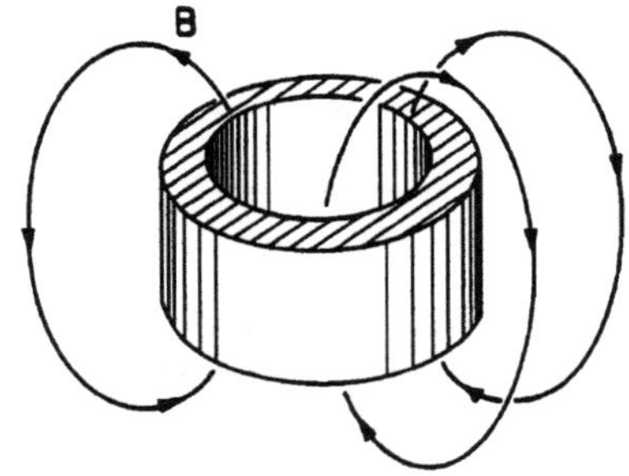

Figure 1.5: A set of magnetic flux lines are trapped in the ring.

as shown in Fig. 1.5. The magnetic moment so generated is found to be maintained by a never-decaying supercurrent flowing around the ring [3].

More delicate experiments [4, 5] showed that the magnetic flux enclosed by the ring is quantized as

$$\Phi = n\Phi_0, \qquad n = 0, 1, 2, ... \tag{1.2}$$

$$\Phi_0 = \frac{h}{2e} = \frac{\pi\hbar}{e} = 2.07 \times 10^{-7} \text{ Gauss cm}^2. \tag{1.3}$$

Φ_0 is called a *flux quantum*. The experimental data obtained by Deaver and Fairbank [4] is shown in Fig. 1.6. The superconductor exhibits a *quantum state* described by a kind of macroscopic wave function.

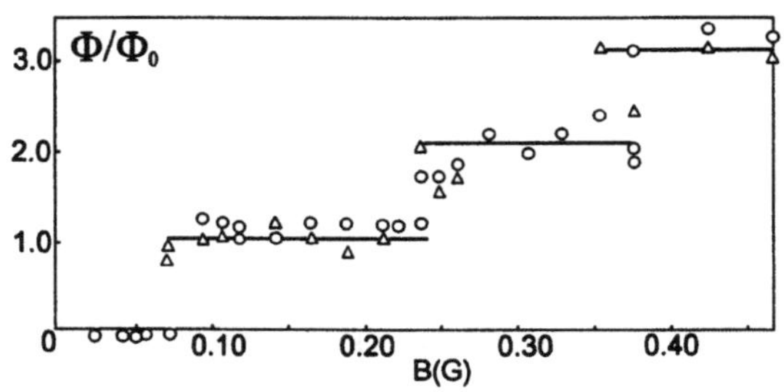

Figure 1.6: The magnetic flux quantization, after Deaver-Fairbank [4]. The two sets of data are shown as Δ and $\circ$.

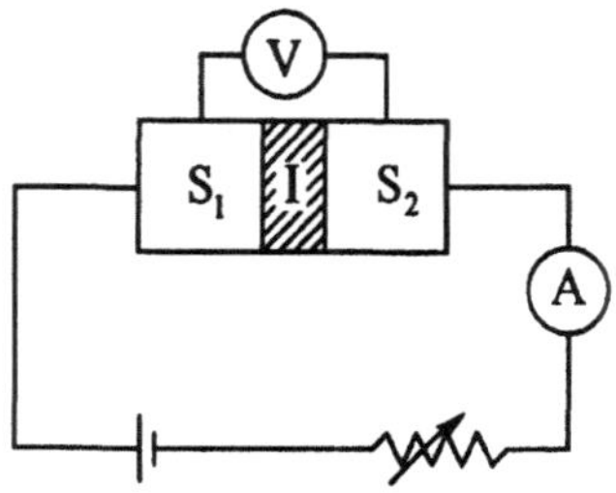

Figure 1.7: Two superconductors, S_1 and S_2, and a Josephson junction I are connected with a battery.

1.1.4 Josephson Effects

Let us take two superconductors separated by an oxide layer of thickness on the order of 10 Å, called a *Josephson junction.* We use this system as part of a circuit including a battery as shown in Fig. 1.7. Above T_c the two superconductors, S_1 and S_2 and the junction I all show potential drops. If temperature is lowered beyond T_c, the potential drops in S_1 and S_2 disappear because of zero resistance. The potential drop across the junction I also disappears! In other words, the supercurrent runs through the junction I with no energy loss. Josephson predicted [6], and later experiments [7] confirmed, this *Josephson tunneling* or *dc Josephson effect.*

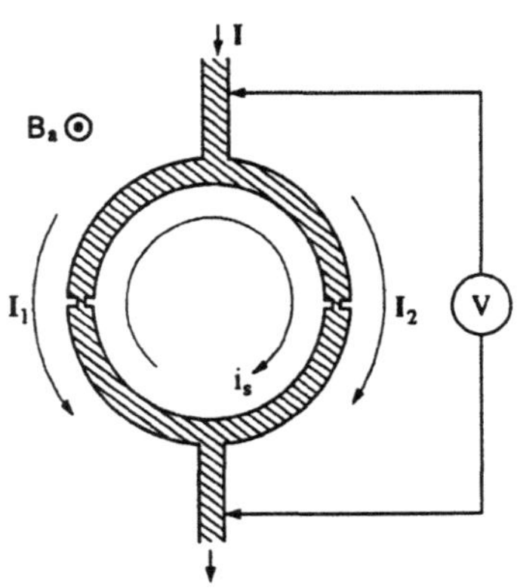

Figure 1.8: Superconducting quantum interference device (SQUID).

We now take a closed loop superconductor containing two similar Josephson junctions and make a circuit as shown in Fig. 1.8. Below T_c, the supercurrent I branches out into I_1 and I_2.

We now apply a magnetic field **B** perpendicular to the loop lying on the paper. The magnetic flux can go through the junctions, and can therefore be changed continuously. The total current I is found to have an oscillatory component:

$$I = I^{(0)} cos(\pi\Phi/\Phi_0), \qquad (I^{(0)} = \text{constant}) \tag{1.4}$$

where Φ is the magnetic flux enclosed by the loop, indicating that the two supercurrents I_1 and I_2, macroscopically separated ($\sim$ 1 mm), interfere just as two laser beams coming from the same source. This is called a *Josephson interference.* A sketch of interference pattern [8] is shown in Fig. 1.9.

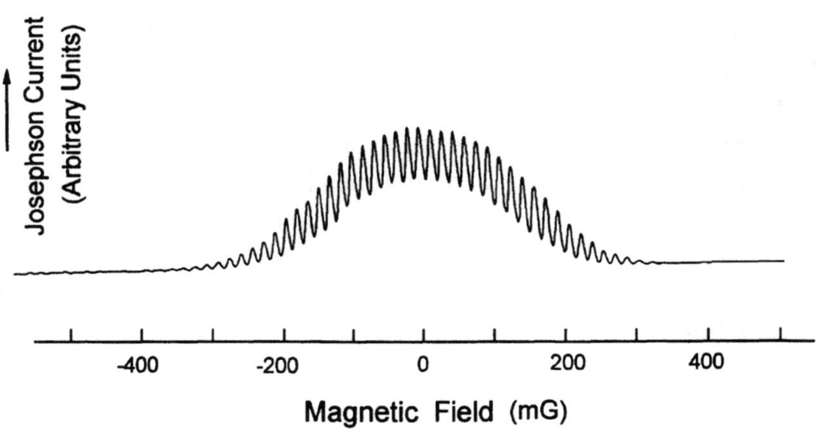

Figure 1.9: Current versus magnetic field, after Jaklevic *et al.* [8].

The circuit in Fig. 1.8 can be used to detect an extremely weak magnetic field. The detector is called the *superconducting quantum interference device* (SQUID).

1.1.5 Energy Gap

If a continuous band of the excitation energy is separated by a finite gap ε_g from the discrete ground-state energy level as shown in Fig. 1.10, this gap can be detected by photo-absorption [9], quantum tunneling [10], heat capacity [11] and other experiments. The energy gap ε_g turns out to be temperature-dependent. The energy gap $\varepsilon_g(T)$ as determined from the tunneling experiments [12] is shown in Fig. 1.11. The energy gap is zero at T_c, and reaches a maximum value $\varepsilon_g(0)$ as temperature approaches toward 0 K.

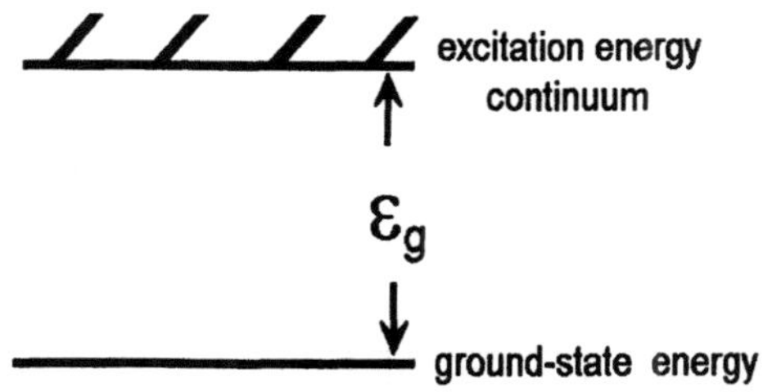

Figure 1.10: Excitation-energy spectrum with a gap.

In true thermodynamic equilibrium, there can be no currents, super or normal. Thus we must deal with a nonequilibrium condition when discussing the basic properties of superconductors, such as zero resistance, flux quantization, and Josephson effects. All of these arise from the supercurrents that dominate the transport and magnetic phenomena. When a superconductor is used to form a circuit with a battery and a steady state is established, all currents passing the superconductor are supercurrents. Normal currents due to the motion of electrons and other charged particles do not show up because no voltage difference can be developed in a homogeneous superconductor.

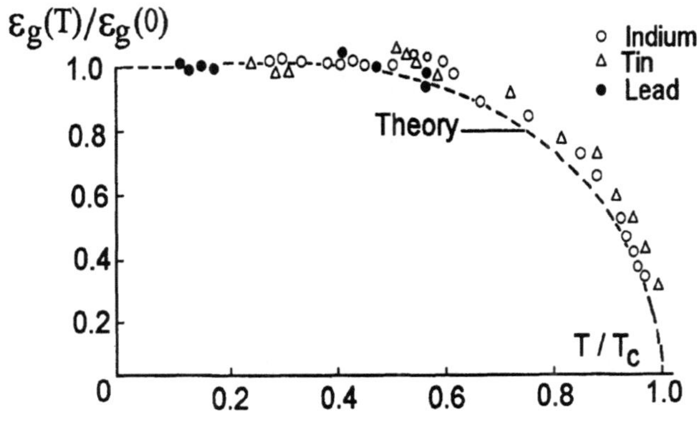

Figure 1.11: The energy gap $\varepsilon_g(T)$ versus temperature, as determined by tunneling experiments, after Giaever and Megerle [12].

1.2 Occurrence of a Superconductor

1.2.1 Elemental Superconductors

Li	Be*											B	C*	N	O	Ne
Na	Mg										T_C= B_C=	Al 1.18 105	Si*	P	S	Ar
K	Ca	Sc	Ti 0.39 100	V 5.38 1420	Cr	Mn	Fe	Co	Ni	Cu	Zn 0.87 53	Ga 1.09 51	Ge*	As	Se*	Kr
Rb	Sr	Y*	Zr 0.54 47	Nb 9.20 1980	Mo 0.92 95	Tc 7.77 1410	Ru 0.51 70	Rh	Pd	Ag	Cd 3.40	In 3.40 293	Sn 3.72 309	Sb*	Te*	Xe
Cs*	Ba*	La 6.00 1100	Hf	Ta 4.48 830	W 0.01 1.07	Re 1.69 198	Os 0.65 65	Ir 0.14 19	Pt*	Au	Hg 4.15 412	Tl 2.39 171	Pb 7.19 803	Bi*	Po	Rn
Fr	Ra	Ac														
			Ce*	Pr	Nd	Pm	Sm	Eu	Gd	Tb	Dy	Ho	Er	Tm	Yb	Lu
			Th 1.36 1.62	Pa 1.4	U 0.68	Np	Pu	Am	Cm	Bk	Cf	Es	Fm	Md	No	Lw

Table 1.1. Superconductivity Parameters of the Elements.
Transition temperature in K and critical magnetic field at 0 K in Gauss.
* denotes superconductivity in thin films or under high pressures.

More than 40 elements have been found to become superconducting. Table 1.1 shows the critical temperature T_c and the critical magnetic fields at 0 K, B_0. Most non-magnetic metals can be superconductors, with notable exceptions being monovalent metals such as Li, Na, K, Cu, Ag, and Au. Some metals can become superconductors under applied pressures and/or in thin films, and these are indicated by asterisks.

1.2.2 Compound Superconductors

Thousands of metallic compounds are found to be superconductors. A selection of compound superconductors with critical temperature T_c are shown in Table 1.2. Note: T_c tends to be higher in compounds than in elements. Nb_3Ge has the highest T_c ($\sim 23\,K$).

Compound	T_c(K)	Compound	T_c(K)
Nb_3Ge	23.0	MoN	12.0
$Nb_3(Al_{0.8}Ge_{0.2})$	20.9	V_3Ga	16.5
Nb_3Sn	18.05	V_3Si	17.1
Nb_3Al	17.5	UCo	1.70
Nb_3Au	11.5	Ti_2Co	3.44
NbN	16.0	La_3In	10.4

Table 1.2. Critical temperatures of selected compounds

Compound superconductors exhibit type II magnetic behavior different from that of type I elemental superconductors. A very weak magnetic field is expelled from the body (the Meissner effect) just as by the type I superconductor. If the field is raised beyond the *lower critical field* H_{c1}, the body allows a partial penetration of the field, still remaining in the superconducting state. A further field increase turns the body to a normal state upon passing the *upper critical field* H_{c2}. Between H_{c1} and H_{c2}, the superconductor is in a *mixed state* in which magnetic flux lines surrounded by supercurrents, called vortices, penetrate the body. The critical fields *versus* temperature are shown in Fig. 1.12. The upper critical field H_{c2} can be very high (20 T $= 2 \times 10^5$ G for Nb_3Sn). Also the critical temperature T_c tends to be high for high-H_{c2} superconductors. These properties make compound superconductors useful for devices and applications.

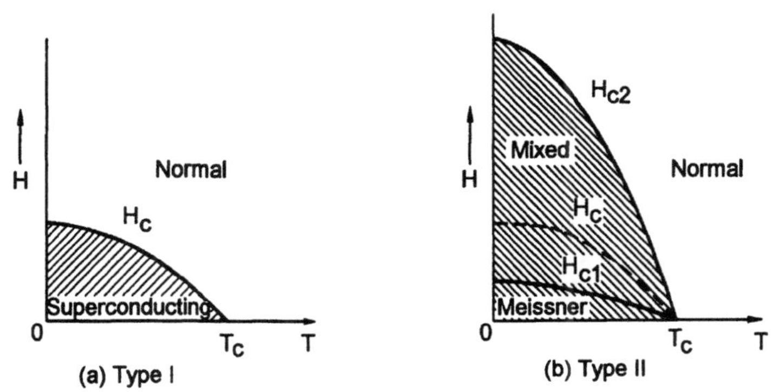

Figure 1.12: Phase diagrams of type I and type II superconductors.

1.2.3 High-T_c Superconductors

In 1986 Bednorz and Müller [14] reported the discovery of the first high-T_c or high-temperature cuprate superconductors, also called high temperature superconductor, (HTSC) ($T_c \sim 30\,\mathrm{K}$) (LaBaCuO). Since then, many investigations have been carried out on the high-T_c superconductors including YBaCuO with $T_c \sim 94\,\mathrm{K}$ [15]. The boiling point of abundantly available and inexpensive liquid nitrogen (N) is 77 K. So the application potential of high-T_c superconductors, which are of type II, appears enormous. The superconducting state of these conductors is essentially the same as that of elemental superconductors.

1.3 Theoretical Background

1.3.1 Metals and Conduction Electrons

A metal is a conducting crystal in which electrical current can flow with little resistance. This electrical current is generated by moving electrons. The electron has mass m and charge, $-e$, which is negative by convention. Their numerical values are $m = 9.1 \times 10^{-28}$ g, $e = 4.8 \times 10^{-10}$ esu $= 1.6 \times 10^{-19}$ C. The electron mass is smaller by about 1837 times than the least massive hydrogen atom. This makes the electron extremely mobile. It also makes the electron's quantum nature more pronounced. The electrons participating in the charge transport, called *conduction electrons*, are those which would have orbited in the outermost shells surrounding the atomic nuclei if the nuclei were separated from each other. Core electrons which are more tightly bound with the nuclei form part of the metallic ions. In a pure crystalline metal, these metallic ions form a relatively immobile array of regular spacing, called a *lattice*. Thus, a metal can be pictured as a system of two components: mobile electrons and relatively immobile lattice ions.

1.3.2 Quantum Mechanics

Superconductivity is a quantum effect manifested on a macroscopic scale. This is most clearly seen by a supercurrent ring with the associated quantized magnetic flux. To interpret this phenomenon, a thorough understanding of quantum theory is essential. Dirac's formulation of quantum theory in his book, *Principles of Quantum Mechanics* [16], is unsurpassed. Dirac's rules that the quantum states are represented by *bra* or *ket* vectors and physical observables by Hermitean operators, are used in the text. There are two distinct quantum effects, the first of which concerns a single particle and the second a system of identical particles.

1.3.3 Heisenberg's Uncertainty Principle

Let us consider a simple harmonic oscillator characterized by the Hamiltonian

$$H = \frac{p^2}{2m} + \frac{kx^2}{2}, \tag{1.5}$$

where m is the mass and k the force constant. The corresponding energy eigenvalues are

$$\varepsilon_n = \hbar\omega_0(n + 1/2), \qquad \omega_0 \equiv (k/m)^{1/2}, \qquad n = 0, 1, 2, \ldots \,. \tag{1.6}$$

The energies are quantized in Eq. (1.6). In contrast the classical energy can be any positive value. The lowest quantum energy $\varepsilon_0 = \hbar\omega_0/2$, called the *energy of zero point motion*, is not zero. It is found that the most stable state of any quantum system is not a state of *static equilibrium* in the configuration of lowest potential energy. It is rather a *dynamic equilibrium* for the zero point motion, which may be characterized by the minimum average total (potential+kinetic) energy under the condition that each coordinate q have a range Δq and the corresponding momentum p has a range Δp, so that the product $\Delta q \Delta p$ satisfies the *Heisenberg uncertainty relation*:

$$\Delta q \Delta p > \hbar/2. \tag{1.7}$$

The most remarkable example of a macroscopic body in dynamic equilibrium is liquid helium (He). This liquid with a boiling point at 4.2 K is known to remain liquid down to 0 K. The zero-point motion of He atoms precludes solidification.

1.3.4 Bosons and Fermions

Electrons are fermions. That is, they are *indistinguishable quantum particles* subject to *Pauli's exclusion principle.* Indistinguishability of the particles is defined by using the permutation symmetry. According to Pauli's principle no two electrons can occupy the same state. Indistinguishable quantum particles not subject to Pauli's exclusion principle are called *bosons.* Bosons can occupy the same state multiply. *Every elementary particle is either a boson or a fermion.* This is known as the *quantum statistical postulate.* Whether an elementary particle is a boson or fermion is related to the magnitude of its spin angular momentum in units of $\hbar$. *Particles with integer spins are bosons while those with half-integer spin are fermions.* This is known as Pauli's *spin-statistics theorem.* According to this theorem and in agreement with all experimental evidence, electrons, protons, neutrons, and μ-mesons, all of which have spin of magnitude $\hbar/2$, are fermions while photons (quanta of electromagnetic radiation) with spin of magnitude $\hbar$, are bosons.

1.3.5 Fermi and Bose Distribution Functions

The average occupation number at state a, denoted by $< N_a >$, for a system of free fermions in equilibrium at temperature T and chemical potential μ, is given by the *Fermi distribution function*:

$$< N_a >= f_F(\varepsilon_a) \equiv \frac{1}{\exp[(\varepsilon_a - \mu)/\mathrm{k}_B T] + 1} \qquad \text{for fermions,} \tag{1.8}$$

where ε_a is the energy associated with the state a. The Boltzmann constant k_B has the following numerical value: $\mathrm{k}_B = 1.38 \times 10^{-16}$ erg/deg $= 1.38 \times 10^{-23}$ JK^{-1}.

The average occupation number at state a for a system of free bosons in equilibrium is given by the *Bose distribution function*:

$$< N_a >= f_B(\varepsilon_a) \equiv \frac{1}{\exp[(\varepsilon_a - \mu)/\mathrm{k}_B T] - 1} \qquad \text{for bosons.} \tag{1.9}$$

1.3.6 Composite Particles

Atomic nuclei are composed of nucleons (protons, neutrons) while atoms are composed of nuclei and electrons. It has been experimentally demonstrated that these composites are indistinguishable quantum particles. According to Ehrenfest-Oppenheimer-Bethe's rule [17] the Center-of-Mass (CM) of a composite moves as a fermion (boson) if it contains an odd (even) number of elementary fermions. Thus He^4 atoms (4 nucleons, 2 electrons) move as bosons and He^3 atoms (3 nucleons, 2 electrons) as fermions. Cooper pairs (two electrons) move as bosons, [see Section 8.2].

1.3.7 Superfluids and Bose-Einstein Condensation

Liquid He^4 (the most abundant isotope) undergoes a superfluid transition at 2.19 K. Below this transition temperature, liquid He^4 exhibits frictionless (zero viscosity) flows remarkably similar to supercurrents. The pioneering experimental works on superfluidity were conducted primarily in the late 1930's. In 1938 Fritz London [18] advanced a view that the superfluid transition in liquid He^4 be interpreted in terms of a B-E condensation [19], where a *finite fraction* of bosons is condensed in the lowest energy state and the rest of bosons have a gas like distribution.

1.3.8 Bloch Electrons and Fermi Liquid Model

In a metal at the lowest temperatures conduction electrons move in a static periodic lattice. Because of the Coulomb interaction among the electrons, the motion of the electrons is correlated. However each electron in a crystal moves in an extremely weak self-consistent periodic field. Combining this result with Pauli's exclusion principle, which applies to electrons with no regard to the interaction, we can obtain the *Fermi Liquid model* of Landau [20] (see Section 3.2). In this model the quantum states for the Bloch electron are characterized by k-vector **k**, zone-number j and energy

$$\varepsilon = E_j(k). \tag{1.10}$$

At 0 K, all of the lowest-energy states are filled with electrons, and there exists a *sharp Fermi surface* represented by

$$E_j(k) = \varepsilon_F, \tag{1.11}$$

where ε_F is the *Fermi energy*. Experimentally, all normal conductors are known to exhibit sharp Fermi surfaces at 0 K. Theoretically the band theory of solids [21] and the microscopic theory of superconductivity are based on this model. The occurrence of superconductors critically depends on the Fermi surface; see Section 7.3.

1.3.9 "Electrons" and "Holes"

"Electrons" ("Holes") in the text are defined as quasiparticles possessing charge e (magnitude) which circulate counterclockwise (clockwise) when viewed

from the tip of the applied magnetic field vector **B**. We use quotation-marked "electron" to distinguish it from the generic electron having the gravitational mass m. This definition is used routinely in semiconductor physics. A "hole" can be regarded as a particle having positive charge, positive mass and positive energy; The "hole" does not, however have the same effective mass m^* (magnitude) as the "electron", so that "holes" are not true antiparticles like positrons. "Electrons" and "holes" are the thermally excited particles with respect to the Fermi energy, and they are closely related to the curvature of the Fermi surface; see, Sections 3.5.

1.3.10 Second Quantization

In second quantization, where *creation and annihilation operators* associated with each quantum state are used, identical particles (bosons or fermions) can be treated simply and naturally. This formulation is indispensable in developing a theory of superconductivity, and for completeness it is reviewed in Appendix A.

1.4 Elements of Kinetic Theory

The understanding of normal conduction is a prerequisite for the superconductivity theory. We discuss the elements of the kinetic theory of electrical conduction.

1.4.1 Electrical Conductivity and Matthiessen's Rule

Let us consider a system of electrons independently moving in a potential field of impurities, which act as scatterers. The impurities are assumed to be distributed uniformly.

Under the action of an electric field **E** pointed along the positive x-axis, a classical electron will move according to Newtons's equation of motion:

$$m\frac{dv_x}{dt} = -eE, \tag{1.12}$$

(in the absence of the impurity potential). Solving this, we obtain

$$v_x = -\frac{e}{m}Et + v_x^0. \tag{1.13}$$

where v_x^0 is the x-component of the initial velocity. For a free electron the velocity v_x can increase indefinitely and leads to infinite conductivity.

In the presence of the impurities, this uniform acceleration will be interrupted by scattering. When the electron hits a scatterer (impurity), the velocity will suffer an abrupt change in direction and grow again following Eq. (1.13) until the electron hits another scatterer. Let us denote the *average time between successive scatterings* or the *mean free time* by τ_f. The average velocity $\langle v_x \rangle$ is then given by

$$\langle v_x \rangle = -(e/m)E\tau_f, \tag{1.14}$$

where we assumed that the electron loses the memory of its preceding motion every time it hits a scatterer, and the average velocity after collision is zero:

$$\langle v_x^0 \rangle = 0. \tag{1.15}$$

The charge current density (average current per unit volume) j is given by

$$j = (\text{charge}) \times (\text{number density}) \times (\text{velocity}) = en\langle v_x \rangle = e^2 n\tau_f E/m, \tag{1.16}$$

where n is the number density of electrons. According to *Ohm's Law*, the current density j is proportional to the applied electric field E when this field is small:

$$\boxed{j = \sigma E.} \tag{1.17}$$

The proportionality factor σ is called the *electrical conductivity.* It represents the facility with which the current flows in response to the electric field. Combining the last two equations, we obtain

$$\boxed{\sigma = e^2 n\tau_f/m} \tag{1.18}$$

This equation is very useful in the qualitative discussion of the electrical transport phenomenon. The inverse mass-dependence law means that the ion contribution to the electric transport in an ionized gas will be smaller by at least three orders of magnitude than the electron contribution. Also note that the conductivity is higher if the number density is greater and/or if the mean free time is greater.

The inverse of the mean free time τ_f:

$$\Gamma = 1/\tau_f \tag{1.19}$$

is called the *scattering rate.* Roughly speaking this represents the mean frequency with which the electron is scattered by impurities. The scattering rate Γ is given by

$$\Gamma = n_I v A, \tag{1.20}$$

where n_I, v, and A are respectively the density of scatterers, the electron speed, and the scattering cross section.

If there is more than one kind of scatterer, the scattering rate may be computed by the addition law:

$$\Gamma = n_1 v A_1 + n_2 v A_2 + ... \equiv \Gamma_1 + \Gamma_2 + ... \; . \tag{1.21}$$

This is often called *Matthiessen's rule.* The total scattering rate is the sum of rates computed separately for each kind of scatterer.

Historically and still today, the analysis of resistance data for a conductor is performed as follows: If the electrons are scattered by impurities and again by phonons (quanta of lattice vibrations), the total resistance will be written as the sum of the resistances due to each separate cause of scattering:

$$R_{total} = R_{impurity} + R_{phonon}. \tag{1.22}$$

This is the original statement of Matthiessen's rule. In further detail, the electron-phonon scattering depends on temperature because of the changing phonon population while the effect of the electron-impurity scattering is temperature-independent. By separating the resistance in two parts, one temperature-dependent and the other temperature-independent, we may apply Matthiessen's rule.

1.4.2 The Hall Effect, "Electrons" and "Holes"

In this section we discuss the Hall effect. As we see later "electrons" and "holes" play very important roles in the microscopic theory of superconductivity. Let us consider a conducting wire connected with a battery. If a magnetic field B is applied, the field penetrates the wire. The Lorentz force

$$\mathbf{F} = q\mathbf{v} \times \mathbf{B}, \tag{1.23}$$

where q is the charge of the carrier, may then affect the electron's classical motion. If so, the picture of the straight line motion of a free electron in kinetic theory has to be modified significantly. If the field (magnitude) B

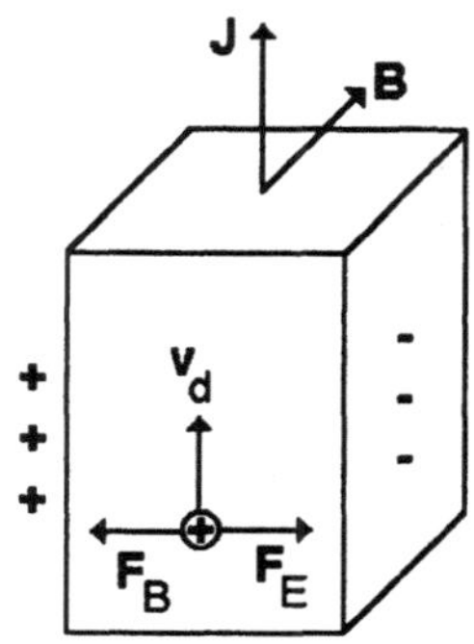

Figure 1.13: The magnetic and electric forces $(\mathbf{F}_B, \mathbf{F}_E)$ are balanced to zero in the Hall effect measurement.

is not too high and the stationary state is considered, the actual physical situation turns out to be much simpler.

Take the case in which the field $\mathbf{B}$ is applied perpendicular to the wire of a rectangular cross-section as shown in Fig. 1.13. Experiments show that a voltage V_c is generated perpendicular to the field $\mathbf{B}$ and the electric current $\mathbf{J}$ such that a steady current flows in the wire apparently unhindered. We may interpret this condition as follows. Let us write the current density $\mathbf{j}$ as

$$\mathbf{j} = qn\mathbf{v}_d, \tag{1.24}$$

where n is the density of conduction electrons and $\mathbf{v}_d$ the *drift velocity*. A charge-carrier having a velocity equal to the drift velocity $\mathbf{v}_d$ is affected by the Lorentz force:

$$\mathbf{F} = q(\mathbf{E}_c + \mathbf{v}_d \times \mathbf{B}), \tag{1.25}$$

where $\mathbf{E}_c$ is the electric field due to the cross voltage V_c. In the geometry shown, only the x-component of the force $\mathbf{F}$ is relevant. If the net force vanishes:

$$q(E_c + v_d B) = 0, \tag{1.26}$$

the carrier can proceed along the wire ($z-$direction) unhindered.

Let us check our model calculation. We define the *Hall coefficient* R_H by

$$R_H \equiv \frac{E_c}{jB}, \tag{1.27}$$

where the three quantities on the rhs can be measured. Using (1.24) and (1.26), we obtain

$$\boxed{R_H = 1/qn.} \tag{1.28}$$

The experimental values for $-qnR_H$ in some metals are given in Table 1.3. For alkali metals the agreement between theory and experiment is nearly perfect. The measured Hall coefficients R_H for most metals are negative. This can be understood by assuming that the charge carriers are "electrons" having a negative charge $q = -e$. There are exceptions, however. As we see in *Table 1.3*, Al, Be and others exhibit positive Hall coefficients. This can be explained only by assuming that in these metals the main charge carriers are "holes" possessing a positive charge $q = +e$. This is a quantum many-body effect. As we shall see later, the existence of "electrons" and "holes" is closely connected with the curvature of the Fermi surface. Non-magnetic metals which have "holes" tend to be superconductors, as will be explained later.

Metal	Valence	$-1/nqR_H$
Li	1	−0.8
Na	1	−1.2
K	1	−1.1
Cu	1	−1.5
Ag	1	−1.3
Au	1	−1.5
Be	2	0.2
Mg	2	0.4
In	3	0.3
Al	3	0.3

Table 1.3. Hall coefficients of selected metals.

Chapter 2

Superconducting Transition

Flux quantization and Josephson interference indicate that the superconducting transition is a Bose-Einstein condensation of independently moving Cooper pairs (pairons) having a linear energy-momentum relation.

2.1 Three Phase Transitions

The superconducting transition is a sharp thermodynamic phase transition similar to the vapor-liquid transition of water and the ferromagnetic transition of a magnet. But there are significant differences between the three transitions.

First, the vapor-liquid transition is a *condensation* in the ordinary space, while the ferromagnetic transition is a condensation in the spin angular momentum space. As we see later, the *superconducting transition is a condensation in the momentum space.*

Second, the cause of the liquid condensation is the intermolecular attraction, which may be represented by the Lenard-Jones potential with a depth ε and a size, see Fig. 2.1. The condensation temperature T_c is of the order of the depth ε:

$$T_c \sim \varepsilon. \qquad \text{(liquid condensation)} \tag{2.1}$$

The correlation of the motion among a number of molecules becomes great near T_c, and their rigorous treatment of the phase transition is extremely difficult. A ferromagnetic transition may be described by the Ising model [1] with an exchange energy J characterizing the alignment tendency of two neighboring spins. The condensation (critical) temperature T_c is of the order

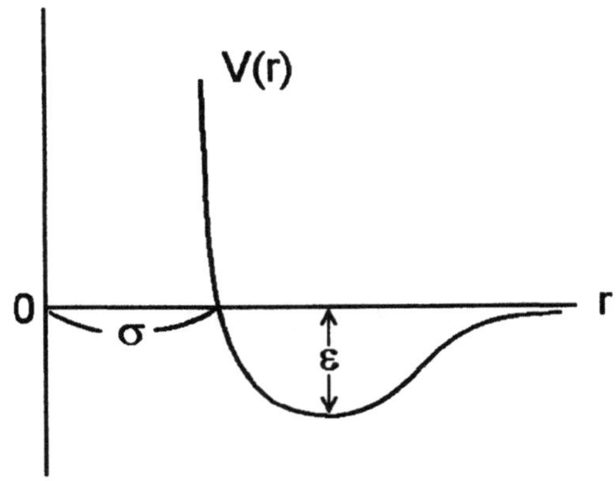

Figure 2.1: The Lennard-Jones potential representing the interaction between atoms is characterized by the potential depth ε and the linear size σ.

of the exchange energy J:

$$T_c \sim J. \qquad \text{(spin condensation)} \tag{2.2}$$

In this case, too, the motional correlation extends over a great number of spins as the system's temperature approaches T_c, making the rigorous treatment of phase transition extremely difficult. In contrast, the superconducting transition is, as we shall see, a Bose-Einstein (B-E) condensation of pairons. The interparticle correlation comes only through the boson nature of the pairons, and the treatment is simple and straightforward. We shall see that the critical temperature T_c depends on the pairon density n.

The three transitions can be distinguished thermodynamically as follows: In the case of water, the gas and liquid phases are both characterized by two independent thermodynamic variables, e. g. the number density n and the temperature T. There is a *latent heat* at the transition point, indicating that there is a finite change in the entropy S at T_c. The *order of phase transition* is defined to be that order of the derivative of the free energy F whose discontinuity appears for the first time. Since $S = -\partial F(T, V)/\partial T$, the vapor-liquid transition is a phase transition of first order.

In contrast the superconducting phase has one peculiar component, called a super part or a *supercondensate*, which dominates the transport and magnetic phenomena, and the other component, called the normal part, which behaves normally [2]. The two components are intermixed in space, but they are distinguished in momentum and move distinctly. In other words, the superconducting phase can be characterized by the normal thermodynamic variables (n, T) and a Ginzburg-Landau wavefunction Ψ (complex order parameter), which represents the quantum state of the supercondensate. [3] The two-fluid (component) model is applicable only to superconductor (and superfluid). The appearance of the supercondensate wave function Ψ below T_c, is somewhat similar to the appearance of a spontaneous magnetization M below T_c in a ferromagnet. The Ψ is complex, while the M is real however. The superconducting transitions at zero magnetic field is of second or higher order, and there is no latent heat.

2.2 Pairons Move as Masless Particles

Let us take a ring-shaped superconductor at 0 K. The *ground state* for a pairon (or any quantum particle) in the absence of electromagnetic fields can be characterized by a *real* wave function $\psi_0(\mathbf{r})$ having no nodes and vanishing at the ring boundary:

$$\psi_0(\mathbf{r}) = \begin{cases} \text{nearly constant} & \text{inside the body} \\ 0 & \text{at the boundary.} \end{cases} \tag{2.3}$$

Such a wave function corresponds to the zero-momentum state, and it cannot carry current. Now consider current-carrying states. There are many nonzero momentum states whose energies are very close to the ground-state energy 0. These states may be represented by the wave functions $\{\psi_n\}$ having a finite number of nodes along the ring. A typical state ψ_n is represented by

$$\psi_n(\mathbf{r}) = u \exp(ip_n x/\hbar), \tag{2.4}$$

$$p_n \equiv \frac{2\pi\hbar n}{L}, \qquad n = \pm 1, \pm 2, \ldots \tag{2.5}$$

where x is the coordinate along the ring of circumference L; the factor u is real and nearly constant inside, and vanishes at the boundary. When a macroscopic ring is considered, the wavefunction ψ_n represents a state having linear momentum p_n along the ring. For small n, the value of $p_n = 2\pi\hbar n/L$ is very small, since L is a macroscopic length. The associated energy eigenvalue is also very small.

Let us now consider the effect of a magnetic field. In flux quantization experiments [4] a minute flux is trapped in the ring, and this flux is maintained by the ring supercurrent (see Fig. 1.5). Experiments (data are summarized in Fig. 1.6), show that the trapped flux is quantized:

$$\Phi = n\Phi_{0,} \qquad \Phi_0 \equiv \Phi_{\text{pairon}} \equiv \frac{h}{2e} \equiv \frac{\pi\hbar}{e}. \tag{2.6}$$

The integers n appearing in (2.5) and (2.6) are the same, which can be seen by applying the Bohr-Sommerfeld quantization rule:

$$\oint pdx = 2\pi\hbar(n + \gamma) \tag{2.7}$$

to the circulating pairons. The *phase* (number) γ is zero for the present ring (periodic) boundary condition.

Flux quantization experiments indicate that a macroscopic supercurrent is generated and maintained around the ring, that the charge carriers in the supercurrent have charge (magnitude) $2e$, and that a great number of pairons participate in the conduction, and that they occupy the same momentum state p_n, implying that pairons move as bosons. Since the steady-state is maintained in the experiments this momentum state p_n corresponds to one of the energy-eigenstates of the pairon, meaning that the energy of a moving pairons is a sole function of its momentum:

$$\varepsilon = \varepsilon(\mathbf{p}). \tag{2.8}$$

In other words the pairons do not interact with each other. Furthermore, since flux quantization is the momentum quantization, the phenomenon is material-independent.

The Josephson interference [5], (see Figs. 1.8 and 1.9), indicate that two supercurrents macroscopically separated up to 1 mm can interfere just as two laser beams coming from the same source. This means that the condensed pairons move with a non-dispersive linear energy-momentum relation:

$$\varepsilon = cp, \qquad c = \text{constant}. \tag{2.9}$$

In fact, supercurrent and laser have a great similarities as shown in Table 2.1. The only significant difference are (i) that pairons have charges and hence respond to electromagnetic fields while photons do not and (ii) that pairons can be stationary while photons must always travel at the speed of light.

pairons	common feature	**photons**
$\varepsilon = v_F p/2$		$\varepsilon = cp = c\hbar k$
(Cooper-Schrieffer relation)		c = ligth speed
	No electromagnetic mass	
	boson	
	diffraction	
coherence range $\sim$ 1 mm	interference	coherence range $\sim$ 1
	self focussing power	
charge $2e$		0 charge

Table 2.1 Analogy Between Supercurrents and Lasers

2.3 The Bose-Einstein Condensation

Electrons are fermions. Pairons, each composed of a pair of electrons, move as bosons. That is, two or more pairons can occupy one and the same *center-of-mass* momentum state. In general the quantum statistics of composites such as pairons, nuclei, atoms, etc. obeys Ehrenfest-Oppenheimer-Bethe's (EOB) rule with respect to the center of mass motion [6]: a composite's motion is fermionic (bosonic) if it contains an odd (even) number of elementary fermions. We stress that the quantum statistics is defined with respect to the occupation possibility at the *center-of-mass* momentum. This theorem applies independently of any interaction and any energy (bound-state or else). No detailed configuration of constituting particles matters. The proof of this theorem is given in Chapter 8 [7].

A system of independently moving particles in equilibrium obeys the Bose or the Fermi distribution law. For free bosons with mass M moving in three dimensions (3D), the condensation temperature T_c is given by [1], [8]

$$T_c = 3.31\,\hbar^2 \mathrm{k}_B^{-1} M^{-1} n^{2/3}, \tag{2.10}$$

where n is the number density of bosons. This formula is applicable to the systems of gaseous Rb^{87} [87 nucleons (N), 37 electrons (e)] and Na [$22N$, $11e$] recently observed by Cornell's [9] and Keterle's [10] groups.

The phase transition in an elemental (and compound) superconductor can also be regarded as a B-E condensation of free pairons. The phonon-exchange attraction can bind any two electrons whose energies are close to the Fermi energy. Hence zero and finite-momentum pairons, are created in a superconductor. In fact, Cooper and Schrieffer [11] showed that pairons moving in 3D have a linear energy momentum relation

$$\varepsilon_q = (1/2)v_F q + w_0, \tag{2.11}$$

where $v_F \equiv (2\varepsilon_F/m^*)^{1/2}$ is the Fermi velocity and w_0 the ground-state energy of a pairon. The condensation temperature T_c for free pairons is given by [12]

$$T_c = 1.01\,\hbar \mathrm{k}_B^{-1} v_F n^{1/3}. \tag{2.12}$$

Note that this formula is independent of the famous BCS formula [13]:

$$2\Delta = 3.53\,\mathrm{k}_B T_c, \tag{2.13}$$

connecting the critical temperature T_c and the zero-temperature quasi-electron energy gap Δ in the weak coupling limit. The Bloch electron (wave packet) in a crystal moves unhindered by the lattice potential. Hence pairons, each made up of two Bloch electrons move freely in the perfect crystal. The pairon size can be estimated by the BCS zero-temperature coherence length [13]

$$\xi_0 = \hbar v_F/\pi\Delta = 0.18\,\hbar v/\mathrm{k}_B T_c. \tag{2.14}$$

After solving Eq. (2.12) for n and introducing the interpairon distance $r_0 \equiv n^{-1/3}$, we obtain

$$r_0 \equiv n^{-1/3} = 1.01\,\hbar v_F/\mathrm{k}_B T_c = 5.61\ \xi_0, \tag{2.15}$$

indicating that the interpairon distance r_0 is several times greater than the pairon size ξ_0. Thus B-E condensation occurs before the picture of free pairons breaks down, and formula (2.12) can be identified as the superconducting (critical) temperature.

In the currently predominant BCS theory the superconducting temperature T_c is identified as the temperature at which the stationary (zero-momentum) pairons break up and disappear in the system. But the heat capacity has a maximum at T_c, which can only be explained in terms of a model in which many particles participate in the phase transition. We take a view that the superconducting transition is a macroscopic change of state with no change in the microscopic constituents (pairons). Our view is supported by experiments, for example, hydrogen molecules H_2 in a gas mixture of (H_2, H) will break up haphazardly with no visible macroscopic change of phase. In the B-E condensation a great number of pairons cooperatively participate and exhibits a macroscopic change of phase. Hence, in our view the pairons do not break up at T_c, and exist above T_c. The current-voltage (I-V) curves in the Giaever tunneling [14] go over smoothly to the straight-line Ohm's law behavior as the temperature is raised through T_c, indicating that both pairons and electrons above T_c contribute to the charge transport.

High-T_c cuprates [15] and organic superconductors [16] have layered lattice structures. The electrical conduction in such a material occurs in a plane perpendicular to the c-axis; the Fermi surface is a right cylinder with its axis parallel to the c-axis. Pairons formed in a two-dimensional (2D) material have a linear energy-momentum relation:

$$\varepsilon = (2/\pi)v_F p + w_0. \tag{2.16}$$

The system of free pairons moving in 2D with the relation (2.16) undergoes a B-E condensation at

$$T_c = 1.24\,\hbar \mathrm{k}_B^{-1} v_F n^{1/2}. \tag{2.17}$$

The condensation of massless bosons in 2D is noteworthy [17] since the B-E condensation of massive bosons ($\varepsilon = p^2/2M$) is known to occur in three dimensions only. This is not a violation of Hohenberg's theorem [18] that there can be no long-range order in two dimensions, which is derived with the assumption of an $f-$sum rule representing the mass conservation law. In fact, no B-E condensation occurs in 2D for finite-mass bosons [17].

High-T_c superconductors above T_c exhibit unusual transport behaviors. This will be explained later with the assumption that there exist two kinds of charge carriers, electrons and pairons, both being scattered by phonons.

Chapter 3

Bloch Electrons

To properly develop a microscopic theory of superconductivity, a deeper understanding of the properties of normal metals than what is provided by the free-electron model is required. Based on the Bloch's theorem, the Fermi liquid model is derived. At 0 K, the normal metal is shown to have a sharp Fermi surface, which is experimentally supported by the fact that the heat capacity is linear in temperature at the lowest temperatures. "Electrons" and "holes", which appear in the Hall effect measurements, are defined in terms of the curvature of the Fermi surface. Newtonian equations of motion for a Bloch electron (wave-packet) are derived and discussed.

3.1 Bloch Theorem

Let us take a monovalent metal such as sodium (Na). The Hamiltonian H of the system may be represented by

$$\begin{aligned} H &= \sum_{j=1}^{N} \frac{p_j^2}{2m} + \sum_{j>k}\sum \frac{k_0 e^2}{|\mathbf{r}_j - \mathbf{r}_k|} + \sum_{\alpha=1}^{N} \frac{\mathbf{P}_\alpha^2}{2M} \\ &\quad + \sum_{\alpha>\gamma}\sum \frac{k_0 e^2}{|\mathbf{R}_\alpha - \mathbf{R}_\gamma|} - \sum_j \sum_\alpha \frac{k_0 e^2}{|\mathbf{r}_j - \mathbf{R}_\alpha|}, \end{aligned} \tag{3.1}$$

$k_0 \equiv (4\pi\varepsilon_0)^{-1}$. The sums on the right hand side (rhs) represent, respectively, the kinetic energy of electrons, the interaction energy among electrons, the kinetic energy of ions, the interaction energy among ions, and the interaction energy between electrons and ions. The metal as a whole is electrically

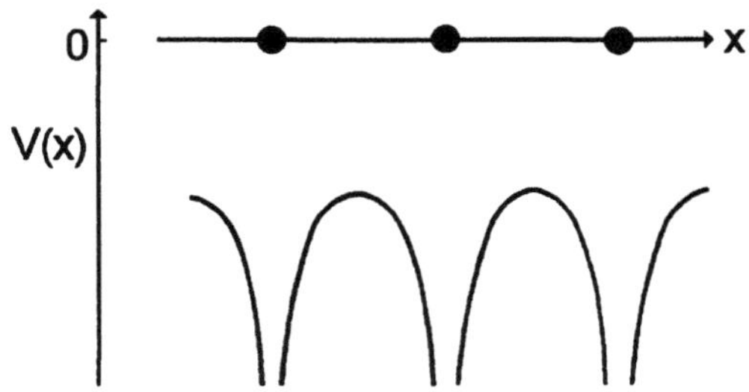

Figure 3.1: A periodic potential in one dimension.

neutral, and hence the number of electrons equals the number of ions. Both numbers are denoted by N.

At the lowest temperatures the ions are almost stationary near the equilibrium lattice points. (Because of quantum zero-point motion, the ions are not at rest even at 0 K. But this fact does not affect the following argument.) The system then can be viewed as the one in which the electrons move in a periodic lattice potential V. The Hamiltonian of this idealized system that now depends on the electron variables only can be written as

$$H = \sum_j \frac{p_j^2}{2m} + \sum_{j>k}\sum \frac{k_0 e^2}{|\mathbf{r}_j - \mathbf{r}_k|} + \sum_j V(\mathbf{r}_j) + C, \tag{3.2}$$

where $V(\mathbf{r})$ represents the lattice potential, and the constant energy C depends on the lattice configuration.

Let us drop the Coulomb interaction energy from (3.2). We then have

$$H = \sum_j \frac{p_j^2}{2m} + \sum_j V(\mathbf{r}_j) + \textit{constant}. \tag{3.3}$$

In Fig. 3.1, we draw a typical lattice potential in one dimension (1D).

For definiteness, we consider an infinite orthorhombic (orc) lattice. We choose a Cartesian frame of coordinates (x, y, z) along the lattice axes. The potential $V(x, y, z) = V(\mathbf{r})$ is *lattice-periodic*:

$$V(\mathbf{r} + \mathbf{R}) = V(\mathbf{r}), \tag{3.4}$$

$$\mathbf{R} \equiv n_1 a\,\mathbf{i} + n_2 b\,\mathbf{j} + n_3 c\,\mathbf{k}, \qquad (n_j : \textit{integers}) \tag{3.5}$$

where (a, b, c) are lattice constants and the vector $\mathbf{R}$ is called the *Bravais lattice vector.*

The Schrödinger equation for an electron is

$$[-\frac{\hbar^2}{2m}\nabla^2 + V(\mathbf{r})]\psi_E(\mathbf{r}) = E\psi_E(\mathbf{r}), \tag{3.6}$$

where E is the energy. Clearly $\psi_E(\mathbf{r}+\mathbf{R})$ also satisfies the same equation. Therefore the values of the wavefunction ψ_E at $\mathbf{r}$ and $\mathbf{r}+\mathbf{R}$ may be different only by a $\mathbf{r}$-independent phase:

$$\psi_E(\mathbf{r}+\mathbf{R}) = e^{i\mathbf{k}\cdot\mathbf{R}}\psi_E(\mathbf{r}), \tag{3.7}$$

where $\mathbf{k} \equiv (k_x, k_y, k_z)$ is called a k-vector. Eq. (3.7) represents a form of the Bloch's theorem [1]. It generates far reaching consequences in theory of conduction electrons.

The three principal properties of the Bloch wavefunctions are:

(A) The probability distribution $P(\mathbf{r})$ is lattice-periodic:

$$P(\mathbf{r}) \equiv |\psi(\mathbf{r})|^2 = P(\mathbf{r}+\mathbf{R}). \tag{3.8}$$

(B) The k-vector $\mathbf{k} = (k_x, k_y, k_z)$ in Eq. (3.7) has the fundamental range:

$$-\pi/a < k_x < \pi/a, \qquad -\pi/b < k_y < \pi/b, \qquad -\pi/c < k_z < \pi/c; \tag{3.9}$$

the end points that form a rectangular box, are called the *Brillouin boundary.*

(C) The energy eigenvalues E have energy gaps, and the allowed energies E can be characterized by the zone number j and the k-vectors:

$$E = \varepsilon_j(\hbar\mathbf{k}) \equiv \varepsilon_j(\mathbf{p}). \tag{3.10}$$

Using Eq. (3.7), we can express the Bloch wave function ψ in the form:

$$\psi_E(\mathbf{r}) \equiv \psi_{j,\mathbf{k}}(\mathbf{r}) = \exp(i\mathbf{k}\cdot\mathbf{r})\, u_{j,\mathbf{k}}(\mathbf{r}), \tag{3.11}$$

$$u_{j,\mathbf{k}}(\mathbf{r}+\mathbf{R}) = u_{j,\mathbf{k}}(\mathbf{r}). \tag{3.12}$$

Equations (3.10)-(3.12) indicate that the Bloch wavefunction $\psi_E(\mathbf{r})$, associated with quantum numbers $(j, \mathbf{k})$, is a plane-wave characterized by k-vector $\mathbf{k}$, angular frequency $\omega \equiv \hbar^{-1}E_j(\mathbf{k})$ and wave train $u_{j,\mathbf{k}}(\mathbf{r})$.

3.2 Fermi Liquid Model

We consider a monovalent metal, whose Hamiltonian H_A is given in Eq. (3.1):

$$H_A = \sum_{j=1}^{N} \frac{p_j^2}{2m} + \sum_{j>k}\sum \frac{k_0 e^2}{|\mathbf{r}_j - \mathbf{r}_k|} + \sum_{\alpha=1}^{N} \frac{P_\alpha^2}{2M}$$

$$+ \sum_{\alpha>\gamma}\sum \frac{k_0 e^2}{|\mathbf{R}_\alpha - \mathbf{R}_\gamma|} - \sum_j \sum_\alpha \frac{k_0 e^2}{|\mathbf{r}_j - \mathbf{R}_\alpha|}. \tag{3.13}$$

The motion of the set of N electrons is correlated because of the interelectronic interaction. If we omit the ionic kinetic energy, interionic and interelectronic Coulomb interaction from Eq. (3.13), we obtain

$$H_B = \sum_j \frac{p_j^2}{2m} + \sum_j V(\mathbf{r}_j) + \text{constant}, \tag{3.14}$$

which characterizes a system of electrons moving in the *bare lattice.*

Since the metal as a whole is neutral, the Coulomb interaction among the electrons, among the ions, and between electrons and ions, all have the same orders of magnitude, and hence they are equally important. We now pick one electron in the system. This electron is in interaction with the system of N ions and $N-1$ electrons, the system (medium) having the net charge $+e$. These other $N-1$ electrons should, in accordance with Bloch's theorem be distributed with the lattice periodicity and all over the crystal in equilibrium. The charge per lattice ion is greatly reduced from e to $N^{-1}e$ because the net charge e of the medium is shared equally by N ions. Since N is a large number, the selected electron moves in an extremely weak *effective lattice potential* V_e as characterized by the model Hamiltonian

$$h_C = \frac{p^2}{2m} + V_e(\mathbf{r}), \qquad V_e(\mathbf{r}+\mathbf{R}) = V_e(\mathbf{r}). \tag{3.15}$$

In other words any chosen electron moves in an environment far different from what is represented by the bare lattice potential V. It moves almost freely in an extremely weak effective lattice potential V_e. This picture was obtained with the aid of Bloch's theorem, and hence it is a result of quantum theory. To illustrate let us examine the same system from the classical point of view. In equilibrium the classical electron distribution is lattice-periodic, so there is one electron near each ion. This electron will not move in the greatly reduced field.

We now assume that electrons move independently in the effective potential field V_e. The total Hamiltonian for the idealized system may then be represented by

$$H_C = \sum_j h_C(\mathbf{r}_j, \mathbf{p}_j) \equiv \sum_j \frac{p_j^2}{2m} + \sum_j V_e(\mathbf{r}_j). \tag{3.16}$$

This Hamiltonian H_C is a far better approximation to the original Hamiltonian H_A than the Hamiltonian H_B. In H_C both interelectronic and interionic Coulomb repulsion are not neglected but are taken into consideration self-consistently. This model is a *one-electron-picture approximation*, but it is hard to improve on by any simple method. The model in fact forms the basis for band theory of electrons.

We now apply Bloch's theorem to the Hamiltonian H_C composed of the kinetic energy and the interaction energy V_e. We then obtain the Bloch energy bands $\varepsilon_j(\hbar\mathbf{k})$ and the Bloch states characterized by band index j and k-vector $\mathbf{k}$. The Fermi-Dirac statistics obeyed by the electrons can be applied to the Bloch electrons with no regard to interaction. This means that there is a certain Fermi energy ε_F for the ground-state of the system. Thus there is a *sharp Fermi surface* represented by

$$\varepsilon_j(\hbar\mathbf{k}) = \varepsilon_F, \tag{3.17}$$

which separates the electron-filled k-space (low-energy side) from the empty k-space (high energy side). The Fermi surface for a real metal in general is complicated in contrast to the free-electron Fermi sphere represented by

$$\frac{p^2}{2m} \equiv \frac{p_x^2 + p_y^2 + p_z^2}{2m} = \varepsilon_F. \tag{3.18}$$

The independent electron model with a sharp Fermi surface at 0 K is called the *Fermi liquid model* of Landau [2]. As we show later, many thermal properties of conductors are dominated by those electrons near the Fermi surface. The shape of the Fermi surface is very important for the occurrence of superconductors. In the following section, we shall examine the Fermi surfaces of some metals.

The Fermi liquid model was obtained in the static lattice approximation in which the motion of the ions is neglected. If the effect of moving ions (phonons) is taken into account, a new model is required. The electron-phonon interaction turns out to be very important in the theory of superconductivity, which will be discussed in Chapter 4.

3.3 The Fermi Surface

Why does a particular metal exist with a particular crystalline state? This is a good question. The answer must involve the composition and nature of the atoms constituting the metal and the interaction between the component particles. To illustrate let us take sodium, which forms a bcc lattice. This monovalent metal may be thought of as an ideal composite system of electrons and ions. The system Hamiltonian may be approximated by H_A in Eq. (3.13), which consists of the kinetic energies of electrons and ions and the Coulomb interaction energies among and between electrons and ions. This is an approximation since the interaction between electron and ion deviates significantly from the ideal Coulomb law at short distances because each ion has core-electrons. At any rate the study of the ground-state energy of the ideal model favors a fcc lattice structure, which is not observed for this metal. If multivalent metals like Pb and Sn are considered, the condition becomes even more complicated, since the core electrons forming part of ions have anisotropic charge distribution. Because of this complexity it is customary in solid state physics to assume the experimentally known lattice structures first, then proceed to study the Fermi surface.

Once a lattice is selected, the Brillouin zone is fixed. For an orc lattice the Brillouin zone is a rectangular box defined by Eqs. (3.9). We now assume a large periodic box of volume

$$V = (N_1\, a)(N_2\, b)(N_3\, c), \qquad N_1, N_2, N_3 \gg 1. \tag{3.19}$$

Let us find the number N of the quantum states in the first Brillouin zone. With the neglect of the spin degeneracy, the number N is equal to the total k-space volume enclosed by the Brillouin boundary divided by unit k-cell volume

$$\left(\frac{2\pi}{a}\right)\left(\frac{2\pi}{b}\right)\left(\frac{2\pi}{c}\right) \div \left(\frac{2\pi}{N_1\, a}\right)\left(\frac{2\pi}{N_2\, b}\right)\left(\frac{2\pi}{N_3\, c}\right) = N_1 N_2 N_3, \tag{3.20}$$

which equals the number of the ions in the normalization volume. It is also equal to the number of the conduction electrons in a monovalent metal.

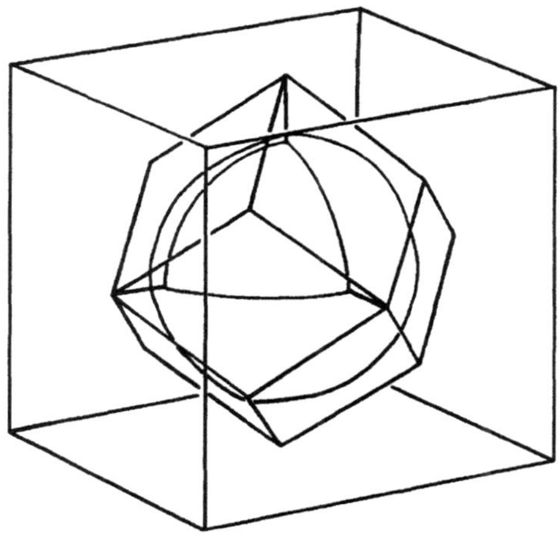

Figure 3.2: The Fermi surface of sodium (bcc) is nearly spherical within the first Brillouin zone.

Thus, *the first Brillouin zone can contain twice (because of spin degeneracy) the number of conduction electrons for the monovalent metal.* This means that at 0 K, half of the Brillouin zone may be filled by electrons. Something similar to this actually happens to alkali metals including Li, Na, K. These metals form bcc lattices. All experiments indicate that the Fermi surface is nearly spherical and entirely within the first Brillouin zone. The Fermi surface of sodium is shown in Fig. 3.2.

The *nearly free electron model* [NFEM] developed by Harrison [3, 4] can predict a Fermi surface for any metal in the first approximation. This model is obtained by applying Heisenberg's uncertainty principle and Pauli's exclusion principle to a solid. Hence it has a general applicability unhindered by the complications due to particle-particle interaction. Briefly in the NFEM, the first Brillouin zone is drawn for a chosen metal. Electrons are filled, starting from the center of the zone, with the assumption of a free-electron dispersion relation. If we apply the NFEM to alkali metals, we simply obtain the Fermi sphere as shown in Fig. 3.2.

Noble metals, including copper (Cu), silver (Ag), and gold (Au), are monovalent fcc metals. The Brillouin zone and Fermi surface of copper are shown in Fig. 3.3. The Fermi surface is far from spherical. Notice that *the Fermi surface approaches the Brillouin boundary at right angles.* This arises from the mirror symmetry possessed by the fcc lattice.

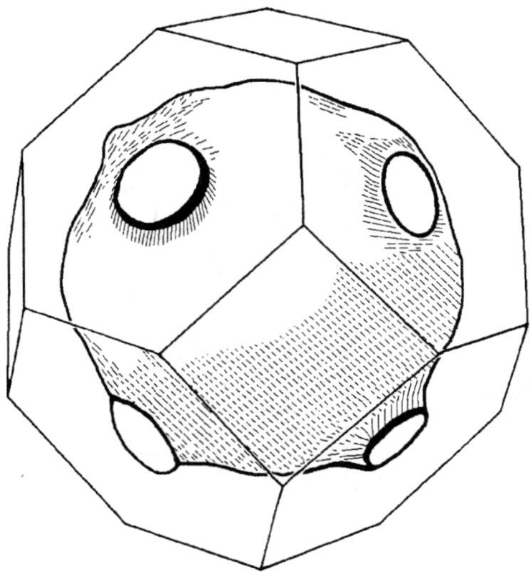

Figure 3.3: The Fermi sphere bulges out in the $\langle 111 \rangle$ direction to make contact with the hexagonal zone faces.

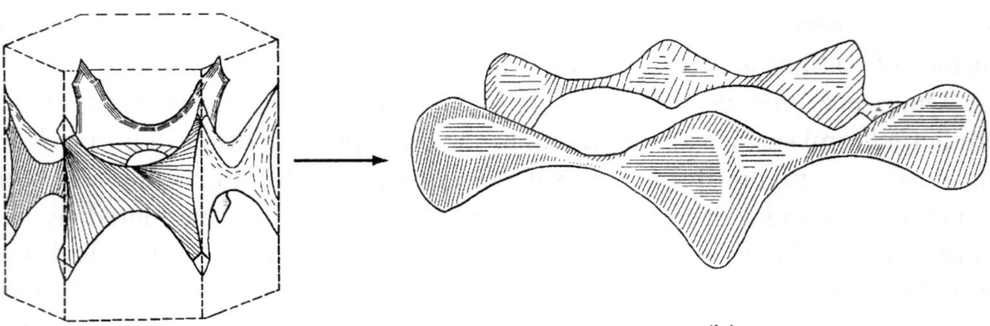

(a) (b)

Figure 3.4: The Fermi surfaces in the second zone for Be. (a) NFEM monster, (b) measured coronet. The coronet encloses unocupied states.

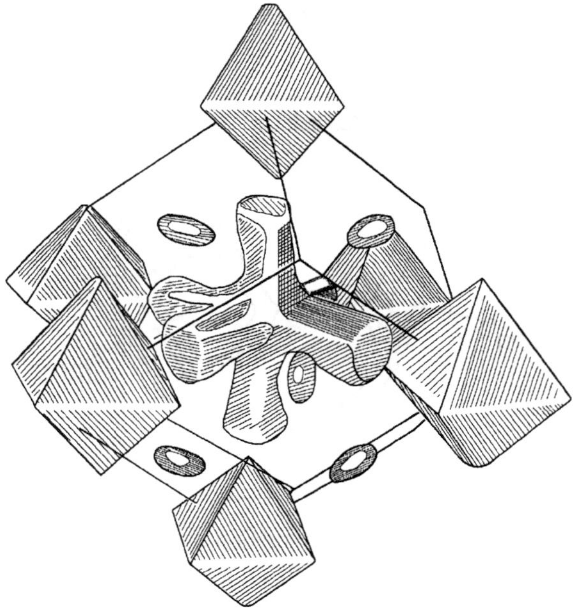

Figure 3.5: The Fermi surface for bcc tungsten. The central figure contains electrons and all other figures contain vacant states, after Schönberg and Gold [6].

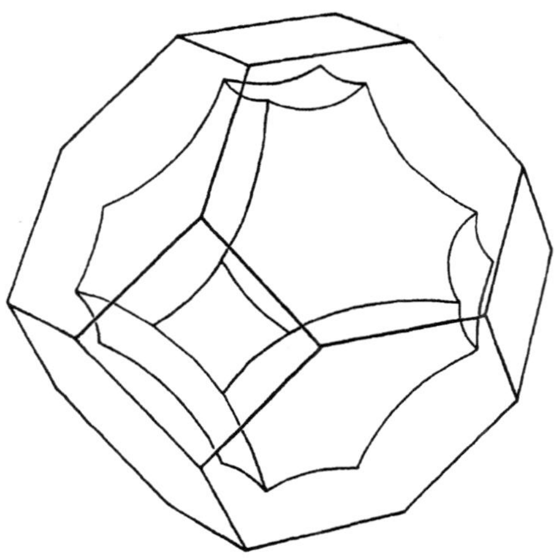

Figure 3.6: The Fermi surface constructed by Harrison's model (NFEM) in the second zone for Al. The convex surface encloses vacant states.

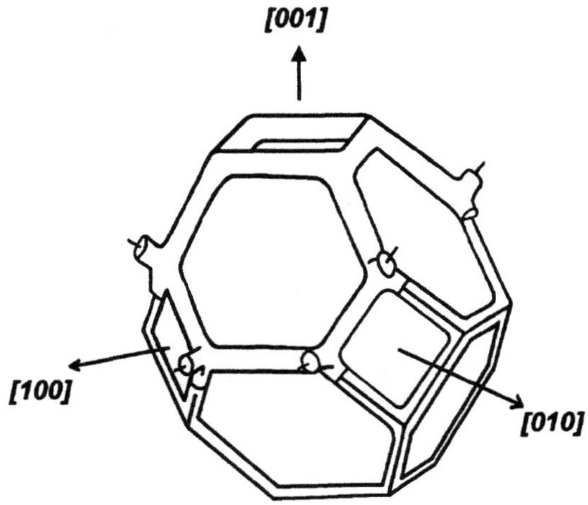

Figure 3.7: The conjectured Fermi surface in the third zone for fcc Pb.

For a divalent metal like calcium (Ca) (fcc), the first Brillouin zone can in principle contain all of the conduction electrons. However, the Fermi surface must approach the zone boundary at right angles, which distorts the ideal configuration considerably. Thus the real Fermi surface for Ca has a set of unfilled corners in the first zone, and the overflow electrons are in the second zone. As a result Ca is a metal, and not an insulator. Besides Ca has "electrons" and "holes". Divalent beryllium (Be) forms a hexagonal closed pack (hcp) crystal. The Fermi surfaces in the second zone (a) constructed in the NFEM and (b) observed [5], are shown in Fig. 3.4. Tungsten (W), which contains *d*-electrons [(xenon) $4f^{14}5d^4 6s^2$] forms a bcc crystal. The conjectured Fermi surface is shown in Fig. 3.5. Let us now consider trivalent aluminum (Al), which forms a fcc lattice. The first Brillouin zone is entirely filled with electrons. The second zone is half filled with electrons, starting with the zone boundary as shown in Fig. 3.6. As another example, we examine the Fermi surface of lead (Pb), which also forms a fcc lattice. Since this metal is quadrivalent and hence has a great number of conduction electrons, the Fermi surface is quite complicated. The conjectured Fermi surface in the third zone is shown in Fig 3.7. For more detailed description of the Fermi surface of metals see standard texts on solid state physics (Refs. [4], [7], and [8]). Al, Be, W and Pb are superconductors, while Na and Cu are not, which will be discussed further in Chapter 8.

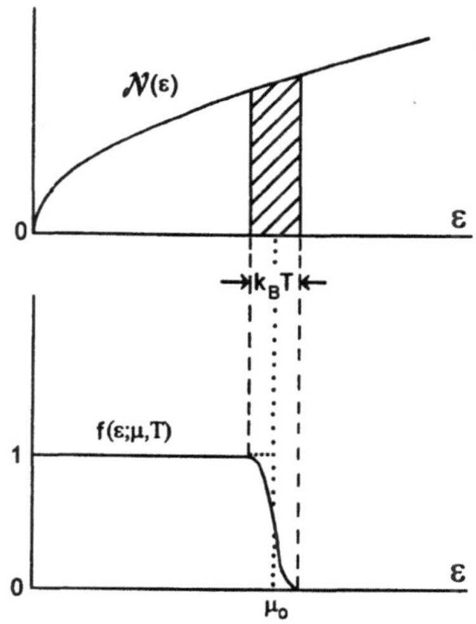

Figure 3.8: The density of states in energy $N(\varepsilon)$ and the Fermi distribution function $f(\varepsilon)$ are drawn as functions of the kinetic energy ε. The change in f is appreciable only near the Fermi energy μ_0 if $k_BT \ll \mu_0$. The shaded area represents approximately the number of thermally excited electrons.

3.4 Heat Capacity and Density of States

The band structures of conduction electrons are quite different from metal to metal. In spite of this, the electronic heat capacities at very low temperatures are all similar, which is shown in this section. We first show that *any normal metal having a sharp Fermi surface has a T-linear heat capacity.* We draw the density of states $\mathcal{N}(\varepsilon)$, and the Fermi distribution function $f(\varepsilon)$ as a function of the kinetic energy ε in Fig. 3.8. The change in $f(\varepsilon)$ is appreciable only near the Fermi energy ε_F. The number of excited electrons, N_x, is estimated by

$$N_x = \mathcal{N}(\varepsilon_F)\mathrm{k}_BT. \tag{3.21}$$

Each thermally excited electron will move up with an extra energy of the order k_BT. The approximate change in the total energy ΔE is given by multiplying these two factors:

$$\Delta E = N_x\,\mathrm{k}_BT = \mathcal{N}(\varepsilon_F)(\mathrm{k}_BT)^2. \tag{3.22}$$

Differentiating this with respect to T, we obtain an expression for the heat capacity:

$$C_V \simeq \frac{\partial}{\partial T}\Delta E = 2\mathrm{k}_B^2 \mathcal{N}(\varepsilon_F)T, \tag{3.23}$$

which indicates the T-linear dependence. This T-dependence comes from the Fermi distribution function. Using the more rigorous calculation, we obtain

$$\boxed{\mathrm{C}_V = (1/3)\pi^2 \mathrm{k}_B^2 \mathcal{N}(\varepsilon_F)\mathrm{T}.} \tag{3.24}$$

The density of states, $\mathcal{N}(\varepsilon_F)$, for any normal metal can be expressed by

$$\mathcal{N}(\varepsilon_F) = \sum_j \frac{2}{(2\pi\hbar)^3} \int d\mathbf{S} \frac{1}{|\boldsymbol{\nabla}\varepsilon_j(\hbar\mathbf{k})|}, \tag{3.25}$$

where the factor 2 is due to the spin degeneracy, and the surface integration is carried out over the Fermi surface represented by

$$\varepsilon_j(\hbar\mathbf{k}) = \varepsilon_F. \tag{3.26}$$

As an illustration, consider a free electron system having the Fermi sphere:

$$\varepsilon = (p_x^2 + p_y^2 + p_z^2)/(2m) \equiv \varepsilon_F. \tag{3.27}$$

The gradient $\boldsymbol{\nabla}\varepsilon(\mathbf{p})$ at any point of the surface has a constant magnitude p_F/m, and the surface integral is equal to $4\pi p_F^2$. Equation (3.25), then, yields

$$\mathcal{N}(\varepsilon_F) = \frac{2V}{(2\pi\hbar)^3}\frac{4\pi p_F^2}{(p_F/m)} = V\frac{2^{1/2}m^{3/2}}{\pi^2\hbar^3}\varepsilon_F^{1/2}. \tag{3.28}$$

As a second example, consider the ellipsoidal surface represented by

$$\varepsilon = \frac{p_x^2}{2m_1} + \frac{p_y^2}{2m_2} + \frac{p_z^2}{2m_3}. \tag{3.29}$$

After elementary calculations, we obtain (Problem 3.4.1).

$$\mathcal{N}(\varepsilon) = V\frac{2^{1/2}}{\pi^2\hbar^3}(m_1 m_2 m_3)^{1/2}\varepsilon^{1/2}, \tag{3.30}$$

which shows that the density of states still grows like $\varepsilon^{1/2}$, but the coefficient depends on the three effective masses (m_1, m_2, m_3).

Problem 3.4.1. (a) Compute the momentum-space volume between the surfaces represented by $\varepsilon = p_x^2 / 2m_1 + p_y^2 / 2m_2 + p_z^2 / 2m_3$, and $\varepsilon + d\varepsilon = p_x'^2/2m_1 + p_y'^2/2m_2 + p_z'^2/2m_3$. By counting the number of quantum states in this volume, obtain Eq. (3.30). (b) Derive Eq. (3.30), starting from the general formula (3.25). Hint: Convert the integral over the ellipsoidal surface into that over a spherical surface.

3.5 Equations of Motion for a Bloch Electron

We discuss, in this section, how conduction electrons respond to the applied electromagnetic fields. [9] Let us recall that in the Fermi liquid model each electron in a crystal moves independently in an extremely weak lattice-periodic effective potential $V_e(\mathbf{r})$:

$$V_e(\mathbf{r} + \mathbf{R}) = V_e(\mathbf{r}). \tag{3.31}$$

We write down the Schrödinger equation:

$$[-\frac{\hbar^2}{2m}\nabla^2 + V(\mathbf{r})]\psi(\mathbf{r}) = E\psi(\mathbf{r}). \tag{3.32}$$

According to Bloch's theorem, the wavefunction ψ satisfies

$$\psi_{j,\mathbf{k}}(\mathbf{r} + \mathbf{R}) = e^{i\mathbf{k}\cdot\mathbf{R}}\psi_{j,\mathbf{k}}(\mathbf{r}). \tag{3.33}$$

The Bravais vector $\mathbf{R}$ can take on only discrete values, and its minimum length can equal the lattice constant a_0. This generates a limitation on the domain in $\mathbf{k}$. For example the values for each k_a $(a = x, y, z)$ for a sc lattice are limited to $(-\pi/a_0, \pi/a_0)$. This means that the Bloch electron's wavelength $\lambda \equiv 2\pi / k$ has a lower bound:

$$\lambda > 2a_{0.}. \tag{3.34}$$

The Bloch electron state is characterized by k-vector $\mathbf{k}$, band index j and energy

$$\varepsilon = \varepsilon_j(\hbar\mathbf{k}) \equiv \varepsilon_j(\mathbf{p}). \tag{3.35}$$

Here we defined the lattice momentum by $\mathbf{p} \equiv \hbar\mathbf{k}$. The energy-momentum or dispersion relation represented by Eq. (3.35) can be probed by transport measurements. A metal is perturbed from the equilibrium condition by an applied electric field; the deviations of the electron distribution from the equilibrium move in the crystal to reach and maintain a stationary state. The deviations, that is, the *localized Bloch wave packets*, should extend over one unit cell or more. This is so because no wave packets constructed from waves whose wavelengths have the lower bounds $2a_0$ can be localized within distances less than a_0.

Dirac demonstrated [9], that for any p-dependence of the kinetic energy $[\varepsilon = \varepsilon_j(\mathbf{p})]$ the center of a quantum wave packet, identified as the position of the corresponding particle, moves in accordance with Hamilton's equations of motion. Hence the Bloch electron representing the wave packet moves classical mechanically under the action of the force averaged over the lattice constants. The lattice force $-\partial V_e/\partial x$ averaged over a unit cell vanishes:

$$\left\langle -\frac{\partial}{\partial x} V_e \right\rangle_{\text{unit cell}} \equiv -a_0^{-3} \int\int dydz \int_0^{a_0} dx \frac{\partial}{\partial x} V_e(x,\, y,\, z) = 0. \tag{3.36}$$

Thus only important forces acting on the Bloch electron are electromagnetic forces.

We now formulate dynamics for the Bloch electron as follows. First, from the quantum principle of *wave-particle duality*, we postulate that

$$(\hbar k_x, \hbar k_y, \hbar k_z) = (p_x,\, p_y,\, p_z) \equiv (p_1,\, p_2,\, p_3) \equiv \mathbf{p}. \tag{3.37}$$

Second, we introduce a model Hamiltonian,

$$H_0(p_1,\, p_2,\, p_3) \equiv \varepsilon_j(\hbar k_1, \hbar k_2, \hbar k_3). \tag{3.38}$$

Third, we generalize our Hamiltonian H to include the electromagnetic interaction energy:

$$H = H_0(\mathbf{p} - q\mathbf{A}) + q\phi, \tag{3.39}$$

where $(\mathbf{A}, \phi)$ are vector and scalar potentials generating electromagnetic fields $(\mathbf{E}, \mathbf{B})$:

$$\mathbf{E} = -\nabla\phi(\mathbf{r}, t) - \frac{\partial \mathbf{A}(\mathbf{r}, t)}{\partial t}, \qquad \mathbf{B} = \nabla \times \mathbf{A}(\mathbf{r}, t), \qquad \mathbf{r} \equiv (x_1, x_2, x_3). \tag{3.40}$$

By using the standard procedures, we then get Hamilton's equations of motion:

$$\dot{\mathbf{r}} \equiv \mathbf{v} = \frac{\partial H}{\partial \mathbf{p}} = \frac{\partial}{\partial \mathbf{p}} H_0(\mathbf{p} - q\mathbf{A}), \quad \dot{\mathbf{p}} = -\frac{\partial H}{\partial \mathbf{r}} = -\frac{\partial H_0}{\partial \mathbf{r}} - q\frac{\partial \phi}{\partial \mathbf{r}}. \tag{3.41}$$

The first equation defines the velocity $\mathbf{v} \equiv (v_1, v_2, v_3)$. Notice that in the zero-field limit these equations are in agreement with the general definition of a group velocity:

$$v_{g,i} \equiv \partial\omega(\mathbf{k})/\partial k_i, \qquad \omega(\mathbf{k}) \equiv \varepsilon(\mathbf{p})/\hbar, \qquad \text{(wave picture)} \tag{3.42}$$

$$v_i \equiv \partial\varepsilon(\mathbf{p})/\partial p_i. \qquad \text{(particle picture)} \tag{3.43}$$

In the presence of magnetic field the first of Eqs. (3.41) gives the velocity $\mathbf{v}$ as a function of $\mathbf{p} - q\mathbf{A}$. Inverting this functional relation, we have

$$\mathbf{p} - q\mathbf{A} = \mathbf{f}(\mathbf{v}). \tag{3.44}$$

Using Eqs. (3.41)-(3.44), we obtain (Problem 3.5.1)

$$\frac{d\mathbf{f}}{dt} = q(\mathbf{E} + \mathbf{v} \times \mathbf{B}). \tag{3.45}$$

Since the vector $\mathbf{f}$ is a function of $\mathbf{v}$, Eq. (3.45) describes how the velocity $\mathbf{v}$ changes by the action of the Lorentz force (the right-hand term).

To see the nature of Eq. (3.45), we take a quadratic dispersion relation represented by

$$\varepsilon = p_1^2/2m_1^* + p_2^2/2m_2^* + p_3^2/2m_3^* + \varepsilon_0, \tag{3.46}$$

where $\{m_j^*\}$ are *effective masses* and ε_0 is a constant. The effective masses $\{m_j^*\}$ may be positive or negative. Depending on their values, the energy surface represented by Eq. (3.46) is ellipsoidal or hyperboloidal. See Fig. 3.9. If the Cartesian axes are taken along the major axes of the ellipsoid, Eq. (3.45) can be written as

$$m_j^* \frac{dv_j}{dt} = q(\mathbf{E} + \mathbf{v} \times \mathbf{B})_j, \tag{3.47}$$

(Problem 3.5.1). These are *Newtonian equations of motion.* Only a set of three effective masses (m_j^*) are introduced. The Bloch electron moves in an anisotropic environment if the effective masses are different.

Let us now go back to the general case. The function $\mathbf{f}$ may be determined from the dispersion relation as follows: Take a point A at the constant-energy surface represented by Eq. (3.35) in the k-space. We choose the positive normal vector to point in the direction in which energy increases. A normal

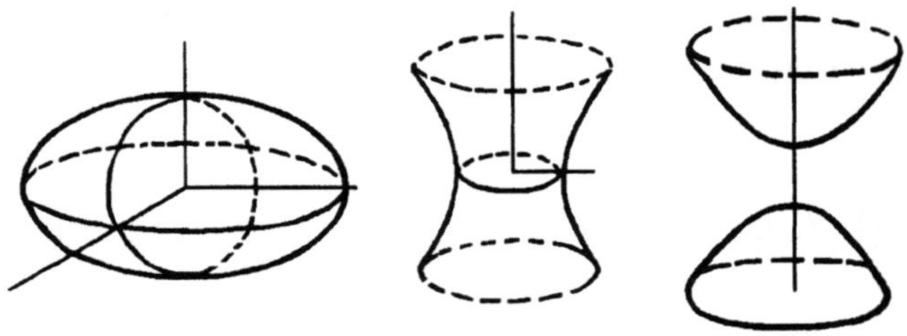

Figure 3.9: (a) Ellipsoid, (b) hyperboloid of one sheet (neck), (c) hyperboloid of two sheets (inverted double caps).

curvature κ is defined as the inverse of the radius of the contact circle at A (in the plane containing the normal vector) times the curvature sign δ_A:

$$\kappa \equiv \delta_A \, R_A^{-1}, \tag{3.48}$$

where δ_A is $+1$ or -1 according to whether the center of the contact circle is on the positive side (which contains the positive normal) or not. In space-surface theory [10], the two planes which contain the greatest and smallest normal curvatures are known to be mutually orthogonal. They are, by construction, orthogonal to the contact plane at A. Therefore, the intersections of these two planes and the contact plane can form a Cartesian set of orthogonal axes with the origin at A, called the *principal axes of curvatures.* By using this property, we define *principal masses* m_i by

$$\frac{1}{m_i} \equiv \frac{\partial^2 \varepsilon}{\partial p_i^2}, \tag{3.49}$$

where dp_i is the differential along the principal axis i. If we choose a Cartesian coordinate system along the principal axes, Eq. (3.45) can be written as

$$\boxed{m_i dv_i/dt = q(\mathbf{E} + \mathbf{v} \times \mathbf{B})_i.} \tag{3.50}$$

Note that these equations are similar to Eqs. (3.47). The principal masses $\{m_i\}$, however, are defined at each point on the constant-energy surface, and hence they depend on $\mathbf{p}$ and $\varepsilon_j(\mathbf{p})$. Let us take a simple example: an ellipsoidal constant-energy surface represented by Eq. (3.46) with all positive m_i^*. At extremal points, e.g. $(p_{1,\max}, 0, 0) = ([2(\varepsilon - \varepsilon_0)m_1^*]^{1/2}, 0, 0)$, the principal axes of curvatures match the major axes of the ellipsoid. Then the principal masses $\{m_i\}$ can simply be expressed in terms of the constant effective masses $\{m_i^*\}$ (Problem 3.5.3).

The proof of the equivalence between (3.45) and (3.50) is carried out as follows. Since f_j are functions of (v_1, v_2, v_3), we obtain

$$\frac{df_i}{dt} = \sum_j \frac{\partial f_i}{\partial v_j}\frac{dv_j}{dt} = \sum_j \left(\frac{\partial v_j}{\partial f_i}\right)^{-1} \frac{dv_j}{dt}. \tag{3.51}$$

The velocities v_i from Eq. (3.41) can be expressed in terms of the first p-derivatives. Thus in the zero-field limit:

$$\frac{\partial v_j}{\partial f_i} \to \frac{\partial^2 \varepsilon}{\partial p_i \partial p_j} \equiv \frac{1}{m_{ij}}, \tag{3.52}$$

which defines the symmetric mass tensor elements $\{m_{ij}\}$. By using Eqs. (3.51) and (3.52), we can re-express Eq. (3.45) as

$$\sum_j m_{ij}\frac{dv_j}{dt} = q(\mathbf{E} + \mathbf{v} \times \mathbf{B})_i\,, \tag{3.53}$$

which is valid in any Cartesian frame of reference. The mass tensor $\{m_{ij}\}$ is real and symmetric, and hence can always be diagonalized by a principal-axes transformation [12]. The principal masses $\{m_i\}$ are given by (3.49) and the principal axes are given by the principal axes of curvature. (q.e.d.)

In Eq. (3.49) the third principal mass m_3 is defined in terms of the second derivative, $\partial^2\varepsilon/\partial p_3^2$, in the energy-increasing (p_3-) direction. The first and second principal masses (m_1, m_2) can be connected with the two principal radii of curvature, $(P_1,\ P_2)$, which by definition equal the inverse principal curvatures (κ_1, κ_2): (Problem 3.5.4.)

$$\frac{1}{m_j} = -\kappa_j v \equiv -\frac{v}{P_j}, \qquad v \equiv |\mathbf{v}|\,, \qquad \frac{1}{P_j} \equiv \kappa_j. \tag{3.54}$$

Eqs. (3.54) is a very useful relation. The signs (definitely) and magnitudes (qualitatively) of the first two principal masses $(m_1,\ m_2)$ can be obtained by a visual inspection of the constant-energy surface, an example of which is the Fermi surface. The sign of the third principal mass m_3 can also be obtained by inspection: the mass m_3 is positive or negative according to whether the center of the contact circle is on the negative or the positive side. For example, the system of free electrons has a spherical constant-energy surface represented by $\varepsilon = p^2/(2m)$ with the normal vector pointing outwards. By inspection the principal radii of curvatures at every point of the surface is negative, and the principal masses (m_1, m_2) are positive and equal to m. The third principal mass m_3 is also positive and equal to m. Eqs. (3.50) were derived from the energy-k relation (3.35) without referring to the Fermi energy. They are valid for all wave vectors $\mathbf{k}$ and all band indices j.

Problem 3.5.1. Assume a quadratic dispersion relation [Eq. (3.46)] and derive Eq. (3.47).

Problem 3.5.2. Assume a general dispersion relation [Eq. (3.38)] and derive Eq. (3.45).

Problem 3.5.3. Consider the ellipsoidal constant-energy surface represented by Eq. (3.46) with all $m_i^* > 0$. At the six extremal points, the principal axes of curvatures match the major axes of the ellipsoid. Demonstrate that the principal masses $\{m_i\}$ at one of these points can be expressed simply in terms of the effective masses $\{m_j^*\}$.

Problem 3.5.4. Verify Eq. (3.54). Consider first the 2D Fermi circle, and then a general case.

Chapter 4

Phonon-Exchange Attraction

The cause of superconductivity is the phonon-exchange attraction between a pair of electrons, which is derived using quantum perturbation theory.

4.1 Phonons and Lattice Dynamics

In the present section, we review a general theory of the heat capacity based on lattice dynamics.

Let us take a crystal composed of N atoms. The potential energy V depends on the configuration of N atoms $(\mathbf{r}_1, \mathbf{r}_2, \cdots, \mathbf{r}_N)$. We regard this energy V as a function of the displacements of the atoms,

$$\mathbf{u}_j \equiv \mathbf{r}_j - \mathbf{r}_j^{(0)}, \tag{4.1}$$

measured from the equilibrium positions $\mathbf{r}_j^{(0)}$. Let us expand the potential $V = V(\mathbf{u}_1, \mathbf{u}_2, \cdots, \mathbf{u}_N) \equiv V(u_{1x}, u_{1y}, u_{1z}, u_{2x}, \cdots)$ in terms of small displacements $\{u_{j\mu}\}$:

$$V = V_0 + \sum_j \sum_{\mu=x,y,z} u_{j\mu} \left[\frac{\partial V}{\partial u_{j\mu}}\right]_0 + \frac{1}{2} \sum_j \sum_\mu \sum_k \sum_\nu u_{j\mu} u_{k\nu} \left[\frac{\partial^2 V}{\partial u_{j\mu} \partial u_{k\nu}}\right]_0 + \cdots, \tag{4.2}$$

where all partial derivatives are evaluated at $\mathbf{u}_1 = \mathbf{u}_2 = \cdots = 0$, which is indicated by subscripts 0. We may set the constant V_0 equal to zero with no loss of rigor. By assumption, the lattice is stable at the equilibrium configuration. Then the potential V must have a minimum, requiring that the first-order derivatives vanish:

$$\left[\frac{\partial V}{\partial u_{j\mu}}\right]_0 = 0. \tag{4.3}$$

For small oscillations we may keep terms of the second order in $u_{j\mu}$ only. We then have

$$V \simeq V' \equiv \sum_j \sum_\mu \sum_k \sum_\nu \frac{1}{2} A_{j\mu k\nu} u_{j\mu} u_{k\nu}, \tag{4.4}$$

$$A_{j\mu k\nu} \equiv \left[\frac{\partial^2 V}{\partial u_{j\mu}\, \partial u_{k\nu}}\right]_0. \tag{4.5}$$

The prime ($'$) indicating the *harmonic approximation* will be dropped hereafter. The kinetic energy of the system is

$$T \equiv \sum_j \frac{m}{2} \dot{r}_j^2 = \sum_j \frac{m}{2} \dot{u}_j^2 \equiv \sum_j \sum_\mu \frac{m}{2} \dot{u}_{j\mu}^2. \tag{4.6}$$

We can now write down the Lagrangian $L \equiv T - V$ as

$$L = \sum_j \sum_\mu \frac{m}{2} \dot{u}_{j\mu}^2 - \sum_j \sum_\mu \sum_k \sum_\nu \frac{1}{2} A_{j\mu k\nu} u_{j\mu} u_{k\nu}. \tag{4.7}$$

This Lagrangian L in the harmonic approximation is quadratic in $u_{j\mu}$ and $\dot{u}_{j\mu}$. According to theory of the *principal-axis transformation* [1], we can in principle transform the Hamiltonian (total energy) $H = T + V$ for the system into the sum of the energies of the normal modes of oscillations:

$$H = \sum_{\kappa=1}^{3N} \frac{1}{2}(P_\kappa^2 + \omega_\kappa^2 Q_\kappa^2), \tag{4.8}$$

where $\{Q_\kappa, P_\kappa\}$ are the normal coordinates and momenta, and ω_κ are characteristic frequencies. Note: there are exactly $3N$ normal modes.

Let us first calculate the heat capacity by means of classical statistical mechanics. This Hamiltonian H is quadratic in canonical variables (Q_κ, P_κ). Hence the equipartition theorem holds. We multiply the average thermal energy for each mode ($\mathrm{k}_B T$) by the number of modes $3N$ and obtain $3N\mathrm{k}_B T$ for the average energy $\langle H \rangle$. Differentiating this with respect to T, we obtain $3N\mathrm{k}_B$ for the heat capacity, in agreement with Dulong-Petit's law. It is interesting to observe that we obtained this result without knowing the actual distribution of normal-mode frequencies. The fact that there are $3N$ normal modes played an important role.

Let us now use quantum theory and calculate the heat capacity based on formula (4.8). The energy eigenvalues of the Hamiltonian H are given by

$$E[\{n_\kappa\}] = \sum_\kappa (\frac{1}{2} + n_\kappa)\hbar\omega_\kappa. \qquad n_\kappa = 0, 1, 2, \cdots . \tag{4.9}$$

We can interpret Eq. (4.9) in terms of *phonons* as follows: the energy of the lattice vibrations is characterized by the set of the numbers of phonons $\{n_\kappa\}$ in the normal modes $\{\kappa\}$. Taking the canonical-ensemble average of Eq. (4.9), we obtain

$$\langle E[\{n\}]\rangle = \sum_\kappa [\frac{1}{2} + \langle n_\kappa\rangle]\hbar\omega_\kappa = \sum_\kappa [\frac{1}{2} + f_0(\hbar\omega_\kappa)]\hbar\omega_\kappa \equiv E(T), \tag{4.10}$$

where

$$f_0(\varepsilon) \equiv \frac{1}{\exp(\varepsilon/\mathrm{k}_B T) - 1} \tag{4.11}$$

is the Planck distribution function.

The normal-modes frequencies $\{\omega_\kappa\}$ depend on the normalization volume V, and they are densely populated for large V. In the bulk limit, we may convert the sum over the normal modes into a frequency integral and obtain

$$E(T) = E_0 + \int_0^\infty d\omega\, \hbar\omega\, f_0(\hbar\omega)\mathcal{D}(\omega), \tag{4.12}$$

$$E_0 \equiv \frac{1}{2}\int_0^\infty d\omega\, \hbar\omega\, \mathcal{D}(\omega), \tag{4.13}$$

where $\mathcal{D}(\omega)$ is the density of states (modes) in angular frequency defined such that

$$\text{number of modes in the interval } (\omega,\, \omega + d\omega) \equiv \mathcal{D}(\omega)d\omega. \tag{4.14}$$

The constant E_0 represents a temperature-independent zero-point energy.

Differentiating $E(T)$ with respect to T, we obtain for the heat capacity

$$C_V = \left[\frac{\partial E}{\partial T}\right]_V = \int_0^\infty d\omega\, \hbar\omega\, \frac{\partial f_0(\hbar\omega)}{\partial T}\mathcal{D}(\omega). \tag{4.15}$$

This expression was obtained under the harmonic approximation only, which is good at very low temperatures.

To proceed further, we have to know the density of normal modes, $\mathcal{D}(\omega)$. To find the set of characteristic frequencies $\{\omega_\kappa\}$ requires solving an algebraic equation of $3N$-th order, and we need the frequency distribution for large N. This is not a simple matter. In fact, a branch of mathematical physics whose principal aim is to find the frequency distribution, is called *lattice dynamics*. Fig. 4.1 represents a result obtained by Walker [2] after the analysis of the X-ray scattering data for aluminum, based on lattice dynamics. Some remarkable features of the curve are

(A) At low frequencies

$$D(\omega) \propto \omega^2. \tag{4.16}$$

(B) There exists a maximum frequency ω_m such that

$$D(\omega) = 0 \qquad \text{for} \quad \omega \geq \omega_m. \tag{4.17}$$

(C) A few sharp peaks exist below ω_m.

The feature **(A)** is common to all crystals. The low frequency modes can be described adequately in terms of longitudinal and transverse elastic waves. This region can be represented very well by the Debye's continuum model, indicated by the broken line. The feature **(B)** is connected with the lattice structure. Briefly, no normal modes of extreme short wavelengths (extreme high frequencies) exist. There is a limit frequency ω_m. Sharp peaks were first predicted by Van Hove [3] on topological grounds, and these peaks are often referred to as *van Hove singularities.* As we will see later, the cause of superconductivity lies in the electron-phonon interaction. The microscopic theory however can be formulated in terms of the generalized BCS Hamiltonian, see Section 7.1, where all phonon variables are eliminated. In this sense the details of lattice dynamics are secondary to our main concern. The following point, however, is noteworthy. All lattice dynamical calculations start with the assumption of a real crystal lattice. For example, to treat aluminum, we start with a fcc lattice having empirically known lattice constants. The equations of motion for a set of ions are solved under the assumption of a periodic lattice-box boundary condition. Thus the k-vectors used in both lattice dynamics and Bloch electron dynamics are the same. The domain of

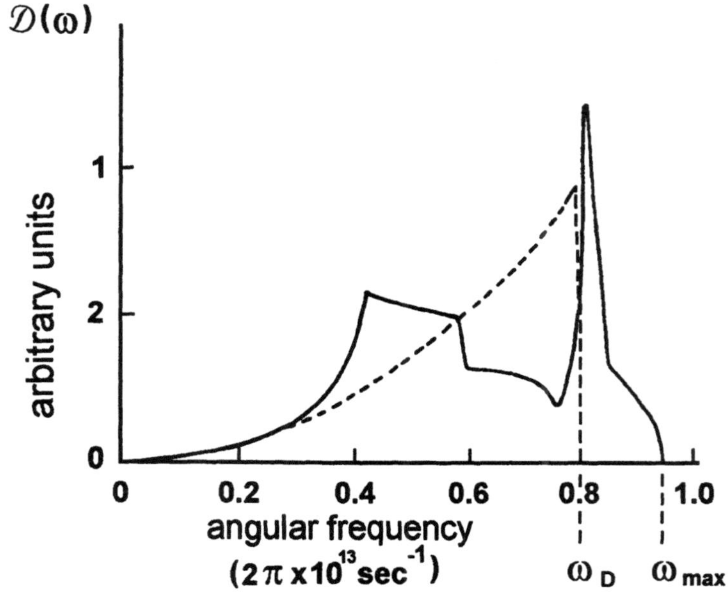

Figure 4.1: The density of normal modes in the angular frequency for aluminum. The solid curve represents the data deduced from X-ray scattering measurements due to Walker [2]. The broken lines indicate the Debye distribution with $\Theta_D = 328\,\text{K}$.

the k-vectors can be restricted to the same first Brillouin zone. Colloquially speaking, phonons (bosons) and electrons (fermions) live together in the same Brillouin zone, which is equivalent to say that electrons and phonons share the same house (crystal lattice). This affinity between electrons and phonons makes the conservation of momentum in the electron-phonon interaction physically meaningful. Thus the fact that the electron-phonon interaction is the cause of superconductivity is not accidental.

4.2 Electron-Phonon Interaction

A crystal lattice is composed of a regular arrays of ions. If the ions move, then the electrons must move in a changing potential field. Fröhlich proposed an interaction Hamiltonian, which is especially suitable for the transport and superconductivity problems. In the present section we derive the Fröhlich

Hamiltonian [4].

For simplicity let us take a simple cubic (sc) lattice. The normal modes of oscillations for a solid are longitudinal and transverse running waves characterized by wave vector $\mathbf{q}$ and frequency ω_q. First, consider the case of a longitudinal wave proceeding in the crystal axis x, which is represented by

$$u_q \exp(-i\omega_q t + i\mathbf{q} \cdot \mathbf{r}) = u_q \exp(-i\omega_q t + iqx), \tag{4.18}$$

where u_q is the displacement in the x-direction. The wavelength $\lambda \equiv 2\pi/q$ ($> 2a_0$) is greater than twice the lattice constant a_0. The case: $\lambda = 12\, a_0$ is shown in Fig. 4.2.

If we imagine a set of parallel plates containing a great number of ions fixed in each plate, we have a realistic picture of the lattice vibration mode. From Fig. 4.2 we see that the density of ions changes in the x-direction. Hence the longitudinal modes are also called the *density-wave* modes. The transverse wave mode can also be pictured from Fig. 4.2 by imagining a set of parallel plates containing a great number of ions fixed in each plate and assuming the transverse displacements of the plates. Notice that this mode generates no charge-density variation.

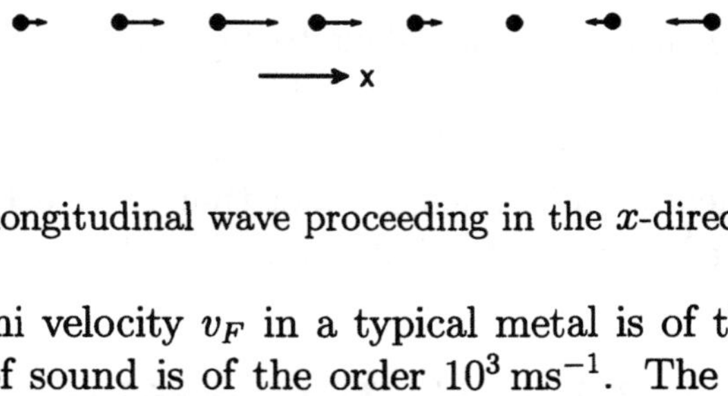

Figure 4.2: A longitudinal wave proceeding in the x-direction; $\lambda = 12\, a_0$.

Now, the Fermi velocity v_F in a typical metal is of the order 10^6 ms^{-1} while the speed of sound is of the order 10^3 ms^{-1}. The electrons are then likely to move quickly to negate any electric field generated by the density variations associated with the lattice wave. Hence the electrons may follow the lattice waves quite easily. Given a traveling normal mode in Eq. (4.18), we may assume an electron density deviation of the form:

$$C_\mathbf{q} \exp(-i\omega_q t + i\mathbf{q} \cdot \mathbf{r}), \tag{4.19}$$

Since electrons follow phonons immediately for all ω_q, the factor $C_\mathbf{q}$ can be regarded as independent of ω_q. We further assume that the deviation is linear in $\mathbf{u}_q \cdot \mathbf{q} = qu_\mathbf{q}$ and again in the electron density $n(\mathbf{r})$. Thus

$$C_\mathbf{q} = A_\mathbf{q} \, qu_\mathbf{q} \, n(\mathbf{r}). \tag{4.20}$$

This is called the *deformation potential approximation.* The dynamic response factor $A_\mathbf{q}$ is necessarily complex since there is a time delay between the field (cause) and the density variation (result). The traveling wave is represented by the exponential form (4.19). Complex conjugation of this equation yields $C_\mathbf{q}^* \exp(i\omega_q t - i\mathbf{q}\cdot\mathbf{r})$. Using this form we can reformulate the electron's response, but the physics must be the same. From this cosideration we obtain (Problem 4.2.1)

$$A_\mathbf{q} = A^*_{-\mathbf{q}}. \tag{4.21}$$

We can express the electron density (field) by

$$n(\mathbf{r}) = \psi^\dagger(\mathbf{r})\psi(\mathbf{r}), \tag{4.22}$$

where $\psi(\mathbf{r})$ and $\psi^\dagger(\mathbf{r})$ are annihilation and creation electron field operators.

We now construct an interaction Hamiltonian H_F, which has the dimension of an energy and which is Hermitean. We propose

$$H = \int d^3r \sum_\mathbf{q} \frac{1}{2}[A_\mathbf{q} \, qu_\mathbf{q} \exp(i\mathbf{q}\cdot\mathbf{r})\psi^\dagger(\mathbf{r})\psi(\mathbf{r}) + h.c.], \tag{4.23}$$

where $h.c.$ denotes the hermitian conjugate. Classically, the displacement $u_\mathbf{q}$ changes, following the harmonic equation of motion:

$$\ddot{u}_\mathbf{q} + \omega_q^2 u_\mathbf{q} = 0. \tag{4.24}$$

Let us write the corresponding Hamiltonian for each mode as

$$H = \frac{1}{2}(p^2 + \omega^2 q^2), \qquad q \equiv u, \qquad p \equiv \dot{q}, \qquad \omega_q \equiv \omega, \tag{4.25}$$

where we dropped the mode index $\mathbf{q}$. If we assume the same Hamiltonian H and the basic commutation relations

$$[q, p] = ih, \qquad [q, q] = [p, p] = 0, \tag{4.26}$$

the quantum description of a harmonic oscillator is complete. The equations of motion are

$$\dot{q} = \frac{1}{i\hbar}[q,\, H] = p, \qquad \dot{p} = \frac{1}{i\hbar}[p,\, H] = -\omega^2 q. \tag{4.27}$$

(Problem 4.2.2). We introduce the dimensionless complex dynamical variables:

$$a^\dagger \equiv (2\hbar\omega)^{-1/2}(p + i\omega q), \qquad a \equiv (2\hbar\omega)^{-1/2}(p - i\omega q). \tag{4.28}$$

Using Eq. (4.27) we obtain

$$\dot{a}^\dagger \equiv (2\hbar\omega)^{-1/2}(-\omega^2 q + i\omega p) = i\omega a^\dagger, \qquad \dot{a} = -i\omega a. \tag{4.29}$$

We can express $(q,\, p)$ in terms of $(a^\dagger,\, a)$:

$$q = -i(\hbar/2\omega)^{1/2}(a^\dagger - a), \qquad p = (\hbar\omega/2)^{1/2}(a^\dagger + a). \tag{4.30}$$

Thus we may work entirely in terms of $(a^\dagger,\, a)$. After straightforward calculations, we obtain [Problem 4.2.3]

$$\begin{aligned} \hbar\omega a^\dagger a &= (2)^{-1}(p + i\omega q)(p - i\omega q) \\ &= (2)^{-1}[p^2 + \omega^2 q^2 + i\omega(qp - pq)] = H - \frac{1}{2}\hbar\omega. \end{aligned}$$

$$\hbar\omega\, a a^\dagger = H + \frac{1}{2}\hbar\omega \tag{4.31}$$

$$a a^\dagger - a^\dagger a \equiv [a,\, a^\dagger] = 1 \tag{4.32}$$

$$H = \frac{1}{2}\hbar\omega(a^\dagger a + a a^\dagger) = \hbar\omega(a^\dagger a + \frac{1}{2}) \equiv \hbar\omega(n + \frac{1}{2}). \tag{4.33}$$

The operators $(a^\dagger,\, a)$ satisfy the Bose commutation rules. We can therefore use second quantization algebras in sections A.1-A.2 and obtain

- Eigenvalues of $n \equiv a^\dagger a :\ n' = 0, 1, 2, \cdots$ [see Eq. (A.1.1)]
- Vacuum ket $|\phi\rangle :\ a\,|\phi\rangle = 0$ [see Eq. (A.1.14)]
- Eigenkets of nT : $|\phi\rangle,\ a^\dagger\,|\phi\rangle$, $(a^\dagger)^2\,|\phi\rangle \cdots$ having the eigenvalues $0, 1, 2, \cdots$ [see Eq. (A.1.16)]

- Eigenvalues of H : $\frac{1}{2}\hbar\omega, \frac{3}{2}\hbar\omega, \frac{5}{2}\hbar\omega, \cdots$

In summary, the quantum Hamiltonian and the quantum states of a harmonic oscillator can be simply described in terms of the bosonic second quantized operators.

We now go back to the case of the lattice normal modes. Each normal mode corresponds to a harmonic oscillator characterized by $(\mathbf{q}, \omega_q)$. The displacements $u_\mathbf{q}$ can be expressed as

$$u_\mathbf{q} = i\left(\frac{\hbar}{2\omega_q}\right)^{1/2}(a_\mathbf{q} - a_\mathbf{q}^\dagger), \tag{4.34}$$

where $(a_\mathbf{q}, a_\mathbf{q}^\dagger)$ are operators satisfying the Bose commutation rules:

$$[a_\mathbf{q}, a_\mathbf{p}^\dagger] \equiv a_\mathbf{q}a_\mathbf{p}^\dagger - a_\mathbf{p}^\dagger a_\mathbf{q} = \delta_{\mathbf{pq}}, \qquad [a_\mathbf{q}, a_\mathbf{p}] = [a_\mathbf{q}^\dagger, a_\mathbf{p}^\dagger] = 0. \tag{4.35}$$

The field operators ψ $(\psi^\dagger)$ can be expanded in terms of the momentum-state electron operators $c_\mathbf{q}$ $(c_\mathbf{q}^\dagger)$:

$$\psi(\mathbf{r}) = \frac{1}{(V)^{1/2}}\sum_\mathbf{q}\exp(i\mathbf{q}\cdot\mathbf{r})c_\mathbf{q}, \qquad \psi^\dagger(\mathbf{r}) = \frac{1}{(V)^{1/2}}\sum_\mathbf{q}\exp(-i\mathbf{q}\cdot\mathbf{r})c_\mathbf{q}^\dagger. \tag{4.36}$$

Using Eqs. (4.20), (4.34) and (4.36), we can reexpress Eq. (4.23) as (Problem 4.2.3):

$$H_F = \sum_\mathbf{k}\sum_\mathbf{q}(V_q\, c_{\mathbf{k}+\mathbf{q}}^\dagger c_\mathbf{k}\, a_\mathbf{q} + h.c.), \qquad V_q \equiv A_q(\hbar/2\omega_q)^{1/2}iq. \tag{4.37}$$

This is the *Fröhlich Hamiltonian.* Electrons describable in terms of $c_\mathbf{k}$'s are now coupled with phonons describable in terms of $a_\mathbf{q}$'s. The term

$$V_q\, c_{\mathbf{k}+\mathbf{q}}^\dagger c_\mathbf{k}\, a_\mathbf{q} \qquad (V_q^*\, c_\mathbf{k}^\dagger c_{\mathbf{k}+\mathbf{q}}\, a_\mathbf{q}^\dagger)$$

can be pictured as an interaction process in which a phonon is absorbed (emitted) by an electron as represented by the Feynman diagram [5] in Fig. 4.3. (a) [(b)]. Note: At each vertex the momentum is conserved. The Fröhlich Hamiltonian H_F is applicable for longitudinal phonons only. As noted earlier, the transverse lattice normal modes generate no charge density variations, making the electron-transverse-phonon interaction negligible.

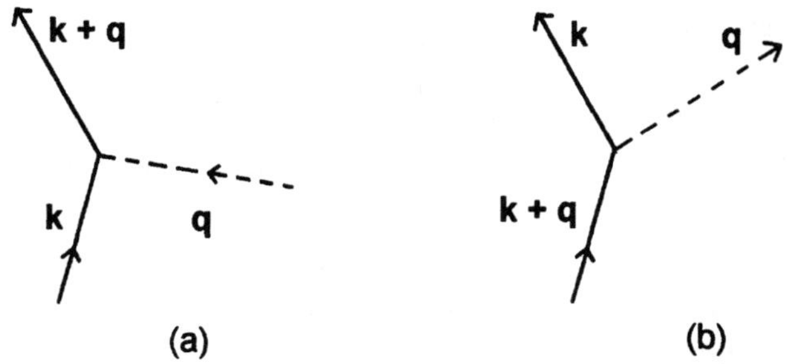

Figure 4.3: Feynman diagrams representing (a) absorption and (b) emission of a phonon by an electron.

Problem 4.2.1. Prove Eq. (4.21).

Problem 4.2.2. Verify Eq. (4.27).

Problem 4.2.3. Verify Eq. (4.31).

Problem 4.2.4. Verify Eq. (4.37).

4.3 Phonon-Exchange Attraction

By exchanging a phonon, two electrons can gain an attraction under a certain condition. In this section we treat this effect by using the many-body perturbation method.

Let us consider an *electron-phonon system* characterized by

$$\begin{aligned} H &= \sum_{\mathbf{k}}\sum_{s} \varepsilon_k c^\dagger_{\mathbf{k}s} c_{\mathbf{k}s} + \sum_{\mathbf{q}} \hbar\omega_q \left(\frac{1}{2} + a^\dagger_{\mathbf{q}} a_{\mathbf{q}}\right) \\ &\quad + \lambda \sum_{\mathbf{k}}\sum_{s}\sum_{\mathbf{q}} (V_q\, a_{\mathbf{q}}\, c^\dagger_{\mathbf{k}+\mathbf{q}s} c_{\mathbf{k}s} + h.c.) \\ &\equiv H_{el} + H_{ph} + \lambda H_F \equiv H_0 + \lambda V, \qquad (V \equiv H_F) \end{aligned} \tag{4.38}$$

where the three sums represent: the total electron kinetic energy (H_{el}), the total phonon energy (H_{ph}), and the Fröhlich interaction Hamiltonian H_F, [see (4.37)].

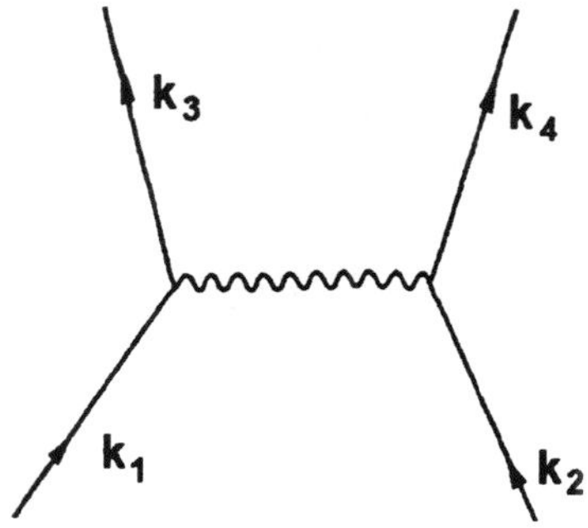

Figure 4.4: The Coulomb interaction represented by the horizontal wavy line generates the change in the momenta of two electrons.

For comparison we consider an *electron gas system* characterized by the Hamiltonian

$$H_c = \sum_{\mathbf{k}} \sum_{s} \varepsilon_k \, c^\dagger_{\mathbf{k}s} c_{\mathbf{k}s} + \frac{1}{2} \sum_{\mathbf{k}_1 s_1} \cdots \sum_{\mathbf{k}_4 s_4} \langle 3,4|\, v_c \,|1,2\rangle \; c^\dagger_4 c^\dagger_3 c_1 c_2 \equiv H_{el} + V_c, \tag{4.39}$$

$$\begin{aligned} \langle 3,4|\, v_c \,|1,2\rangle &\equiv \langle \mathbf{k}_3 s_3, \mathbf{k}_4 s_4 \mid v_c \mid \mathbf{k}_1 s_1, \mathbf{k}_2 s_2 \rangle \\ &= \frac{4\pi e^2 k_0}{V} \frac{1}{\mathbf{q}^2} \delta_{\mathbf{k}_1+\mathbf{k}_2, \mathbf{k}_3+\mathbf{k}_4} \delta_{\mathbf{k}_1-\mathbf{k}_3, \mathbf{q}} \, \delta_{s_3 s_1} \delta_{s_4 s_2}. \end{aligned} \tag{4.40}$$

The elementary interaction process can be represented by a diagram in Fig. 4.4. The wavy horizontal line represents the instantaneous Coulomb interaction v_c. The net momentum of a pair of electrons is conserved:

$$\mathbf{k}_1 + \mathbf{k}_2 = \mathbf{k}_3 + \mathbf{k}_4, \tag{4.41}$$

as seen by the appearance of the Kronecker's delta in Eq. (4.40). Physically, the Coulomb force between a pair of electrons is an internal force, and hence it cannot change the net momentum.

We wish to find an *effective* Hamiltonian v_e between a pair of electrons generated by a phonon exchange. If we look for this v_e in the second order in the coupling constant λ, the likely candidates may be represented by two Feynman diagrams in Fig. 4.5. Here, the time is measured upward. (Historically, Feynman represented the elementary interaction processes in the Dirac Picture (DP) by diagrams. Such a diagram representation is very popular and widely used in quantum field theory [5].) In the diagrams in Fig. 4.5, we follow the motion of two electrons. We may therefore consider a system of two electrons and obtain the effective Hamiltonian v_e through a study of the evolution of two-body density operator ρ_2. For brevity we shall hereafter drop the subscript 2 on ρ indicating two-body system.

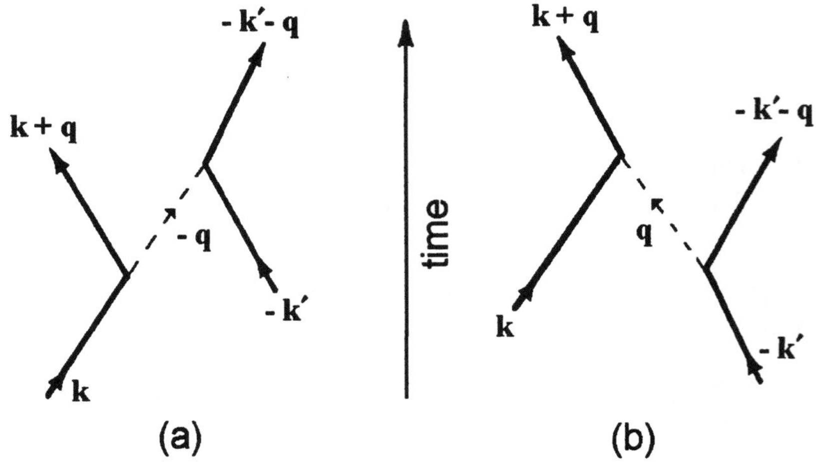

Figure 4.5: A one-phonon-exchange process generates the change in the momenta of two electrons similar to that caused by the Coulomb interaction.

The system-density operator $\rho(t)$ changes in time, following the quantum Liouville equation:

$$i\hbar\frac{\partial\rho(t)}{\partial t} = [H,\, \rho] \equiv \mathcal{H}\rho. \tag{4.42}$$

We assume the Hamiltonian H in Eq. (4.38), and study the time evolution of $\rho(t)$, using quantum many-body perturbation theory. We sketch only the important steps; more detailed calculations were given in Fujita-Godoy's book 1,[6].

Let us introduce a quantum Liouville operators

$$\mathcal{H} \equiv \mathcal{H}_0 + \lambda V, \tag{4.43}$$

which generates a commutator upon acting on ρ, see Eq. (4.42). We assume that the initial-density operator ρ_0 for the combined electron-phonon system can be written as

$$\rho_0 = \rho_{\mathrm{electron}}\, \rho_{\mathrm{phonon}}, \tag{4.44}$$

which is reasonable at 0 K, where there are no real phonons and only virtual phonons are involved in the dynamical processes. We can then choose

$$\rho_{\text{phonon}} = |0\rangle\langle 0|, \tag{4.45}$$

where $|0\rangle$ is the vacuum-state ket for phonons:

$$a_{\mathbf{q}}|0\rangle = 0 \qquad \text{for any } \mathbf{q}. \tag{4.46}$$

The phonon vacuum average will be denoted by an upper bar or angular brackets:

$$\overline{\rho}(t) \equiv \langle 0|\rho(t)|0\rangle \equiv \langle \rho(t)\rangle_{av}. \tag{4.47}$$

Using a time-dependent perturbation theory and taking a phonon-average, we obtain from Eq. (4.42)

$$\frac{\partial \overline{\rho}(t)}{\partial t} = -\frac{\lambda^2}{\hbar^2}\int_0^t d\tau \, \langle \mathcal{V}\exp(-i\tau\hbar^{-1}\mathcal{H}_0)\mathcal{V}\,\rho(t-\tau)\rangle_{av}. \tag{4.48}$$

In the weak-coupling approximation, we may calculate the phonon-exchange effect to the lowest (second) order in λ so that

$$\lambda^2\langle \mathcal{V}\exp(-i\tau\hbar^{-1}\mathcal{H}_0)\mathcal{V}\,\rho(t-\tau)\rangle = \lambda^2\langle \mathcal{V}\exp(-i\tau\hbar^{-1}\mathcal{H}_0)\mathcal{V}\rangle_{av}\,\overline{\rho}(t-\tau). \tag{4.49}$$

In the Markoffian approximation we may replace $\overline{\rho}(t-\tau)$ by $\overline{\rho}(t)$ and take the upper limit t of the τ-integration to ∞. Using these two approximations, we obtain from Eq. (4.48)

$$\frac{\partial \overline{\rho}(t)}{\partial t} = i\lambda^2\hbar^{-1}\lim_{a\to 0}\langle \mathcal{V}(\mathcal{H}_0 - ia)^{-1}\mathcal{V}\rangle_{av}\,\overline{\rho}(t), \qquad a > 0. \tag{4.50}$$

Let us now take momentum-state matrix elements of Eq. (4.50). The lhs is

$$\frac{\partial}{\partial t}\langle \mathbf{k}_1 s_1, \mathbf{k}_2 s_2|\rho(t)|\mathbf{k}_3 s_3, \mathbf{k}_4 s_4\rangle \equiv \frac{\partial}{\partial t}\rho(1,2;3,4,t), \tag{4.51}$$

where we dropped the upper bar indicating the phonon average. The rhs requires more sophisticated computations since the Liouville operators $(\mathcal{V}, \mathcal{H}_0)$ are involved. After lengthy but straightforward calculations, we obtain from Eq. (4.50)

$$\begin{aligned}\frac{\partial}{\partial t}\rho(1,2;3,4,t) \;=\; & \sum_{\mathbf{k}_5 s_5}\sum_{\mathbf{k}_6 s_6} -i\hbar^{-1}[\langle 1,2|v_e|5,6\rangle\,\rho_2(5,6;3,4,t) \\ & - \langle 5,6|v_e|3,4\rangle\,\rho_2(1,2;5,6,t)]\end{aligned} \tag{4.52}$$

$$\langle 3,4|\, v_e \,|1,2\rangle \equiv |V_q|^2 \frac{\hbar\omega_q}{(\varepsilon_3 - \varepsilon_1)^2 - \hbar^2\omega_q^2} \delta_{\mathbf{k}_1+\mathbf{k}_2,\, \mathbf{k}_3+\mathbf{k}_4} \delta_{\mathbf{k}_3-\mathbf{k}_1,\mathbf{q}}\, \delta_{s_3 s_1} \delta_{s_4 s_2}. \quad (4.53)$$

Kronecker's delta $\delta_{\mathbf{k}_1+\mathbf{k}_2,\, \mathbf{k}_3+\mathbf{k}_4}$ in (4.53) means that the net momentum is conserved, since the phonon-exchange interaction is an internal interaction.

For comparison, consider the electron-gas system. The two-electron density matrix ρ_c for this system changes, following

$$\begin{aligned} \frac{\partial}{\partial t}\rho_c(1,2;3,4,t) &= \sum_{\mathbf{k}_5 s_5}\sum_{\mathbf{k}_6 s_6} -i\hbar^{-1}[\langle 1,2|\, v_c \,|5,6\rangle\, \rho_c(5,6;3,4,t) \\ &\quad - \langle 5,6|\, v_c \,|3,4\rangle\, \rho_c(1,2;5,6,t)] \end{aligned} \quad (4.54)$$

which is of the same form as Eq. (4.52). The only differences are in the interaction matrix elements. Comparison between Eqs. (4.40) and (4.53) yields

$$\frac{4\pi e^2 k_0}{V}\frac{1}{q^2} \qquad \text{(Coulomb interaction),} \quad (4.55)$$

$$|V_q|^2 \frac{\hbar\omega_q}{(\varepsilon_{\mathbf{k}_1+\mathbf{q}} - \varepsilon_{\mathbf{k}_1})^2 - \hbar^2\omega_q^2} \qquad \text{(phonon-exchange interaction).} \quad (4.56)$$

In our derivation, the weak-coupling and the Markoffian approximation were used. The Markoffian approximation is justified in the steady state condition in which the effect of the duration of interaction can be neglected. The electron mass is four orders of magnitude smaller than the lattice-ion mass, and hence the coupling between the electron and ionic motion must be small by the mass mismatch. Thus expression (4.56) is highly accurate for the effective phonon-exchange interaction at 0 K. This expression has remarkable features. First, it depends on the phonon energy $\hbar\omega_q$. Second, it depends on the electron energy difference $\varepsilon_{\mathbf{k}_1+\mathbf{q}} - \varepsilon_{\mathbf{k}_1}$ before and after the transition. Third, if

$$|\varepsilon_{\mathbf{k}_1+\mathbf{q}} - \varepsilon_1| < \hbar\omega_q, \quad (4.57)$$

the effective interaction is *attractive*. Fourth, the atraction is greatest when $\varepsilon_{\mathbf{k}_1+\mathbf{q}} - \varepsilon_{\mathbf{k}_1} = 0$, that is, when the phonon momentum $\mathbf{q}$ is parallel to the constant energy (Fermi) surface. A bound electron-pair may be formed due to the phonon-exchange attraction as demonstrated by Cooper [7], which will be discussed in the following chapter.

Chapter 5

Quantum Statistical Theory

In a quantum statistical theory one starts with a reasonable Hamiltonian and derive everything from this, following step-by-step calculations. Only Heisenberg's equation of motion (quantum mechanics), Pauli's exclusion principle (quantum statistics), and Boltzmann's statistical principle (grand canonical ensemble theory) are assumed.

5.1 Theory of Superconductivity

The major superconducting properties were enumerated in section 1.1. The purpose of a microscopic theory is to explain all of these from first principles, starting with a reasonable Hamiltonian. Besides, one must answer basic questions such as:

- What causes superconductivity? The answer is the phonon-exchange attraction. We have discussed this interaction in Chapter 4. It generates Cooper pairs [1], called pairons for short, under certain conditions.
- Why do impurities that must exist in any superconductor not hinder the supercurrent? Why is the supercurrent stable against an applied voltage? Why does increasing the magnetic field destroy the superconducting state?
- Why does the supercurrent flow only in the surface layer? Why does the supercurrent dominate the normal current in the steady state?

- What is the supercondensate whose motion generates the supercurrent? How does magnetic-flux quantization arise? Josephson interference indicates that two supercurrents can interfere macroscopically just as two lasers from the same source. Where does this property come from?

- Below the critical temperature T_c, there is a profound change in the behavior of the electrons by the appearance of a temperature-dependent energy gap $\Delta(T)$. This was shown by Bardeen Cooper and Schrieffer (BCS) in their classic work [2]. What is the cause of the energy gap? Why does the energy gap depend on the temperature? Can the gap $\Delta(T)$ be observed directly?

- Phonons can be exchanged between any electrons at all time and at all temperatures. The phonon-exchange attraction can bound a pair of quasi-electrons to form moving (or excited) pairons. What is the energy of excited pairons? How do the moving pairons affect the low temperature behavior of the superconductor?

- All superconductors behave alike below T_c. Why does the law of corresponding states work here? Why is the supercurrent temperature- and material-independent?

- What is the nature of the superconducting transition? Does the transition depend on dimensionality?

- About half of all elemental metals are superconductors. Why does sodium remain normal down to 0 K? What is the criterion for superconductivity? What is the connection between superconductivity and band structures?

- Compound, organic, and high-T_c superconductors in general show type II magnetic behaviors. Why do they behave differently compared with type I elemental superconductors?

- All superconductors exhibit five basic properties: (1) zero resistance, (2) Meissner effect, (3) flux quantization, (4) Josephson effects and (5) gaps in the elementary excitation energy spectra. Can a quantum statistical theory explain all types of superconductors in a unified manner?

- Below 2.2 K, liquid He^4 exhibits a superfluid phase in which the superfluid can flow without a viscous resistance, the flow property remarkably similar to the supercurrent. Why and how this similarity arise?

5.2 The Bardeen-Cooper-Schrieffer Theory

In 1957 BCS published a classic paper [2] which is regarded as one of the most important theoretical works in the twentieth century. The Nobel physics prize in 1972 was shared by Bardeen, Cooper and Schrieffer for this work.

We shall briefly review this theory.

In spite of the Coulomb interaction among electrons there exists a sharp Fermi surface for the normal state of a conductor, as described by the Fermi liquid model. The phonon exchange attraction can bound pairs of electrons near the Fermi surface within a distance (energy) equal to Planck's constant $\hbar$ times the Debye frequency ω_D. The electron pairs having antiparallel spins and charge (magnitude) $2e$ are called *Cooper pairs (pairons)*. Cooper pair and pairons both denote the same entity. When we emphasize the quasi-particle aspect rather than the two electron composition aspect, we use the term pairon more often. Under the two conditions, we may write a Hamiltonian H in the form:

$$
\begin{aligned}
H &= \sum_{\substack{\mathbf{k} \\ \varepsilon_k>0}}\sum_{s} \varepsilon_k\, c^\dagger_{\mathbf{k}s} c_{\mathbf{k}s} + \sum_{\substack{\mathbf{k} \\ \varepsilon_k<0}}\sum_{s} |\varepsilon_k|\; c_{\mathbf{k}s} c^\dagger_{\mathbf{k}s} \\
&\quad + \frac{1}{2}\sum_{\mathbf{k}_1}\sum_{s_1}\cdots\sum_{\mathbf{k}_4}\sum_{s_4} \langle 1,2|\, U\, |3,4\rangle\; c^\dagger_1 c^\dagger_2\, c_4 c_3, \qquad (5.1)
\end{aligned}
$$

where $\varepsilon_{k_1} \equiv \varepsilon_1$ is the kinetic energy of a free electron measured relative to the Fermi energy ε_F, and $c^\dagger_{\mathbf{k}_1 s_1}\,(c_{\mathbf{k}_1 s_1}) \equiv c^\dagger_1\,(c_1)$ are creation (annihilation) operators satisfying the Fermi anticommutation rules:

$$
\{c_{\mathbf{k}s},\, c^\dagger_{\mathbf{k}'s'}\} = \delta_{\mathbf{k},\mathbf{k}'}\,\delta_{s,s'}\,, \qquad \{c_{\mathbf{k}s},\, c_{\mathbf{k}'s'}\} = \{c^\dagger_{\mathbf{k}s},\, c^\dagger_{\mathbf{k}'s'}\} = 0. \qquad (5.2)
$$

The first (second) sum on the rhs of (5.1) represents the total kinetic energy of "electrons" with positive ε_k ("holes" with negative ε_k). The matrix element $\langle 1,2|\, U\, |3,4\rangle$ denotes the net interaction arising from the virtual exchange of a phonon and the Coulomb repulsion between electrons. Specifically, we assume that

$$\langle 1,2|\,U\,|3,4\rangle = \begin{cases} -V_0 V^{-1}\delta_{\mathbf{k}_1+\mathbf{k}_2,\,\mathbf{k}_3+\mathbf{k}_4}\delta_{s_1,\,s_3}\delta_{s_2,\,s_4} & \text{if } |\varepsilon_m| < \hbar\omega_D \\ 0 & \text{otherwise,} \end{cases} \tag{5.3}$$

where V_0 is a constant (energy).

Starting with this Hamiltonian (5.1), they obtained an expression W for the ground-state energy

$$W = \hbar\omega_D \mathcal{N}(0) w_0 = N_0 w_0, \tag{5.4}$$

where

$$w_0 = \frac{-2\hbar\omega_D}{\exp[2/v_0\mathcal{N}(0)] - 1} \tag{5.5}$$

is the pairon ground-state energy, and

$$N_0 \equiv \hbar\omega_D \mathcal{N}(0) \tag{5.6}$$

is the total number of pairons and $\mathcal{N}(0)$ is the density of states per spin at the Fermi energy. In the variational calculation of the ground-state energy BCS found that the *unpaired electrons*, often called the *quasi-electrons*, not joining the ground pairons which form the supercondensate, have the energy

$$E_k = (\Delta^2 + \varepsilon_k^2)^{1/2}. \tag{5.7}$$

The energy constant Δ, called the quasi-electron energy gap, in Eq. (5.7) is greatest at 0 K and decreases to zero as temperature is raised to the critical temperature T_c. BCS further showed that the energy gap at 0 K, $\Delta(T = 0) \equiv \Delta_0$ and the critical temperature T_c are related (in the weak coupling limit) by

$$2\Delta_0 = 3.53\,\mathrm{k}_B T_c. \tag{5.8}$$

These findings are among the most important results obtained in the BCS theory. A large body of theoretical and experimental work followed several years after the BCS theory. By 1964 the general consensus was that the BCS theory is an essentially correct theory of superconductivity [3].

5.3 Remarks

BCS assumed the Hamiltonian in Eq. (5.1) containing "electron" and "hole" kinetic energies. They also assumed a spherical Fermi surface. These two assumptions however contradict each other. If a Fermi sphere whose inside (outside) is filled with electrons is assumed, there are "electrons" ("holes") only as we saw in Section 3.5. Besides this logical inconsistency, if a free electron model having a spherical Fermi surface is assumed, the question why only certain metals are superconductors cannot be answered. We must incorporate the band structures of electrons more explicitly. We shall later, in Chapter 7, discuss a generalization of the BCS Hamiltonian.

Chapter 6

Cooper Pairs (Pairons)

In 1956 Cooper demonstrated [1] that, however weak the attraction may be, two electrons just above the Fermi sea could be bound. The binding energy is greatest if the two electrons have opposite momenta ($\mathbf{p}$, $-\mathbf{p}$) and antiparallel spins ($\uparrow, \downarrow$). The lowest bound energy w_0 is

$$w_0 = \frac{-2\hbar\omega_D}{\exp[2/\mathcal{N}(0)\, v_0] - 1}, \tag{6.1}$$

where ω_D is the Debye frequency, v_0 a positive constant characterizing the attraction, and $\mathcal{N}(0)$ the electron density of states per spin at the Fermi energy. If electrons having nearly opposite momenta ($\mathbf{p}$, $-\mathbf{p}+\mathbf{q}$) are paired, the binding energy is less than $|w_0|$. For small q, which represents the net momentum (magnitude) of a Cooper pair, the energy momentum relation is

$$w_q = w_0 + cq < 0, \tag{6.2}$$

with $c/v_F = 1/2\ (2/\pi)$ for 3 (2) D, and $v_F \equiv (2\varepsilon_F/m^*)^{1/2}$ is the Fermi velocity. Equations (6.1) and (6.2) play very important roles in the theory of superconductivity. We shall derive them in this chapter.

6.1 The Cooper Problem

We saw in Section 4.3 that two electrons near the Fermi surface can gain attraction by exchanging a phonon. This attraction can generate bound states for the electron pair. We shall look for the ground state energy of the Cooper pair (pairon). We anticipate that the energy is lowest for the pairon

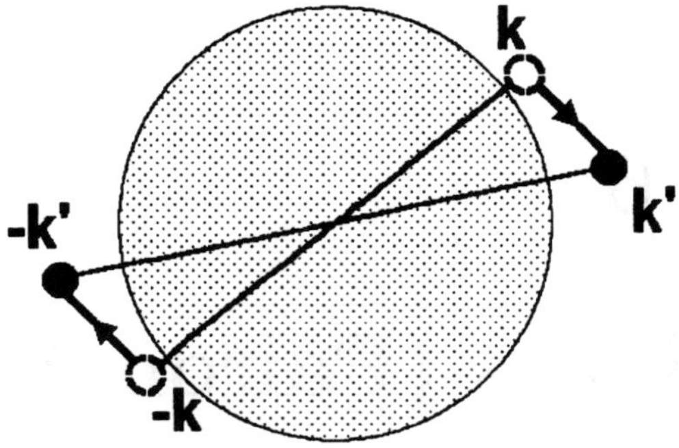

Figure 6.1: A stationary Cooper pair having zero net momentum.

with zero net momentum, that is, a stationary pairon. Moving pairons will be considered in the following section.

We consider a 2D motion. This will simplify the concept and calculations. The 3D case can be treated similarly. Take two electrons just above the Fermi surface (circle), one electron having momentum **k** and up spin and the other having momentum $-\mathbf{k}$ and down spin, see Fig. 6.1. We measure the energy relative to the Fermi energy ε_F:

$$\varepsilon(|\mathbf{k}|) \equiv \varepsilon_k = \frac{k^2 - k_F^2}{2m}. \tag{6.3}$$

The sum of the kinetic energies of the two electrons is $2\varepsilon_k$. By exchanging a phonon, the pair's momenta change from $(\mathbf{k}, -\mathbf{k})$ to $(\mathbf{k}', -\mathbf{k}')$. This process lowers the energy of the pair.

Following Cooper, we write down the energy-eigenvalue equation, postponing its microscopic derivation to Section 6.4: [1]

$$w_0 A(\mathbf{k}) = 2\varepsilon_k A(\mathbf{k}) - \frac{1}{(2\pi\hbar)^2} v_0 \int' d^2k' \, A(\mathbf{k}'), \tag{6.4}$$

where w_0 is the pairon energy, and $A(\mathbf{k})$ the wave function; the prime on the integral sign means the restriction:

$$0 < \varepsilon_k < \hbar\omega_D. \tag{6.5}$$

Eq. (6.4) can be solved simply as follows. Consider the last integral:

$$C \equiv \frac{1}{(2\pi\hbar)^2} v_0 \int' d^2k' \, A(\mathbf{k}'), \tag{6.6}$$

which is a constant. Assume that the energy w_0 is negative:

$$w_0 < 0. \tag{6.7}$$

Then, $2\varepsilon_k - w_0 = 2\varepsilon_k + |w_0| > 0$. After rearranging the terms in Eq. (6.4) and dividing the result by $2\varepsilon_k + |w_0|$, we obtain

$$A(\mathbf{k}) = \frac{1}{2\varepsilon_k + |w_0|} C. \tag{6.8}$$

Substituting this expression into Eq. (6.6) and dropping the common factor C, we obtain

$$1 = \frac{1}{(2\pi\hbar)^2} v_0 \int' d^2k \frac{1}{2\varepsilon_k + |w_0|}. \tag{6.9}$$

By introducing the density of states at the Fermi energy, $\mathcal{N}(0)$, we can evaluate the k-integral as follows: (Problem 6.1.1)

$$1 = v_0 \mathcal{N}(0) \int_0^{\hbar\omega_D} d\varepsilon \frac{1}{2\varepsilon + |w_0|} = \frac{1}{2} v_0 \mathcal{N}(0) \ln[(2\hbar\omega_D + |w_0|)/|w_0|]$$

or

$$w_0 = \frac{-2\hbar\omega_D}{\exp[2/v_0\mathcal{N}(0)] - 1}. \tag{6.10}$$

We thus found a *negative energy* for the stationary pairon. The v_0-dependence of the energy w_0 is noteworthy. Since $\exp(2/x)$ cannot be expanded in powers of $x = v_0\mathcal{N}(0)$, the energy w_0 cannot be obtained by a perturbation (v_0)-expansion method. We note that formula (6.10) holds for any dimensions.

Problem 6.1.1. Verify Eq. (6.10).

6.2 Moving Pairons

The phonon exchange attraction is in action for any pair of electrons near the Fermi surface. Such pairs can in general have net momenta and hence move. They are called *moving pairons.* The energy w_q of a moving pairon can be obtained from a generalization of Eq. (6.4):

$$\begin{aligned} w_q\, a(\mathbf{k}, \mathbf{q}) &= [\varepsilon(|\mathbf{k} + \mathbf{q}/2|) + \varepsilon(|-\mathbf{k} + \mathbf{q}/2|)]\, a(\mathbf{k}, \mathbf{q}) \\ &\quad - \frac{1}{(2\pi\hbar)^2} v_0 \int' d^2k' a(\mathbf{k}', \mathbf{q}), \end{aligned} \tag{6.11}$$

which is Cooper's equation, Eq. (1) of his 1956 Physical Review Letter [1]. We note that the net momentum $\mathbf{q}$ is a constant of motion, which arises from the fact that the phonon exchange is an internal process and hence cannot change the net momentum. The pair wavefunctions $a(\mathbf{k}, \mathbf{q})$ are coupled with respect to the other variable $\mathbf{k}$, meaning that the exact (or energy-eigenstate) pairon wave functions are superpositions of the pair wave functions $a(\mathbf{k}, \mathbf{q})$.

We note that Eq. (6.11) is reduced to Eq. (6.4) in the small-q limit. The latter equation was solved earlier. Using the same technique we obtain from Eq. (6.11)

$$1 = \frac{v_0}{(2\pi\hbar)^2} \int' d^2k \, [\varepsilon(|\mathbf{k} + \mathbf{q}/2|) + \varepsilon(|-\mathbf{k} + \mathbf{q}/2|) + |w_q|]^{-1}. \tag{6.12}$$

We now assume a free-electron model, whose Fermi surface is a circle of radius (momentum)

$$k_F \equiv (2m_1\varepsilon_F)^{1/2}, \tag{6.13}$$

where m_1 represents the effective mass.

The prime on the k-integral means the restriction: $0 < \varepsilon(|\mathbf{k} + \mathbf{q}/2|)$, $\varepsilon(|-\mathbf{k} + \mathbf{q}/2|) < \hbar\omega_D$. We may choose the x-axis along $\mathbf{q}$ as shown in Fig. 6.2. We assume a small q and keep terms up to the first order in q. The k-integral can then be expressed by (Problem 6.2.1.)

$$\frac{(2\pi\hbar)^2}{v_0} = 4 \int_0^{\pi/2} d\theta \int_{k_F + \frac{1}{2}q\cos\theta}^{k_F + k_D - \frac{1}{2}q\cos\theta} \frac{k\,dk}{|w_q| + (k^2 - k_F^2)/m_1}$$

$$= 2m_1 \int_0^{\pi/2} d\theta \ln \left| \frac{|w_q| + 2\hbar\omega_D - v_F\, q\cos\theta}{|w_q| + v_F\, q\cos\theta} \right|, \tag{6.14}$$

$$k_D \equiv m_1\omega_D\hbar k_F^{-1}, \tag{6.15}$$

where we retained the linear term in (k_D/k_F) only.

After performing the θ-integration, we obtain (Problem 6.2.2.)

$$w_q = w_0 + (2/\pi)\, v_F\, q, \tag{6.16}$$

where w_0 is given by Eq. (6.10). This result was first obtained (but unpublished) by Cooper. It is recorded in Schrieffers' book [2], Eq. (2-15). As

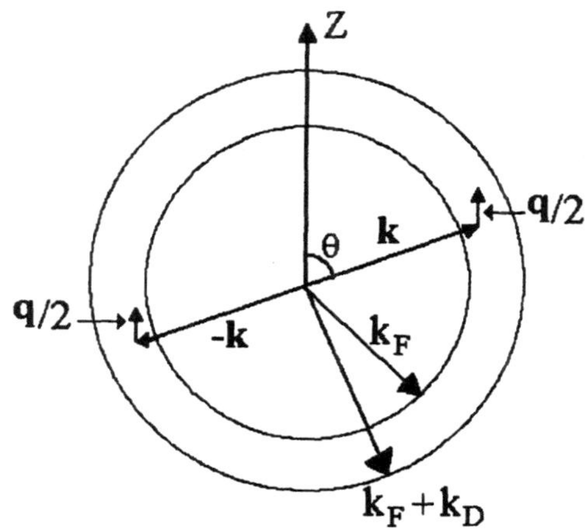

Figure 6.2: The range of the integration variables (k, θ) is restricted to the shell of thickness k_D.

expected, the zero-momentum pairon has the lowest energy. The excitation energy is continuous with no energy gap. The energy w_q increases *linearly* with momentum q for small q rather than quadratically. This fact arises since the density of states is strongly reduced with increasing momentum q, and this behavior dominates the q^2 increase of the kinetic energy. *Pairons move like massless particles* with a common speed $2v_F/\pi$. This linear dispersion relation plays a vital role in the B-E condensation of pairons. See Chapter 9.

Such a linear relation is valid for pairons moving in any dimension (D). However the coefficients slightly depend on the dimensions; in fact

$$w_q = w_0 + cq, \tag{6.17}$$

where $c/v_F = 1/2$, $2/\pi$ and 1 for 3, 2 and 1 D. (see Problem 6.2.3.).

Problem 6.2.1. Verify (6.14). Use the diagram in Fig. 6.2.

Problem 6.2.2. Derive (6.16).

Problem 6.2.3. Derive an energy-momentum relation [Eq.(6.17)] for 3D with the assumption of a Fermi sphere. Use a diagram similar to that in Fig. 6.2.

6.3 Energy-Eigenvalue Problem for a Quasi-particle

In the Cooper problem we considered two electrons above the Fermi sea of electrons with the Fermi sphere represented by

$$\frac{p^2}{2m} = \frac{1}{2m}(p_x^2 + p_y^2 + p_z^2) = \varepsilon_F. \tag{6.18}$$

It is not strictly a two-electron problem. Due to the Pauli exclusion principle, the two electrons are forbidden to go into the states whose energies are less than the Fermi energy ε_F. Thus the role of the electrons which form the Fermi sea is passive.

We wish to show that the Cooper equation is an example of the energy-eigenvalue equation for a quasi-particle. We do this in two steps. In this section we set up the energy-eigenvalue problem for a quasiparticle. Using this technique we derive the Cooper equation in Section 6.4. As a preliminary, we consider an electron characterized by the Hamiltonian

$$h(x, p) = \frac{p^2}{2m} + u(x). \tag{6.19}$$

We may set up the eigenvalue equations for the position x, the momentum p, and the Hamiltonian H as follows:

$$x\,|x'\rangle = x'\,|x'\rangle, \tag{6.20}$$

$$p\,|p'\rangle = p'\,|p'\rangle, \tag{6.21}$$

$$h\,|\varepsilon_\nu\rangle = \varepsilon_\nu\,|\varepsilon_\nu\rangle \equiv \varepsilon_\nu\,|\nu\rangle, \tag{6.22}$$

where x', p', and ε_ν are eigenvalues.

By multiplying Eq. (6.22) from the left by $\langle x|$, we obtain

$$h(x, -i\hbar d/dx)\phi_\nu(x) = \varepsilon_\nu\phi_\nu(x) \tag{6.23}$$

$$\phi_\nu(x) \equiv \langle x|\nu\rangle. \tag{6.24}$$

Eq. (6.23) is just the Schrödinger energy-eigenvalue equation, and $\phi_\nu(x)$ the familiar quantum wave function. If we know with certainty that the system is in the energy eigenstate ν, we can choose a density operator ρ_1 to be

$$\rho_1 \equiv |\nu\rangle\langle\nu|, \qquad \langle\nu|\nu\rangle = 1. \tag{6.25}$$

Let us now consider $\mathrm{tr}\{|\nu\rangle\langle x|\,\rho_1\}$, which can be transformed as follows:

$$\mathrm{tr}\{|\nu\rangle\langle x|\,\rho_1\} = \sum_\alpha \langle\alpha|\,\nu\rangle\langle x|\,\nu\rangle\langle\nu|\,\alpha\rangle = \sum_\alpha \langle x|\,\nu\rangle\langle\nu|\,\alpha\rangle\langle\alpha|\,\nu\rangle = \langle x|\,\nu\rangle$$

or

$$\mathrm{tr}\{|\nu\rangle\langle x|\,\rho_1\} = \langle x|\,\rho_1\,|\nu\rangle = \phi_\nu\,(x). \tag{6.26}$$

This means that the wave function $\phi_\nu(x)$ can be regarded as a *mixed* representation of the density operator ρ_1 in terms of the states (ν, x). In a parallel manner, we can show (Problem 6.3.1) that the wave function in the momentum space $\phi_\nu(p) \equiv \langle p|\,\nu\rangle$ can be regarded as a mixed representation of ρ_1 in terms of energy-state ν and momentum-state p:

$$\phi_\nu(p) = \mathrm{tr}\{|\nu\rangle\langle p|\,\rho_1\} = \langle p|\,\rho_1\,|\nu\rangle. \tag{6.27}$$

In analogy with (6.26) we introduce a *quasi-wavefunction* $\Psi_\nu(p)$ through

$$\Psi_\nu(p) \equiv \mathrm{Tr}\{\psi_\nu^\dagger a_p \rho\}, \tag{6.28}$$

where $\psi_\nu^\dagger$ is the energy-state creation operator, a_p the momentum-state annihilation operator, and ρ a many-body-system density operator that commutes with the Hamiltonian:

$$[\rho,\, H] = 0. \tag{6.29}$$

This is the necessary condition that ρ be a stationary density operator, which is seen at once from the quantum Liouville equation:

$$i\hbar\frac{\partial\rho}{\partial t} = [H,\, \rho]. \tag{6.30}$$

Let us consider a system for which the total Hamiltonian H is the sum of single-electron energies h:

$$H = \sum_j h^{(j)}. \tag{6.31}$$

For example, the single-electron Hamiltonian h may contain the kinetic energy and the lattice potential energy. We assume that the Hamiltonian H does not depend on the time explicitly.

In second quantization the Hamiltonian H can be represented by

$$H = \sum_a \sum_b \langle\alpha_a|\,h\,|\alpha_b\rangle\,\eta_a^\dagger\eta_b \equiv \sum_a \sum_b h_{ab}\eta_a^\dagger\eta_b, \tag{6.32}$$

where η_a ($\eta_a^\dagger$) are annihilation (creation) operators and satisfying the Fermi anticommutation rules:

We calculate the commutator $[H, \psi_a^\dagger]$. In such a commutator calculation, the following identities are very useful

$$[A, BC] = [A, B]C + B[A, C], \qquad [AB, C] = A[B, C] + [A, C]B, \tag{6.33}$$

$$[A, BC] = \{A, B\}C - B\{A, C\}, \qquad [AB, C] = A\{B, C\} - \{A, C\}B. \tag{6.34}$$

Note: the negative signs on the right-hand terms in Eqs. (6.34) occur when the cyclic order is destroyed for the case of the anticommutator: $\{A, B\} \equiv AB + BA$. We obtain after simple calculation (Problem 6.3.2)

$$[H, \psi_\nu^\dagger] = \varepsilon_\nu \psi_\nu^\dagger = \sum_\mu \psi_\mu^\dagger h_{\mu\nu}. \tag{6.35}$$

We multiply Eq. (6.35) by $a_p\rho$ from the right, take a many-body trace and obtain

$$\sum_\mu \Psi_\mu(p)\, h_{\mu\nu} = \varepsilon_\nu \Psi_\nu(p), \tag{6.36}$$

which is formally identical with the Schrödinger energy-eigenvalue equation for the one-body problem: (Problem 6.3.3)

$$\sum_\mu \phi_\mu(p)\, h_{\mu\nu} = \varepsilon_\nu \phi_\nu(p), \qquad \phi_\nu(p) \equiv \langle p | \nu \rangle\,. \tag{6.37}$$

The quasi-wave-function $\Psi_\nu(p)$ can be regarded as a mixed representation of the *one-body density operator* n in terms of the states (ν, p) (Problem 6.3.4):

$$\Psi_\nu(p) = \langle p | \, n \, | \nu \rangle. \tag{6.38}$$

The operator n is defined through

$$\mathrm{Tr}\{\eta_b \rho \eta_a^\dagger\} \equiv \langle \alpha_b | \, n \, | \alpha_a \rangle \equiv n_{ba}. \tag{6.39}$$

These n_{ba} are called b-a elements of the *one-body density matrix.*

We reformulate Eq. (6.36) for a later use. Using Eqs. (6.29), (6.35) and the following general property

$$\mathrm{Tr}\{AB\rho\} = \mathrm{Tr}\{\rho AB\} = \mathrm{Tr}\{B\rho A\}. \tag{6.40}$$

(cyclic permutation under the trace), we obtain (Problem 6.3.5)

$$\mathrm{Tr}\{[H,\,\psi_\nu^\dagger]\,a_p\rho\} = \mathrm{Tr}\{\psi_\nu^\dagger\,[a_p,\,H]\,\rho\} = \varepsilon_\nu\Psi_\nu(p), \tag{6.41}$$

whose complex-conjugate is

$$\varepsilon_\nu\Psi_\nu^*(p) = \mathrm{Tr}\left\{[H, a_p^\dagger]\,\psi_\nu\rho]\right\}. \tag{6.42}$$

Either Eq. (6.41) or Eq. (6.42) can be used to formulate the energy-eigenvalue problem. If we choose the latter, we may proceed as follows:

1. Given H in the momentum space, compute $[H,\,a_p^\dagger]$; the result can be expressed as a linear function of $a^\dagger$;

2. Multiply the result by $\psi_\nu\rho$ from the right, and take a trace; the result is a linear function of Ψ^*;

3. Use Eq. (6.42); the result is a linear homogeneous equation for Ψ^*, a standard form of the energy eigenvalue equation.

Problem 6.3.1. Verify Eq. (6.27).

Problem 6.3.2. Derive Eq. (6.35).

Problem 6.3.3. Derive Eq. (6.37) from Eq. (6.22).

Problem 6.3.4. Prove Eq. (6.38).

Problem 6.3.5. Verify Eq. (6.41).

6.4 Derivation of the Cooper Equation

Second-quantized operators for a pair of electrons are introduced as

$$B_{12}^\dagger \equiv B_{\mathbf{k}_1\uparrow\mathbf{k}_2\downarrow}^\dagger \equiv c_1^\dagger c_2^\dagger, \qquad B_{34} = c_4c_3. \tag{6.43}$$

Odd-numbered electrons carry up spins $\uparrow$ and even-numbered carry down spins $\downarrow$. The commutators among B and $B^\dagger$ can be computed using the Fermi anticommutation relations, and they are given by (Problem 6.4.1)

$$[B_{12},\,B_{34}] \equiv B_{12}B_{34} - B_{34}B_{12} = 0, \tag{6.44}$$

$$B_{12}^2 = B_{12}B_{12} = 0, \tag{6.45}$$

$$[B_{12},\ B_{34}^{\dagger}] = \begin{cases} 1-n_1-n_2 & \text{if} \quad k_1 = k_3,\ k_2 = k_4 \\ c_2 c_4^{\dagger} & \text{if} \quad k_1 = k_3,\ k_2 \neq k_4 \\ c_1 c_3^{\dagger} & \text{if} \quad k_1 \neq k_3,\ k_2 = k_4 \\ 0 & \text{otherwise,} \end{cases} \tag{6.46}$$

where

$$n_1 \equiv c_{\mathbf{k}_1\uparrow}^{\dagger} c_{\mathbf{k}_1\uparrow}, \qquad n_2 \equiv c_{\mathbf{k}_2\downarrow}^{\dagger} c_{\mathbf{k}_2\downarrow} \tag{6.47}$$

are the number operators for electrons.

Let us now introduce the relative and net (CM) momenta $(\mathbf{k}, \mathbf{q})$ such that

$$\mathbf{k} \equiv \frac{1}{2}(\mathbf{k}_1 - \mathbf{k}_2), \qquad \mathbf{q} \equiv \mathbf{k}_1 + \mathbf{k}_2; \qquad \mathbf{k}_1 = \mathbf{k} + \frac{1}{2}\mathbf{q}, \qquad \mathbf{k}_2 = -\mathbf{k} + \frac{1}{2}\mathbf{q}. \tag{6.48}$$

We may alternatively represent the pair operators by

$$B'_{\mathbf{kq}} \equiv B_{\mathbf{k}_1\uparrow\mathbf{k}_2\downarrow} \equiv c_{-\mathbf{k}+\mathbf{q}/2\downarrow} c_{\mathbf{k}+\mathbf{q}/2\uparrow}, \qquad B'^{\dagger}_{\mathbf{kq}} \equiv c^{\dagger}_{\mathbf{k}+\mathbf{q}/2\uparrow} c^{\dagger}_{-\mathbf{k}+\mathbf{q}/2\downarrow}. \tag{6.49}$$

The prime on $B_{\mathbf{kq}}$ will be dropped hereafter. In the k-q representation the commutation relations can be re-expressed as

$$[B_{\mathbf{k},\mathbf{q}},\ B_{\mathbf{k}'\mathbf{q}'}] = 0, \tag{6.50}$$

$$[B_{\mathbf{kq}}]^2 = 0, \tag{6.51}$$

$$[B_{\mathbf{k},\mathbf{q}},\ B^{\dagger}_{\mathbf{k}'\mathbf{q}'}] = \begin{cases} 1 - n_{\mathbf{k}+\mathbf{q}/2\uparrow} - n_{-\mathbf{k}+\mathbf{q}/2\downarrow} & \text{if } \mathbf{k} = \mathbf{k}' \text{ and } \mathbf{q} = \mathbf{q}' \\ c_{-\mathbf{k}+\mathbf{q}/2\downarrow} c^{\dagger}_{-\mathbf{k}'+\mathbf{q}'/2\downarrow} & \text{if } \mathbf{k}+\mathbf{q}/2 = \mathbf{k}'+\mathbf{q}'/2 \text{ and} \\ & \quad -\mathbf{k}+\mathbf{q}/2 \neq -\mathbf{k}'+\mathbf{q}'/2 \\ c_{\mathbf{k}+\mathbf{q}/2\uparrow} c^{\dagger}_{\mathbf{k}'+\mathbf{q}'/2\uparrow} & \text{if } \mathbf{k}+\mathbf{q}/2 \neq \mathbf{k}'+\mathbf{q}'/2 \text{ and} \\ & \quad -\mathbf{k}+\mathbf{q}/2 = -\mathbf{k}'+\mathbf{q}'/2 \\ 0 & \text{otherwise.} \end{cases} \tag{6.52}$$

If we drop the "hole" contribution from the original BCS Hamiltonian in (5.1), we obtain the *Cooper Hamiltonian*:

$$H_C = \sum_{\substack{\mathbf{k} \\ \epsilon_k > 0}} \sum_{s} \epsilon_k\, c^{\dagger}_{\mathbf{k}s} c_{\mathbf{k}s} - v_0 {\sum_{\mathbf{k}}}' {\sum_{\mathbf{k}'}}' {\sum_{\mathbf{q}'}}' B^{\dagger}_{\mathbf{kq}} B_{\mathbf{k}'\mathbf{q}}, \tag{6.53}$$

where we used (5.3); $v_0 \equiv V_0/V^{-1}$; the prime on the summation means the restriction:

$$0 < \varepsilon(|\mathbf{k}+\mathbf{q}/2|), \quad \varepsilon(|-\mathbf{k}+\mathbf{q}/2|) < \hbar\omega_D. \tag{6.54}$$

Our Hamiltonian H_C in Eq. (6.53) can be expressed in terms of pair operators (B, $B^\dagger$):

$$\begin{aligned} H_C &= \sum_{\mathbf{k}}{}' \sum_{\mathbf{q}}{}' [\varepsilon(|\mathbf{k}+\mathbf{q}/2|) + \varepsilon(|-\mathbf{k}+\mathbf{q}/2|]\, B^\dagger_{\mathbf{kq}} B_{\mathbf{kq}} \\ &\quad - \sum_{\mathbf{k}}{}' \sum_{\mathbf{k}'}{}' \sum_{\mathbf{q}'}{}' v_0 B^\dagger_{\mathbf{kq}} B_{\mathbf{k}'\mathbf{q}}. \end{aligned} \tag{6.55}$$

Using Eqs. (6.52) and (6.51), we obtain (Problem 6.4.2)

$$\begin{aligned} [H_C,\, B^\dagger_{\mathbf{kq}}] &= [\varepsilon(|\mathbf{k}+\mathbf{q}/2|) + \varepsilon(|-\mathbf{k}+\mathbf{q}/2|]\, B^\dagger_{\mathbf{kq}} \\ &\quad - v_0 \sum_{\mathbf{k}'}{}' B^\dagger_{\mathbf{k}'\mathbf{q}} (1 - n_{\mathbf{k}+\mathbf{q}/2\uparrow} - n_{-\mathbf{k}+\mathbf{q}/2\downarrow}). \end{aligned} \tag{6.56}$$

If we represent the energies of pairons by w_ν and the associated pair annihilation operators by ϕ_ν, we can in principle re-express H_C as

$$H_C = \sum_\nu w_\nu\, \phi^\dagger_\nu \phi_\nu, \tag{6.57}$$

which is similar to Eq. (6.32) with the only difference that here we deal with pair energies and pair-state operators. We multiply Eq. (6.56) by $\phi_\nu \rho_{gc}$ from the right and take a grand-ensemble trace. After using Eq. (6.42), we obtain

$$\begin{aligned} w_\nu\, a_{\mathbf{kq}} &\equiv w_q\, a_{\mathbf{kq}} = [\varepsilon(|\mathbf{k}+\mathbf{q}/2|) + \varepsilon(|-\mathbf{k}+\mathbf{q}/2|]\, a_{\mathbf{kq}} \\ &\quad - v_0 \sum_{\mathbf{k}'}{}' \langle B^\dagger_{\mathbf{k}'\mathbf{q}} (1 - n_{\mathbf{k}+\mathbf{q}/2\uparrow} - n_{-\mathbf{k}+\mathbf{q}/2\downarrow}) \phi_\nu \rangle \\ a_{\mathbf{kq},\nu} &\equiv \mathrm{TR}\{B^\dagger_{\mathbf{kq}}\, \phi_\nu \rho_{gc}\} \equiv a_{\mathbf{kq}}. \end{aligned} \tag{6.58}$$

The energy w_ν can be characterized by $\mathbf{q}$: $w_\nu \equiv w_q$. In other words, excited pairons have net momentum $\mathbf{q}$ and energy w_q. We shall omit the subindex ν in the pairon wavefunction: $a_{\mathbf{kq},\nu} \equiv a_{\mathbf{kq}}$. The angular brackets mean the grand-canonical-ensemble average:

$$\langle A \rangle \equiv \mathrm{TR}\left\{A\, \phi_\nu \rho_{gc}\right\} \equiv \frac{\mathrm{TR}\{A \exp(\alpha N - \beta H)\}}{\mathrm{TR}\{\exp(\alpha N - \beta H)\}}. \tag{6.59}$$

In the bulk limit: $N \to \infty$, $V \to \infty$ while $n \equiv N/V =$ finite, where N represents the number of electrons, k-vectors become continuous. Denoting the wavefunction in this limit by $a(\mathbf{k},\mathbf{q})$ and using a factorization approximation, we obtain from Eq. (6.58)

$$\begin{aligned} w_{\mathbf{q}}\, a(\mathbf{k},\mathbf{q}) &= [\varepsilon(|\mathbf{k}+\mathbf{q}/2|) + \varepsilon(|-\mathbf{k}+\mathbf{q}/2|]\, a(\mathbf{k},\mathbf{q}) - \frac{v_0}{(2\pi\hbar)^3}\int' d^3k'\, a(\mathbf{k}',\mathbf{q}) \\ &\quad \times \{1 - f_F\,[\varepsilon(|\mathbf{k}+\mathbf{q}/2|)] - f_F\,[\varepsilon(|-\mathbf{k}+\mathbf{q}/2|]\}, \end{aligned} \quad (6.60)$$

$$\langle n_p \rangle = \frac{1}{\exp(\beta\varepsilon_p)+1} \equiv f_F(\varepsilon_p). \quad (6.61)$$

The factorization is justified since the coupling between electrons and pairons is weak.

In the low temperature limit ($T \to 0$ or $\beta \to \infty$),

$$f_F(\varepsilon_p) \to 0, \qquad (\varepsilon_p > 0). \quad (6.62)$$

We then obtain

$$w_{\mathbf{q}}\, a(\mathbf{k},\mathbf{q}) = [\varepsilon(|\mathbf{k}+\mathbf{q}/2|) + \varepsilon(|-\mathbf{k}+\mathbf{q}/2|]\, a(\mathbf{k},\mathbf{q}) - \frac{v_0}{(2\pi\hbar)^3}\int' d^3k'\, a(\mathbf{k}',\mathbf{q}), \quad (6.63)$$

which is identical with Cooper's equation (6.11).

In the above derivation we obtained the Cooper equation in the zero-temperature limit. Thus the energy of the pairon, w_q, is temperature-independent. This low temperature approximation is adequate for the description of superconductors below T_c. But as we will see later the pairons exist above T_c. It is then important to know the validity of the low-temperature approximation.

This problem can simply be answered by examining Eq. (6.60). The energy $w_q(T)$ is now temperature-dependent. The numerical solution of Eq. (6.60) yields that $w_q(T)$ can still be written in the form:

$$w_q(T) = w_0(T) + cq \quad (6.64)$$

with $|w_0(T)|$ decreasing smoothly over a few Θ_D as long as $\Theta_D \ll T_F$.

In the present many-electron calculations the picture of two electrons above the Fermi sea of electrons is not used. All electrons are treated equally. Our derivation gives a firm foundation to the linear dispersion relation for a moving pairon.

Problem 6.4.1. Derive Eqs. (6.44)-(6.46).

Problem 6.4.2 Derive Eq. (6.56).

Chapter 7

Superconductors at 0 K

We construct a generalized BCS Hamiltonian which contains the kinetic energies of "electrons" and "holes", and the pairing Hamiltonian arising from phonon-exchange attraction and Coulomb repulsion. We follow the original BCS theory to construct a many-pairon ground state and find a ground state energy: $W = (1/2)\hbar\omega_D \mathcal{N}(0)(w_1 + w_2)$, where w_1 and w_2 are respectively the ground state energies of "electron"(1) and "hole"(2) pairons. Energy gaps Δ_j are found in the quasi-electron excitation spectra: $E_k^{(j)} \equiv (\varepsilon_k^2 + \Delta_j^2)^{1/2}$.

7.1 The Generalized BCS Hamiltonian

BCS assumed a Hamiltonian containing "electron" and "hole" kinetic energies and a pairing interaction [1]. They also assumed a spherical Fermi surface. But if we assume a free electron model, we cannot explain why only some, and not all, metals are superconductors. We must incorporate the band structures of electrons explicitly. In this section we set up and discuss a *generalized BCS Hamiltonian.*[2] We assume that

- In spite of the Coulomb interaction, there exists a sharp Fermi surface at 0 K for the normal state of a conductor (the Fermi liquid model [3]).
- The phonon exchange can bound Cooper pairs [4] near the Fermi surface within a distance (energy) equal to Planck's constant $\hbar$ times the Debye frequency ω_D.
- "Electrons" and "holes" have different effective masses (magnitude).

- The pairing interaction strengths V_{ij} among *and* between "electron" (1) and "hole" (2) pairons are different so that

$$\langle 1,2;i \mid U \mid 3,4;j\rangle = \begin{cases} -V_{ij}V^{-1}\delta_{\mathbf{k}_1+\mathbf{k}_2,\,\mathbf{k}_3+\mathbf{k}_4}\delta_{s_1,s_3}\delta_{s_2,s_4} & \text{if } |\varepsilon_m| < \hbar\omega_D \\ 0 & \text{otherwise.} \end{cases} \tag{7.1}$$

"Electrons" and "holes" are different quasiparticles, which will be denoted distinctly. Let us introduce the vacuum state $|\phi_1\rangle$, creation and annihilation operators, $(c^{(1)\dagger}, c^{(1)})$, and the number operators $n^{(1)}_{\mathbf{k}s}$ for "electrons" $(\varepsilon_k > 0)$:

$$c^{(1)}_{\mathbf{k}s} \equiv c_{\mathbf{k}s}, \quad c^{(1)\dagger}_{\mathbf{k}s} = c^{\dagger}_{\mathbf{k}s}, \quad n^{(1)}_{\mathbf{k}s} \equiv c^{(1)\dagger}_{\mathbf{k}s}c^{(1)}_{\mathbf{k}s}, \quad c^{(1)}_{\mathbf{k}s}\,|\phi_1\rangle = 0. \tag{7.2}$$

For "holes" $(\varepsilon_k < 0)$, we introduce the vacuum state $|\phi_2\rangle$, creation and annihilation operators, $(c^{(2)\dagger}, c^{(2)})$, and the number operators $n^{(2)}_{\mathbf{k}s}$ as follows:

$$c^{(2)}_{\mathbf{k}s} \equiv c^{\dagger}_{\mathbf{k}s}, \quad c^{(2)\dagger}_{\mathbf{k}s} = c_{\mathbf{k}s}, \quad n^{(2)}_{\mathbf{k}s} \equiv c^{(2)\dagger}_{\mathbf{k}s}c^{(2)}_{\mathbf{k}s}, \quad c^{(2)}_{\mathbf{k}s}\,|\phi_2\rangle = 0. \tag{7.3}$$

Hereafter we adopt Dirac's convention that a "hole" has a positive mass m_2, a positive energy $\varepsilon^{(2)}_k \equiv |\varepsilon_k|$ and the positive charge e. At 0 K there are only *zero-momentum pairons* of the lowest energy. The ground state Ψ for the system can then be described in terms of a *reduced* Hamiltonian:

$$\begin{aligned} H_0 \;=\; & \sum_{\mathbf{k}}\sum_{s}\varepsilon^{(1)}_k n^{(1)}_{\mathbf{k}s} + \sum_{\mathbf{k}}\sum_{s}\varepsilon^{(2)}_k n^{(2)}_{\mathbf{k}s} - \sum_{\mathbf{k}}{}'\sum_{\mathbf{k}'}{}'[v_{11}b^{(1)\dagger}_{\mathbf{k}}b^{(1)}_{\mathbf{k}'} \\ & + v_{12}b^{(1)\dagger}_{\mathbf{k}}b^{(2)\dagger}_{\mathbf{k}'} + v_{21}b^{(2)}_{\mathbf{k}}b^{(1)}_{\mathbf{k}'} + v_{22}b^{(2)}_{\mathbf{k}}b^{(2)\dagger}_{\mathbf{k}'}], \end{aligned} \tag{7.4}$$

where $v_{ij} \equiv V^{-1}V_{ij}$, and $b^{(j)}$ are pair annihilation operators defined by

$$b^{(1)}_{\mathbf{k}} \equiv c^{(1)}_{-\mathbf{k}\downarrow}c^{(1)}_{\mathbf{k}\uparrow}, \qquad b^{(2)}_{\mathbf{k}} \equiv c^{(2)}_{\mathbf{k}\uparrow}c^{(2)}_{-\mathbf{k}\downarrow}; \tag{7.5}$$

the primes on the last summation symbols indicate the restriction that

$$\begin{aligned} 0 < \varepsilon^{(1)}_k \equiv \varepsilon_k < \hbar\omega_D \quad & \text{for "electrons"} \\ 0 < \varepsilon^{(2)}_k \equiv |\varepsilon_k| < \hbar\omega_D \quad & \text{for "holes".} \end{aligned} \tag{7.6}$$

For the sake of arguments, let us drop the interaction Hamiltonian altogether in Eq. (7.4). We then have the first two sums representing the kinetic energies of "electrons" and "holes". These energies $(\varepsilon^{(1)}_k,\ \varepsilon^{(2)}_k)$ are positive by

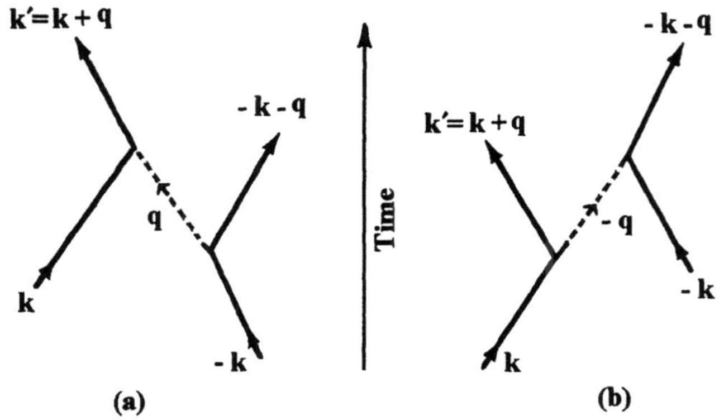

Figure 7.1: Two Feynman diagrams representing a phonon exchange between two electrons.

definition. Then the lowest energy of this system, called the Bloch system, is zero, and the corresponding ground state is characterized by zero "electrons" and zero "holes". This state will be called the *physical vacuum state.* In the theoretical developments in this chapter, we look for the ground state of the generalized BCS system whose energy is negative.

We now examine the physical meaning of the interaction strengths v_{ij}. Consider part of the interaction terms in Eq. (7.4):

$$-v_{11}\, b_{\mathbf{k}'}^{(1)\dagger} b_{\mathbf{k}}^{(1)}, \qquad -v_{22}\, b_{\mathbf{k}'}^{(2)} b_{\mathbf{k}}^{(2)\dagger}. \tag{7.7}$$

The first term generates a transition of the electron pair from $(\mathbf{k}\uparrow, -\mathbf{k}\downarrow)$ to $(\mathbf{k}'\uparrow, -\mathbf{k}'\downarrow)$. This transition is represented by the k-space diagram in Fig. 6.1. Such a transition may be generated by the emission of a virtual phonon with momentum $\mathbf{q} = \mathbf{k}' - \mathbf{k}\,(-\mathbf{q})$ by the down (up)-spin "electron" and subsequent absorption by the up (down)-spin "electron" as shown in Fig. 7.1 (a) and (b). Note: These two processes are distinct, but yield the same net transition. As we saw earlier the phonon exchange generates an attractive change of states between two "electrons" whose energies are nearly the same. The Coulomb interaction generates a repulsive correlation. The effect of this interaction can be included in the strength v_{11}. Similarly, the exchange of a phonon induces a change of states between two "holes", and it is represented by the second term in Eq. (7.7). The exchange of a phonon can also pair create or pair annihilate "electron" ("hole") pairons, called $-(+)$pairons, and the effects of these processes are represented by

$$-v_{12}\, b_{\mathbf{k}'}^{(1)\dagger} b_{\mathbf{k}}^{(2)\dagger}, \qquad -v_{21}\, b_{\mathbf{k}}^{(1)} b_{\mathbf{k}'}^{(2)}. \tag{7.8}$$

These two processes are indicated by k-space diagrams (a) and (b) in Fig. 7.2.

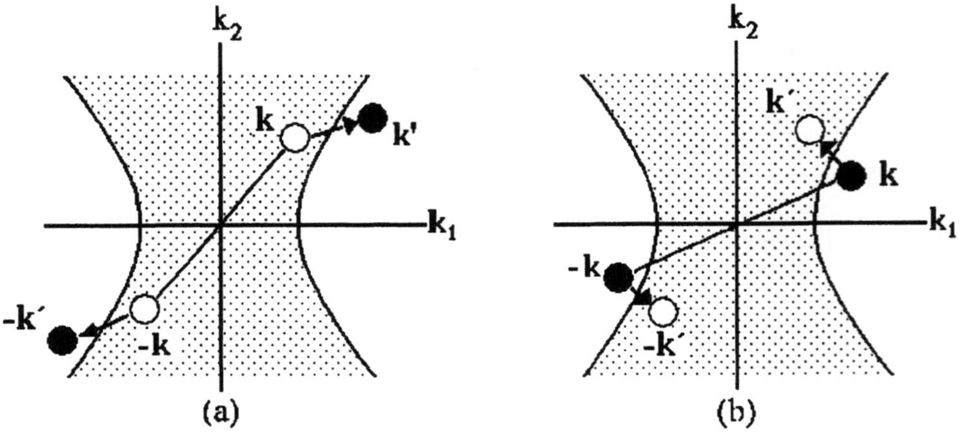

Figure 7.2: k-space diagrams representing (a) pair creation of ground pairons and (b) pair annihilation.

The same processes can be represented by Feynman diagrams in Fig. 7.3. We assume that the time flows upwards, and that "electrons" ("holes") proceed in the positive (negative) time directions. A phonon is electrically neutral; hence the total charge before and after the phonon exchange must be the same. The interaction Hamiltonians in Eqs. (7.7) and (7.8) all conserve charge. For type I elemental superconductors, we additionally assume that there are only acoustic phonons having a linear energy-momentum relation:

$$\varepsilon = c_s q, \tag{7.9}$$

where c_s is the sound speed.

The interaction strength v_{ij} can be the same for all (i,j) if the phonon exchange only is considered. By including the effect of the Coulomb repulsion, only the values of the strengths $v_{11} = v_{22}$ are reduced, (see below for the explanation); hence the following inequalities hold:

$$v_{11} = v_{22} < v_{12} = v_{21}. \tag{7.10}$$

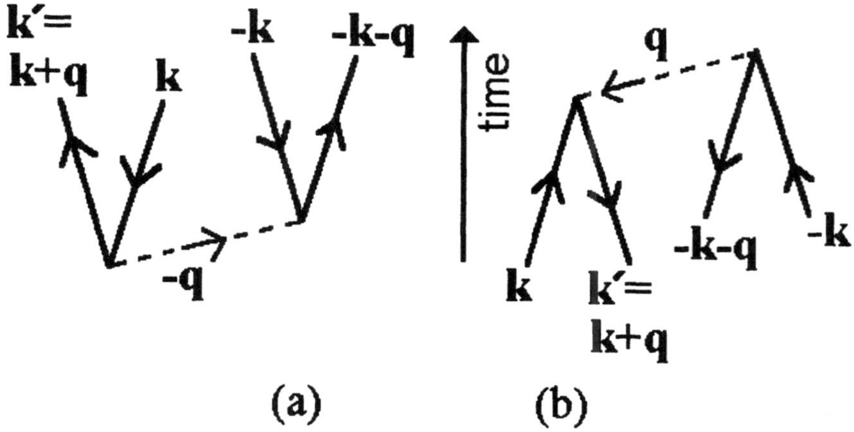

Figure 7.3: Feynman diagrams representing: (a) pair-creation of $\pm$ ground pairons from the physical vacuum, and (b) pair annililation.

The two forces, which are generated by the exchange of longitudinal acoustic phonons and photons are both long ranged. But the speeds of light and sound differ by a factor on the order of 10^5. Hence for the same k, the energies of light and sound are different by the same factor 10^5. It then follows that the photon-exchange interaction acts at a much shorter k-distance than the phonon-exchange interaction. This means that only the phonon-exchange attraction can pair-create or pair-annihilate $\pm$ pairons by emission and absorption of a virtual phonon with momentum $\mathbf{q}$ crossing over the Fermi surface. In other words exchange of a photon with an extremely small q can generate a correlation among pairons, while exchange of an acoustic phonon not only generates a correlation among pairons but also pair-creates or pair-annihilates $\pm$ pairons. This difference generates the inequalities (7.10). The zero-temperature coherence length ξ_0, which is a measure of the Cooper pair size, is on the order of 10^4 Å. This can be accounted for by the phonon-exchange attraction of a range of the order ξ_0, which is much greater than the lattice constant. The bare Coulomb repulsion has an even greater range, but this force can be screened by the motion of other electrons in the conductor. Therefore the effect of the Coulomb repulsion for the Cooper pair separated by 10^4 Å, is completely negligible. This means that for type I superconductors the interaction strengths are all equal to each other:

$$v_{11} = v_{22} = v_{12} \equiv v_0. \qquad \text{(type I)} \tag{7.11}$$

This assumption simplifies the algebra considerably. In this and the following chapters, we discuss mainly the case of elemental superconductors. In high-T_c superconductors, the interaction strengths v_{ij} are not equal because the Coulomb repulsion is not negligible and inequalities (7.10) hold. This case will be discussed separately in detail in Chapter 18. The theoretical treatments for both cases are not dissimilar. We often develop our theory without assuming Eq. (7.11) and derive general results; we then take the limit: $v_{11} = v_{22} \to v_{12}$ in the final expressions. A further discussion on the nature of the generalized BCS Hamiltonian will be given in Section 7.3.

7.2 The Ground State

In this section we look for the ground state of the generalized BCS system. At 0 K there are only $\pm$ ground pairons. The ground state Ψ for the system may then be constructed based on the reduced Hamiltonian H_0 in (7.4), which can be reexpresed in terms of pairon operators b's only: (Problem 7.2.1)

$$\begin{aligned} H_0 &= \sum_{\mathbf{k}} 2\varepsilon_k^{(1)}\, b_{\mathbf{k}}^{(1)\dagger} b_{\mathbf{k}}^{(1)} + \sum_{\mathbf{k}} 2\varepsilon_k^{(2)}\, b_{\mathbf{k}}^{(2)\dagger} b_{\mathbf{k}}^{(2)} - \sum_{\mathbf{k}}{}' \sum_{\mathbf{k}'}{}' [v_{11} b_{\mathbf{k}}^{(1)\dagger} b_{\mathbf{k}'}^{(1)} \\ &\quad + v_{12} b_{\mathbf{k}}^{(1)\dagger} b_{\mathbf{k}'}^{(2)\dagger} + v_{21} b_{\mathbf{k}}^{(2)} b_{\mathbf{k}'}^{(1)} + v_{22} b_{\mathbf{k}}^{(2)} b_{\mathbf{k}'}^{(2)\dagger}]. \end{aligned} \tag{7.12}$$

We calculate the commutation relations for pair operators and obtain (Problem 7.2.2.)

$$\begin{aligned} [b_{\mathbf{k}}^{(j)}, b_{\mathbf{k}'}^{(i)}] \equiv b_{\mathbf{k}}^{(j)} b_{\mathbf{k}'}^{(i)} - b_{\mathbf{k}'}^{(i)} b_{\mathbf{k}}^{(j)} = 0, \qquad [b_{\mathbf{k}}^{(j)}]^2 \equiv b_{\mathbf{k}}^{(j)} b_{\mathbf{k}}^{(j)} = 0, \\ [b_{\mathbf{k}}^{(j)}, b_{\mathbf{k}'}^{(i)\dagger}] = (1 - n_{\mathbf{k}\uparrow}^{(j)} - n_{-\mathbf{k}\downarrow}^{(j)})\delta_{\mathbf{k},\mathbf{k}'}\, \delta_{j,i}. \end{aligned} \tag{7.13}$$

Following BCS [1], we assume that the normalized ground state ket $|\Psi\rangle$ can be written as

$$|\Psi\rangle \equiv \prod_{\mathbf{k}}{}' \frac{1 + g_{\mathbf{k}}^{(1)} b_{\mathbf{k}}^{(1)\dagger}}{(1+ \mid g_{\mathbf{k}}^{(1)} \mid^2)^{1/2}} \prod_{\mathbf{k}'}{}' \frac{1 + g_{\mathbf{k}'}^{(2)} b_{\mathbf{k}'}^{(2)\dagger}}{(1+ \mid g_{\mathbf{k}'}^{(2)} \mid^2)^{1/2}} |0\rangle. \tag{7.14}$$

Here the ket $|0\rangle$ by definition satisfies

$$c_{\mathbf{k}s}^{(j)} |0\rangle \equiv c_{\mathbf{k}s}^{(j)} |\phi_1\rangle\, |\phi_2\rangle = 0. \tag{7.15}$$

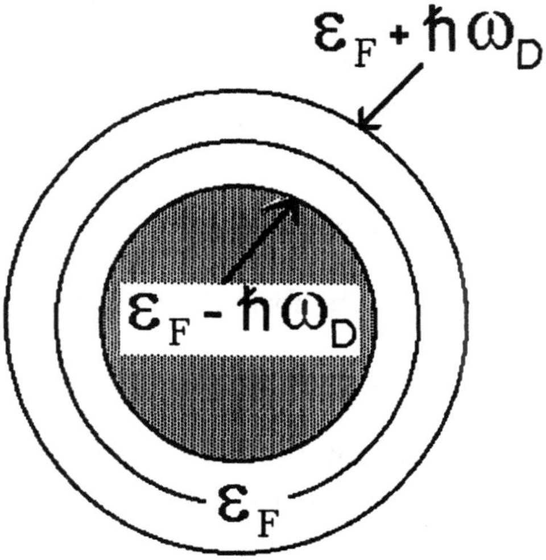

Figure 7.4: The k-space shell in 2D where pairons of both charge types are generated and intercorrelated.

It represents the *physical vacuum state* for "electrons" and "holes", that is, the ground state of the Bloch system with no "electrons" and no "holes" present. In Eq. (7.14) the product-variables $\mathbf{k}$ (and $\mathbf{k}'$) extend over the region of the momenta whose associated energies are bounded: $0 < \varepsilon_k^{(1)}, \varepsilon_k^{(2)} < \hbar\omega_D$, and this is indicated by the primes on the product symbols. The k-space shell in which pairons are generated and intercorrelated are shown in Fig. 7.4. Since $[b_{\mathbf{k}}^{(j)\dagger}]^2 = 0$, [see Eq. (7.13)], only two terms appear for each $\mathbf{k}\,(\mathbf{k}')$. The quantity $|\,g_{\mathbf{k}}^{(j)}|^2$ represents the probability that the pair states $(\mathbf{k}\uparrow, -\mathbf{k}\downarrow)$ are occupied. By expanding the product, we can see that the BCS ground state $|\Psi\rangle$ contains the zero-pairon state $|0\rangle$, one-pairon states $b_{\mathbf{k}}^{(j)\dagger}\,|0\rangle$, two-pairon states $b_{\mathbf{k}}^{(i)\dagger}b_{\mathbf{k}}^{(j)\dagger}\,|0\rangle\,, \ldots$. The physical meaning of the state $|\Psi\rangle$ will further be discussed in the paragraph after Eq. (7.28). The ket $|\Psi\rangle$ is normalized such that

$$\langle\Psi|\Psi\rangle = 1. \tag{7.16}$$

In the case where there is only one state $\mathbf{k}$ in the product, we obtain

$$\langle\Psi|\Psi\rangle = \langle 0|\,\frac{1+g_{\mathbf{k}}^{(1)}b_{\mathbf{k}}^{(1)}}{(1+\mid g_{\mathbf{k}}^{(1)}\mid^2)^{1/2}}\cdot\frac{1+g_{\mathbf{k}}^{(1)}b_{\mathbf{k}}^{(1)\dagger}}{(1+\mid g_{\mathbf{k}}^{(1)}\mid^2)^{1/2}}\,|0\rangle = 1.$$

The general case can be worked out similarly. (Problem 7.2.3.)

Since the ground state wavefunction has no nodes, we may choose $g_{\mathbf{k}}^{(j)}$ to be non-negative with no loss of rigor: $g_{\mathbf{k}}^{(j)} \geq 0$. We now determine $\{g_{\mathbf{k}}^{(j)}\}$ such that the ground state energy

$$W \equiv \langle\Psi| H_0 |\Psi\rangle \tag{7.17}$$

has a minimum value. This may be formulated by the extremum condition:

$$\delta W \equiv \delta \langle\Psi| H_0 |\Psi\rangle = 0. \tag{7.18}$$

The extremum problem meant by (7.18) with respect to the variation in g's can more effectively be solved by working with variations in the real probability amplitudes u's and v's defined by

$$u_{\mathbf{k}}^{(j)} \equiv [1+g_{\mathbf{k}}^{(j)2}]^{-1/2}, \qquad v_{\mathbf{k}}^{(j)} \equiv g_{\mathbf{k}}^{(j)}[1+g_{\mathbf{k}}^{(j)2}]^{-1/2}, \qquad u_{\mathbf{k}}^{(j)2}+v_{\mathbf{k}}^{(j)2} = 1. \tag{7.19}$$

The normalized ket $|\Psi\rangle$ can then be expressed by

$$|\Psi\rangle \equiv \prod_{\mathbf{k}}{}' (u_{\mathbf{k}}^{(1)} + v_{\mathbf{k}}^{(1)} b_{\mathbf{k}}^{(1)\dagger}) \prod_{\mathbf{k}'}{}' (u_{\mathbf{k}'}^{(2)} + v_{\mathbf{k}'}^{(2)} b_{\mathbf{k}'}^{(2)\dagger}) |0\rangle . \tag{7.20}$$

The energy W can be written from Eq. (7.17) as (Problem 7.2.4.)

$$W = \sum_{\mathbf{k}}{}' 2\varepsilon_k^{(1)} v_{\mathbf{k}}^{(1)2} + \sum_{\mathbf{k}'}{}' 2\varepsilon_{k'}^{(2)} v_{\mathbf{k}'}^{(2)2} - \sum_{\mathbf{k}}{}' \sum_{\mathbf{k}'}{}' \sum_i \sum_j v_{ij}\, u_{\mathbf{k}}^{(i)} v_{\mathbf{k}}^{(i)} u_{\mathbf{k}'}^{(j)} v_{\mathbf{k}'}^{(j)} . \tag{7.21}$$

Taking the variations in v's and u's, and noting that $u_{\mathbf{k}}^{(j)} \delta u_{\mathbf{k}}^{(j)} + v_{\mathbf{k}}^{(j)} \delta v_{\mathbf{k}}^{(j)} = 0$, we obtain from Eqs. (7.18) and (7.21) (Problem 7.2.5.)

$$2\varepsilon_k^{(j)} u_{\mathbf{k}}^{(j)} v_{\mathbf{k}}^{(j)} - (u_{\mathbf{k}}^{(j)2} - v_{\mathbf{k}}^{(j)2}) \sum_{\mathbf{k}'}{}' [v_{j1} u_{\mathbf{k}'}^{(1)} v_{\mathbf{k}'}^{(1)} + v_{j2} u_{\mathbf{k}'}^{(2)} v_{\mathbf{k}'}^{(2)}] = 0. \tag{7.22}$$

To simply treat these equations subject to Eqs. (7.19), we introduce a set of energy-parameters:

$$\Delta_{\mathbf{k}}^{(j)}, \qquad E_k^{(j)} \equiv (\varepsilon_k^{(j)2} + \Delta_{\mathbf{k}}^{(j)2})^{1/2}$$

such that

$$u_{\mathbf{k}}^{(j)2} - v_{\mathbf{k}}^{(j)2} = \varepsilon_k^{(j)} / E_k^{(j)}, \qquad u_{\mathbf{k}}^{(j)} v_{\mathbf{k}}^{(j)} = \Delta_{\mathbf{k}}^{(j)} / 2E_k^{(j)} . \tag{7.23}$$

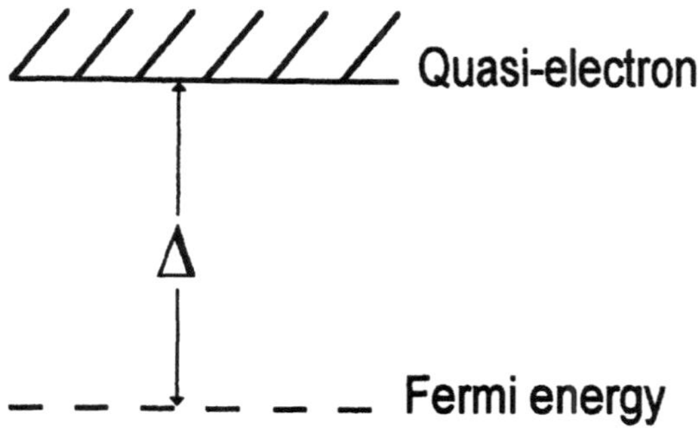

Figure 7.5: Quasi-electrons have an energy gap Δ relative to the Fermi energy.

(Problem 7.2.6.). Then, Eqs. (7.22) can be reexpressed as

$$\Delta_{\mathbf{k}}^{(j)} = \sum_{\mathbf{k}'}{}' \sum_{i=1}^{2} v_{ij} \frac{\Delta_{\mathbf{k}'}^{(i)}}{2E_{\mathbf{k}'}^{(i)}}. \tag{7.24}$$

Since the rhs of Eqs. (7.24) does not depend on **k**, the "energy gaps"

$$\Delta_{\mathbf{k}}^{(j)} \equiv \Delta_j \tag{7.25}$$

are independent of **k**. Hence we can simplify Eqs. (7.24) to

$$\Delta_j = \sum_{\mathbf{k}'}{}' \sum_{i} v_{ij} \frac{\Delta_i}{2E_{\mathbf{k}'}^{(i)}}. \tag{7.26}$$

These are called *generalized energy gap equations.* As we shall see later, $E_k^{(j)}$ is the energy of an unpaired electron or (quasi-electron). Quasi-electrons have energy gaps $\Delta^{(j)}$ relative to the Fermi energy as shown in Fig. 7.5. Notice that there are in general "electron" and "hole" energy gaps, (Δ_1, Δ_2).

Using Eqs. (7.23)-(7.26), we calculate the energy W and obtain (Problem 7.2.7.)

$$\begin{aligned} W &\equiv \sum_{\mathbf{k}}{}' \sum_{j} 2\varepsilon_k^{(j)} v_{\mathbf{k}}^{(j)2} - \sum_{\mathbf{k}}{}' \sum_{\mathbf{k}'}{}' \sum_{i} \sum_{j} v_{ij} u_{\mathbf{k}}^{(i)} v_{\mathbf{k}}^{(i)} u_{\mathbf{k}'}^{(j)} v_{\mathbf{k}'}^{(j)} \\ &= \sum_{\mathbf{k}}{}' \sum_{j} \{\varepsilon_k^{(j)} [1 - \frac{\varepsilon_k^{(j)}}{E_k^{(j)}}] - \frac{\Delta_j^2}{2E_k^{(j)}}\}. \end{aligned} \tag{7.27}$$

In the bulk limit the sums over **k** are converted into energy integrals, yielding

$$W = \sum_{j=1}^{2} \mathcal{N}_j(0) \int_0^{\hbar\omega_D} d\varepsilon \, [\varepsilon - \frac{\varepsilon^2}{(\varepsilon^2 + \Delta_j^2)^{1/2}} - \frac{\Delta_j^2}{2(\varepsilon^2 + \Delta_j^2)^{1/2}}]. \qquad (7.28)$$

The ground state $|\Psi\rangle$ from Eq. (7.14) is a superposition of many-pairon states. Each component state can be reached from the physical vacuum state $|0\rangle$ by pair-creation and/or pair-annihilation of $\pm$ pairons and pair-state change through a succession of phonon exchanges. Since the phonon-exchange processes, as represented by Eq. (7.8), can pair-create (or pair-annihilate) $\pm$ pairons simultaneously from the physical vacuum, the supercondensate is composed of equal numbers of $\pm$ pairons. We can see from Fig. 7.3 that the maximum numbers of $+(-)$ pairons are given by $(1/2)\hbar\omega_D\mathcal{N}_1(0)$ $[(1/2)\hbar\omega_D\mathcal{N}_2(0)]$. We must then have

$$\mathcal{N}_1(0) = \mathcal{N}_2(0) \equiv \mathcal{N}(0), \qquad (7.29)$$

which appears to be unrealistic, but it will be justified by the assumption that the supercondensate is generated only on part of the Fermi surface, (see Section 7.3.8).

Using Eq. (7.29), we obtain from Eq. (7.28) (Problem 7.2.8.)

$$W = \frac{1}{2} N_0 (w_1 + w_2), \qquad w_i \equiv \hbar\omega_D\{1 - [1 + (\Delta_i/\hbar\omega_D)^2]^{1/2}\} < 0. \qquad (7.30)$$

We thus find that the ground state energy of the generalized BCS system is negative, that is, the energy is *lower* than that of the Bloch system. Further note that the binding energy $|w_i|$ per pairon may in general be different for different charge types.

Let us now find Δ_j from the gap equations (7.26). In the bulk limit, these equations are simplified to

$$\begin{aligned} \Delta_j &= \frac{1}{2} v_{j1} \mathcal{N}(0) \int_0^{\hbar\omega_D} d\varepsilon \frac{\Delta_1}{(\varepsilon^2 + \Delta_1^2)^{1/2}} + \frac{1}{2} v_{j2} \mathcal{N}(0) \int_0^{\hbar\omega_D} d\varepsilon \frac{\Delta_2}{(\varepsilon^2 + \Delta_2^2)^{1/2}} \\ &= \frac{v_{j1}}{2} \mathcal{N}(0) \Delta_1 \sinh^{-1}(\hbar\omega_D/\Delta_1) + \frac{v_{j2}}{2} \mathcal{N}(0) \Delta_2 \sinh^{-1}(\hbar\omega_D/\Delta_2). \qquad (7.31) \end{aligned}$$

For type I superconductors, we assume that the interaction strengths v_{ij} are all equal to each other: $v_{11} = v_{12} = v_{22} \equiv v_0$. We see from Eqs. (7.31) that "electron" and "hole" energy gaps coincide:

$$\Delta_1 = \Delta_2 \equiv \Delta. \qquad (7.32)$$

Equations (7.26) are then reduced to a single equation:

$$\Delta = \sum_{\mathbf{k}'}{}' \sum_i v_0 \frac{\Delta}{2E_{\mathbf{k}'}^{(i)}}, \tag{7.33}$$

which is called the BCS *energy gap equation.* After dropping the common factor Δ and taking the bulk limit, we obtain (Problem 8.5.7.) $1 = v_0\mathcal{N}(0)\sinh^{-1}(\hbar\omega_D/\Delta)$ or

$$\Delta = \frac{\hbar\omega_D}{\sinh[1/v_0\mathcal{N}(0)]}. \tag{7.34}$$

Equation (7.34) is similar to Eq. (6.10) for the binding energy of the ground pairon $|w_0|$. The exponential factor, however, is a little different. The factor $\exp[1/v_0\mathcal{N}(0)]$ appears in Eq. (7.34) as opposed to the factor $\exp[2/v_0\mathcal{N}(0)]$ in Eq. (6.10).

We now substitute Eq. (7.34) into Eq. (7.30) and calculate the ground state energy. After straightforward calculations, we obtain (Problem 7.2.10.)

$$W = \frac{-2\mathcal{N}(0)\hbar^2\omega_D^2}{\exp[2/v_0\mathcal{N}(0)] - 1} \quad (= N_0 w_0). \tag{7.35}$$

Equations (7.34) and (7.35) are the famous BCS formulas for the energy gap and the ground state energy, respectively. They correspond to Eqs. (2.40) and (2.42) of the original paper [1]. We stress that these results are *exact,* without assuming the weak coupling limit ($v_0 \to 0$), and that they were obtained from the reduced BCS Hamiltonian H_0.

Problem 7.2.1. Verify Eq. (7.12).

Problem 7.2.2. Derive Eqs. (7.13).

Problem 7.2.3. Verify Eq. (7.16). Hint: Assume that there are only two k-states in the product. If successful, then treat the general case.

Problem 7.2.4. Derive Eq. (7.21).

Problem 7.2.5. Derive Eq. (7.22).

Problem 7.2.6. Check the consistency of Eqs. (7.19) and (7.23). Use the identity: $(u^2 + v^2)^2 - (u^2 - v^2)^2 = 4u^2v^2$.

Problem 7.2.7. Verify Eq. (7.27).

Problem 7.2.8. Verify Eq. (7.30).

Problem 7.2.9. Verify Eq. (7.34).

Problem 7.2.10. Derive Eq. (7.35).

7.3 Discussion

We have uncovered several significant features of the ground state of the generalized BCS system.

7.3.1 The Nature of the Reduced Hamiltonian

The reduced BCS Hamiltonian H_0 in Eq. (7.4) has a different character from the normal starting Hamiltonian for a metal, which is composed of interacting electrons and ions. Bardeen Cooper and Schrieffer envisioned that there are only zero-momentum pairons at 0 K. Only the basic ingredients to build up pairons are introduced and incorporated in the BCS Hamiltonian. Both "electrons" and "holes" are introduced from the outset. These particles are the elementary excitations in the normal state.

7.3.2 Binding Energy per Pairon

We may rewrite Eq. (7.35) for the ground state energy in the form:

$$W = N_0 w_0, \quad N_0 = \hbar\omega_D \mathcal{N}(0), \qquad w_0 = \frac{-2\hbar\omega_D}{\exp[2/v_0\mathcal{N}(0)] - 1}, \tag{7.36}$$

which can be interpreted as follows: The greatest total number of pairons generated consistent with the BCS Hamiltonian is equal to $\hbar\omega_D\mathcal{N}(0) = N_0$. Each pairon contributes a binding energy $|w_0|$. This energy $|w_0|$ can be measured directly by quantum tunneling experiments as we shall see in Chapter 11. Our interpretation of the ground state energy is quite natural, but it is distinct from that of the BCS theory, where the energy gap Δ is regarded as a measure of the binding energy. Our calculations does not support this view, see section 7.3.4.

7.3.3 Critical Field B_c and Binding Energy $|w_0|$

By the Meissner effect a superconductor expels a weak magnetic field **B** from its interior. The magnetic energy stored is higher in proportion to B^2 *and* the excluded volume than that for the uniform B-flux configuration. For a macroscopic superconductor the difference in the energy is given by

$$\frac{VB^2}{2\mu_0}. \tag{7.37}$$

If this energy exceeds the difference of the energy between super and normal conductors, $W_S - W_N$, which is equal to $|W_0|$, the superconducting state should break down. The minimum magnetic field B_c that destroys the superconducting state is the *critical field* at 0 K, $B_c(0) \equiv B_0$. We therefore obtain

$$|W_S - W_N| = |W_0| = N_0 |w_0| = \frac{1}{2} V B_0^2 \mu_0^{-1}, \tag{7.38}$$

which gives a rigorous relation between $|w_0|$ and B_0.

7.3.4 The Energy Gap

In the process of obtaining the ground state energy W by the variational calculation, we derived energy-gap equations, Eq. (7.26) which contain the energy parameters

$$E_k^{(j)} \equiv (\varepsilon_k^{(j)2} + \Delta_j^2)^{1/2}. \tag{7.39}$$

The fact that $E_k^{(j)}$ represents the energy of a *quasi-electron*, can be seen as follows [5]. The quasiparticle energy is defined to be the total excitation energy of the system when an extra particle is added to the system. From Eq. (7.21) we see that by negating the pair state $(\mathbf{k}\uparrow, -\mathbf{k}\downarrow)$, the energy is increased by

$$\begin{aligned} & -2\varepsilon_k^{(1)} v_{\mathbf{k}}^{(1)2} + 2\{\sum_{\mathbf{k}'}{}' [v_{11} u_{\mathbf{k}'}^{(1)} v_{\mathbf{k}'}^{(1)} + v_{12} u_{\mathbf{k}'}^{(2)} v_{\mathbf{k}'}^{(2)}]\} u_{\mathbf{k}}^{(1)} v_{\mathbf{k}}^{(1)} \\ = & -2\varepsilon_k^{(1)} v_{\mathbf{k}}^{(1)2} + 2\Delta_1 u_{\mathbf{k}}^{(1)} v_{\mathbf{k}}^{(1)}, \end{aligned} \tag{7.40}$$

where we used Eqs. (7.23) and (7.24). To this we must add the energy $\varepsilon_k^{(1)}$ of the added "electron". Thus the total excitation energy $\Delta\varepsilon$ is

$$\Delta\varepsilon = \varepsilon_k^{(1)} [1 - 2v_{\mathbf{k}}^{(1)2}] + 2\Delta_1 u_{\mathbf{k}}^{(1)} v_{\mathbf{k}}^{(1)} = E_k^{(1)}. \tag{7.41}$$

Thus, the unpaired electron has the energy $E_k^{(1)}$ as shown in Fig. 7.5. The validity domain for the above statement is $0 < \varepsilon_k^{(1)} < \hbar\omega_D$.

7.3.5 The Energy Gap Equations

The reduced Hamiltonian H_0 was expressed in terms of pairon operators b's only as in Eq. (7.12). The ground state Ψ in Eq. (7.14) contains b's only. Yet in the energy-gap equations, which result from the extremum condition for the ground state energy, the energies of quasi-electron, $E_k^{(j)}$, appear unexpectedly. Generally speaking the physics is lost in the variational calculation. We shall derive the gap equations from a different angle by using the equation-of-motion method in Chapter 10.

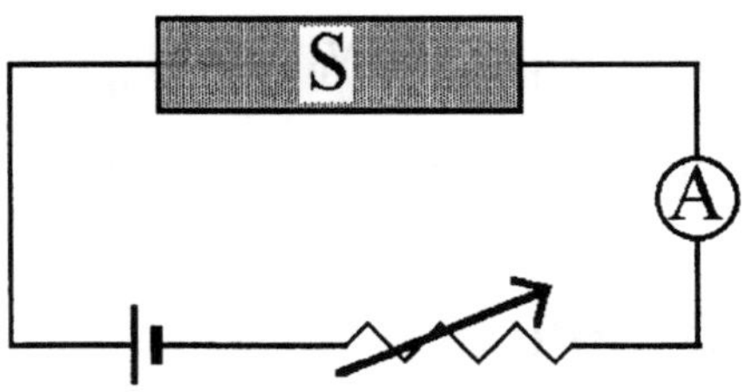

Figure 7.6: A circuit containing a superconductor (S), battery and resistance.

7.3.6 Neutral Supercondensate

The supercondensate composed of equal numbers of $\pm$pairons is electrically *neutral*. This neutrality explains the stability against a weak electric field because no Lorentz electric force can be exerted on the supercondensate. This stability is analogous to that of a stationary excited atomic state, say, the $2p$-state of a neutral hydrogen atom.

A neutral supercondensate is supported by experiments. If a superconducting wire S is used as part of a circuit connected to a battery, as shown in Fig. 7.6. the wire S, having no resistance, generates no potential drop. If a low-frequency AC voltage is applied to it, its response becomes more complicated. But the behavior can be accounted for if we assume that it has a normal component with a finite resistance and a super part. This is the *two fluid model* [6]. The super part, or supercondensate, decreases with rising temperature and vanishes at T_c. The normal part may be composed of any charged elementary excitations including quasi-electrons and excited pairons. At any rate, analyses of all experiments indicate that the *supercondensate is not accelerated by the electric force.* This must be so, otherwise the supercondensate would gain energy without limit since the supercurrent is slowed down by neither impurities nor phonons, and a stationary state would never have been observed in the circuit.

7.3.7 Cooper Pairs (Pairons)

The concept of pairons is inherent in the BCS theory, which is most clearly seen in the reduced Hamiltonian H_0, expressed in terms of pairon operators b's only. The direct evidence for the fact that a Cooper pair is a bound quasi-particle having charge (magnitude) $2e$ comes from flux quantization experiments.

7.3.8 Formation of a Supercondensate and Occurrence of Superconductors

We now discuss the formation of a supercondensate based on the band structures of electrons and phonons. Let us first take lead (Pb), which forms an fcc lattice and which is a superconductor. This metal is known to have a neck-like hyperboloidal Fermi surface represented by [7]

$$E = \frac{p_1^2}{2m_1} + \frac{p_2^2}{2m_2} + \frac{p_3^2}{2m_3}, \qquad (m_1, m_2, m_3) = (1.18, 0.244, -8.71)m. \quad (7.42)$$

We postulate that the supercondensate composed of $\pm$ ground pairons is generated near the "necks". The electron transitions are subject to Pauli's exclusion principle, and hence creating pairons require a high degree of symmetry in the Fermi surface. A typical way of generating pairons of both charge types by one phonon exchange near the neck is shown in Fig. 7.3. Only part of "electrons" and "holes" near the specific part of the Fermi surface are involved in the formation of the supercondensate. The numbers of $\pm$ pairons, which are mutually equal by construction, may both then be represented by $\hbar_D\omega\mathcal{N}(0)/2$, which justifies Eq. (7.29). Next take aluminum (Al), which is also a known fcc superconductor. Its Fermi surface contains inverted double caps. Acoustic phonons with small momenta may generate a supercondensate near the inverted double caps. Supercondensation occurs independently of the lattice structure. Beryllium (Be) forms a hcp crystal. Its Fermi surface in the second zone, shown in Fig. 3.4, has necks. Thus, Be is a superconductor. Tungsten (W) is a bcc metal, and its Fermi surface, shown in Fig. 3.5, has necks. This metal is also a superconductor. In summary type I elemental superconductors should have hyperboloidal Fermi surfaces favorable for the creation of $\pm$ pairons mediated by small-momentum phonons. All of the elemental superconductors whose Fermi surfaces are known appear to satisfy this condition.

To test further let us consider a few more examples. A monovalent metal, such as sodium (Na), has a nearly spherical Fermi surface within the first Brillouin zone. Such a metal cannot become superconducting at any temperature since it does not have "holes" to begin with; it cannot have $+$ pairons and therefore cannot form a neutral supercondensate. A monovalent fcc metal like Cu has a set of necks at the Brillouin boundary. This neck is forced by the inversion symmetry of the lattice (see Fig. 3.3). The region of the hyperboloidal Fermi surface may be more severely restricted than those necks (unforced) in Pb. Thus this metal may become superconducting at extremely low temperatures, which is not ruled out but unlikely.

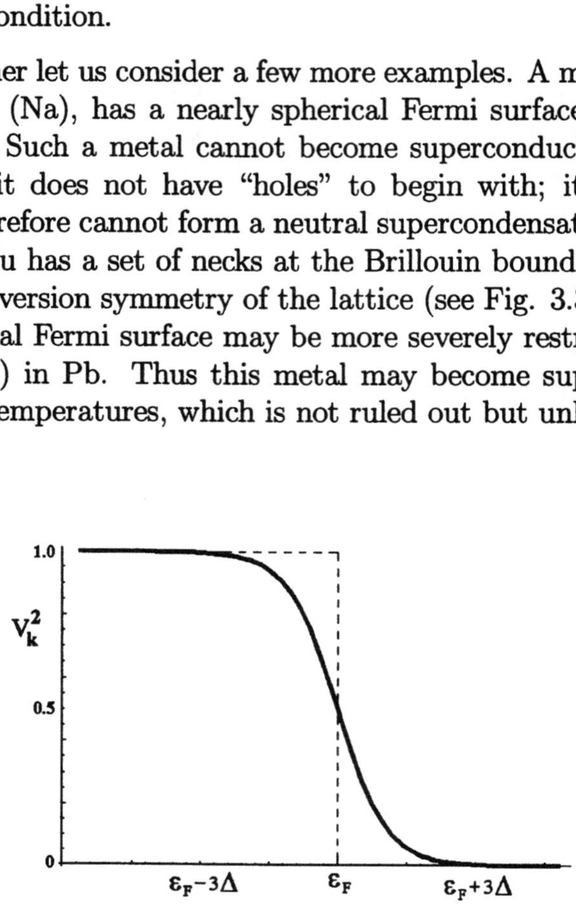

Figure 7.7: The behavior of $v_{\mathbf{k}}^2$ near the Fermi energy.

7.3.9 Blurred Fermi Surface

In Sections 3.2 - 3.3, we saw that a normal metal has a sharp Fermi surface at 0 K. This fact manifests itself in the T-linear heat capacity universally observed at the lowest temperatures. The T-linear law is in fact the most important support for the Fermi liquid model. For a superconductor the Fermi surface is not sharp everywhere. To see this, let us solve Eq. (7.23) with respect to $u_{\mathbf{k}}^2$ and $v_{\mathbf{k}}^2$ in the BCS limits. We obtain

$$v_{\mathbf{k}}^2 = \frac{1}{2}[1 - \varepsilon_k/(\varepsilon_k^2 + \Delta^2)^{1/2}], \qquad u_{\mathbf{k}}^2 = \frac{1}{2}[1 + \varepsilon_k/(\varepsilon_k^2 + \Delta^2)^{1/2}]. \qquad (7.43)$$

Fig. 7.7 shows a general behavior of $v_{\mathbf{k}}^2$ near the Fermi energy. For the normal state $\Delta = 0$, there is a sharp boundary at $\varepsilon_k = 0$; but for a finite Δ, the quantity $v_{\mathbf{k}}^2$ drops off to zero over a region of the order $2 \sim 3\Delta$. This $v_{\mathbf{k}}^2$ represents the probability that the virtual electron pair at $(\mathbf{k} \uparrow, -\mathbf{k} \downarrow)$ participates in the formation of the supercondensate. It is *not* the probability that either electron of the pair occupies the state $\mathbf{k}$. Still, the diagram indicates the nature of the changed electron distribution in the ground state. The supercondensate is generated only near the necks and/or inverted double caps. Thus these parts of the Fermi surface are blurred or fuzzy.

Chapter 8

Quantum Statistics of Composites

8.1 Ehrenfest-Oppenheimer-Bethe's Rule

Experiments indicate that every quantum particle in nature moves either as a boson or as a fermion [1]. This statement, applied to elementary particles, is known as the *quantum statistical postulate* (or *principle*). Bosons (fermions), by definition, can (cannot) multiply occupy one and the same quantum-particle state. Spin and isopin (charge), which are part of particle state variables are included in the definition of the state. Electrons (e) and nucleons (protons p, neutrons n) are examples of elementary fermions [1, 2]. Composites such as deuterons (p, n) tritons $(p, 2n)$, hydrogen H (p, e) are indistinguishable and obey quantum statistics. According to Ehrenfest-Oppenheimer-Bethe (EOB) rule [3, 4] a composite is fermionic (bosonic) if it contains an odd (even) number of elementary fermions. Let us review the arguments leading to EOB's rule as presented in Bethe-Jackiw's book [4]. Take a composite of two identical fermions and study the symmetry of the wavefunction for two composites, which has four particle-state variables, two for the first composite and two for the second one. Imagine that the exchange between the two composites is carried out particle by particle. Each exchange of fermions (bosons) changes the wavefunction by the factor -1 $(+1)$. In the present example, the sign changes twice and the wavefunction is therefore unchanged. If a composite contains different types of particle as in the case of H, the symmetry of the wavefunction is deduced by the interchange within each type. We shall see later that these arguments are incomplete. We note that Feynman used these arguments to deduce that Cooper pairs [5] (pairons) are bosonic [6]. The symmetry of the many-particle wavefunction and the quantum statistics for elementary particles are one-to-one [1]. A set of elementary fermions (bosons) can be described in terms of creation and annihilation operators satisfying the Fermi anticommutation (Bose commutation) rules, see Eqs. (8.2) and (8.24). But no one-to one correspondence exists for composites since composites by construction have extra degrees of freedom. Wavefunctions and second-quantized operators are important auxiliary quantum variables but they are not observables in Dirac's sense [1]. We must examine the observable occupation numbers for the study of the quantum statistics of composites. In the present chapter we shall show that EOB's rule applies to the Center-of-Mass (CM) motion of composites.

8.2 Two-Particle Composites

Let us consider two-particle composites. There are four important cases represented by (**A**) electron-electron (pairon), (**B**) electron-proton (hydrogen H), (**C**) nucleon-pion, and (**D**) boson-boson.

(**A**) **Identical fermion composite**. Second-quantized operators for a pair of electrons are defined by [8]

$$B_{12}^{\dagger} \equiv B_{\mathbf{k}_1\mathbf{k}_2}^{\dagger} \equiv c_{\mathbf{k}_1}^{\dagger} c_{\mathbf{k}_2}^{\dagger} \equiv c_1^{\dagger} c_2^{\dagger}, \qquad B_{34} = c_4 c_3, \tag{8.1}$$

where $c_{\mathbf{k}_1}^{\dagger} (c_{\mathbf{k}_1}) \equiv c_1^{\dagger}(c_1)$ are creation (annihilation) operators (spins indices omitted) satisfying the Fermi anticommutation rules:

$$\{c_1,\ c_2^{\dagger}\} \equiv c_1 c_2^{\dagger} + c_2^{\dagger} c_1 = \delta_{\mathbf{k}_1\mathbf{k}_2}, \qquad \{c_1,\ c_2\} = 0. \tag{8.2}$$

The commutators among B and $B^{\dagger}$ can be computed by using (8.2) and are given by [8]

$$[B_{12},\ B_{34}] \equiv B_{12}B_{34} - B_{34}B_{12} = 0, \qquad (B_{12})^2 = 0, \tag{8.3}$$

$$[B_{12},\ B_{34}^{\dagger}] = \begin{cases} 1 - n_1 - n_2 & \text{if} \quad k_1 = k_3,\ k_2 = k_4 \\ c_2 c_4^{\dagger} & \text{if} \quad k_1 = k_3,\ k_2 \neq k_4 \\ c_1 c_3^{\dagger} & \text{if} \quad k_1 \neq k_3,\ k_2 = k_4 \\ 0 & \text{otherwise,} \end{cases} \tag{8.4}$$

where

$$n_j = c_j^\dagger c_j \qquad (j = 1, 2) \tag{8.5}$$

represent the number operators for electrons. Using Eqs. (8.1)-(8.5) and

$$n_{12} \equiv B_{12}^\dagger B_{12}, \tag{8.6}$$

we obtain

$$n_{12}^2 = B_{12}^\dagger (1 - n_1 - n_2 + B_{12}^\dagger B_{12}) B_{12} = n_{12}. \tag{8.7}$$

Hence

$$(n_{12}^2 - n_{12}) \mid n_{12}^{'}\rangle = (n_{12}^{'2} - n_{12}^{'}) \mid n_{12}^{'}\rangle = 0, \qquad \mid n_{12}^{'}\rangle \neq 0,$$

yielding

$$n_{12}^{'} = 0 \text{ or } 1. \tag{8.8}$$

Let us now introduce the relative and net (or CM) momenta $(\mathbf{k}, \mathbf{q})$ such that

$$\mathbf{k} \equiv \frac{1}{2}(\mathbf{k}_1 - \mathbf{k}_2), \qquad \mathbf{q} \equiv \mathbf{k}_1 + \mathbf{k}_2; \qquad \mathbf{k}_1 = \mathbf{k} + \frac{1}{2}\mathbf{q}, \qquad \mathbf{k}_2 = -\mathbf{k} + \frac{1}{2}\mathbf{q}. \tag{8.9}$$

We may alternatively represent the pair operators by

$$B_{\mathbf{kq}}^{'} \equiv c_{-\mathbf{k}+\frac{1}{2}\mathbf{q}} c_{\mathbf{k}+\frac{1}{2}\mathbf{q}} \equiv B_{12}, \qquad B_{\mathbf{kq}}^{'\dagger} \equiv c_{\mathbf{k}+\frac{1}{2}\mathbf{q}}^\dagger c_{-\mathbf{k}+\frac{1}{2}\mathbf{q}}^\dagger. \tag{8.10}$$

The prime on $B_{\mathbf{kq}}$ will be dropped hereafter. From Eq. (8.8) we deduce that the number operator in the k-q representation.

$$n_{\mathbf{kq}} \equiv B_{\mathbf{kq}}^\dagger B_{\mathbf{kq}}, \tag{8.11}$$

has eigenvalues 0 or 1:

$$n_{\mathbf{kq}}^{'} = 0 \text{ or } 1. \tag{8.12}$$

The total number of a system of pairons, N, is represented by

$$N \equiv \sum_{\mathbf{k}_1}\sum_{\mathbf{k}_2} n_{12} = \sum_{\mathbf{k}}\sum_{\mathbf{q}} n_{\mathbf{kq}} = \sum_{\mathbf{q}} n_{\mathbf{q}}, \tag{8.13}$$

where

$$n_{\mathbf{q}} \equiv \sum_{\mathbf{k}} n_{\mathbf{kq}} = \sum_{\mathbf{k}} B_{\mathbf{kq}}^\dagger B_{\mathbf{kq}} \tag{8.14}$$

represents the number of pairons having net momentum $\mathbf{q}$. From Eqs. (8.12)-(8.14) we see that the eigenvalues of the number operator $n_{\mathbf{q}}$ can be non-negative integers. To explicitly see this property, we introduce

$$B_{\mathbf{q}} \equiv \sum_{\mathbf{k}} B_{\mathbf{kq}}, \tag{8.15}$$

and obtain, after using Eqs. (8.2)-(8.5) and (8.10),

$$[B_{\mathbf{q}},\ n_{\mathbf{q}}] = \sum_{\mathbf{k}} (1 - n_{\mathbf{k}+\frac{1}{2}\mathbf{q}} - n_{-\mathbf{k}+\frac{1}{2}\mathbf{q}}) B_{\mathbf{kq}} = B_{\mathbf{q}}, \qquad [n_{\mathbf{q}},\ B_{\mathbf{q}}^{\dagger}] = B_{\mathbf{q}}^{\dagger}. \tag{8.16}$$

Although the occupation number $n_{\mathbf{q}}$ is not connected with $B_{\mathbf{q}}$ as $n_{\mathbf{q}} \neq B_{\mathbf{q}}^{\dagger} B_{\mathbf{q}}$, the eigenvalues $n'_{\mathbf{q}}$ of $n_{\mathbf{q}}$ satisfying Eqs. (8.16) can be shown straightforwardly to yield [1]

$$n'_{\mathbf{q}} = 0, 1, 2, \ldots . \tag{8.17}$$

with the eigenstates

$$|0\rangle,\ |1\rangle = B_{\mathbf{q}}^{\dagger} |0\rangle,\ |2\rangle = B_{\mathbf{q}}^{\dagger} B_{\mathbf{q}}^{\dagger} |0\rangle, \ldots \tag{8.18}$$

This is important. We illustrate it by taking a one-dimensional motion. The pairon occupation-number states may be represented by drawing quantum cells in the (k, q) space. From Eq. (8.12) the number n'_{kq} are limited to 0 or 1, see Fig. 8.1. The number of pairons characterized by net momentum q only, n'_{q}, is the sum of the numbers of pairs at column q, and clearly it is zero or a positive integer.

In summary, pairons with both $\mathbf{k}$ and $\mathbf{q}$ specified are subject to the Pauli exclusion principle, see Eq. (8.12). Yet, the occupation numbers $n'_{\mathbf{q}}$ of pairons having CM momentum $\mathbf{q}$ are $0, 1, 2, \ldots$, see Eq. (8.17). Note that our results Eqs. (8.8), (8.12) and (8.17) are obtained by using the pair commutators (8.3)-(8.4). Further note that our result (8.17) does not follow from consideration of the symmetry of the wavefunction, the symmetry arising from Eqs. (8.3) only. Eq. (8.4) is needed to prove Eq. (8.16). The fact that the quantum statistics depends on whether we specify $(\mathbf{k}, \mathbf{q})$ or $\mathbf{q}$ alone, arises because a composite by construction has more degrees of freedom than an elementary particle (electron). Only with respect to the CM motion pairons are bosonic and can multiply occupy the same momentum state $\mathbf{q}$. We say in short that pairons move as bosons.

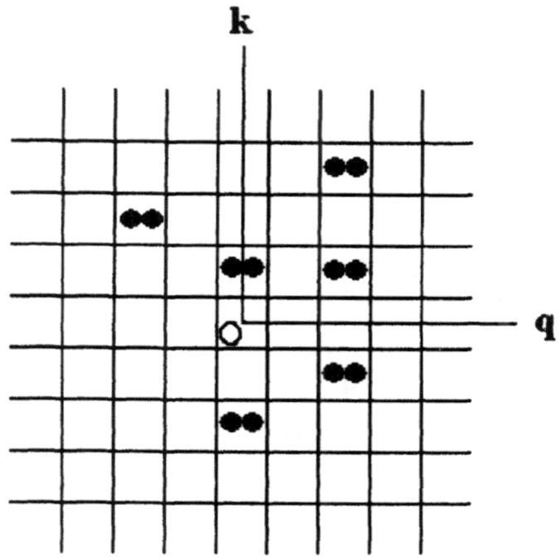

Figure 8.1: The number representation of many electron-pairs in the (k, q) space.

(B) Different-fermion composite. The quantum state for two *distinguishable* particles $(1, 2)$ can be represented by

$$| \mathbf{k}_a^{(1)}, \mathbf{k}_b^{(2)} \rangle \equiv | \mathbf{k}_a^{(1)} \rangle | \mathbf{k}_b^{(2)} \rangle. \tag{8.19}$$

We may represent the state $| \mathbf{k}_a^{(j)} \rangle$ for the particle j by specifying a set of occupation numbers $(n_a^{'(j)}, n_b^{'(j)}, ...)$ with the restriction that each $n_a^{'(j)}$ can take on a value either 0 or 1 and only one member of the set takes the value 1. These numbers $n_a^{'(j)}$ can be represented by [10]

$$n_a^{(j)} \equiv \eta_a^{(j)\dagger} \eta_a^{(j)}, \tag{8.20}$$

where creation (annihilation) operators $\eta_a^{(j)\dagger}$ $(\eta_a^{(j)})$ satisfy the Fermi anticommutation rules (8.2). The quantum state for a many-electron-many-proton system may be represented by a generalization of Eq. (8.19), the direct product of a many-electron (antisymmetric) state and a many-proton (antisymmetric) state. Such states can be described in terms of second-quantized operators $c's$ (electrons) and $a's$ (protons), both satisfying the anticommutation rules (8.2). Following Dirac [1], we postulate that observables for different particles commute:

$$[n_a^{(1)},\ n_b^{(2)}] = 0, \tag{8.21}$$

based on which we may choose such that $c's$ and $a's$ anticommute with each other:

$$\{c,\ a^{\dagger}\} = \{c,\ a\} = 0. \tag{8.22}$$

Pair operators are defined by

$$B_{12}^{\dagger} \equiv a_1^{\dagger} c_2^{\dagger}, \qquad B_{34} \equiv c_4 a_3. \tag{8.23}$$

We study the number operator $n_{\mathbf{q}}^{(H)}$ defined in the form (8.14) and show by means of Eq. (8.16) that the eigenvalues of $n_{\mathbf{q}}^{(H)}$ are $0, 1, 2, ...$ That is, hydrogens H move as bosons.

(C) Fermion-boson composite. We define pair operators in the form (8.23) with boson operators ($b's$) satisfying the Bose commutation rules:

$$[b_1,\ b_2^{\dagger}] \equiv b_1 b_2^{\dagger} - b_2^{\dagger} b_1 = \delta_{k_1 k_2}, \qquad [b_1,\ b_2] = 0. \tag{8.24}$$

Fermion and boson operators mutually anticommute, which is in accord with Eq. (8.21). We obtain

$$\{B_{12},\ B_{34}\} \equiv B_{12}B_{34} + B_{34}B_{12} = 0 \tag{8.25}$$

$$\{B_{12},\ B_{34}^{\dagger}\} = \begin{cases} 1 + n_1 - n_2 & \text{if } k_1 = k_3, k_2 = k_4 \\ c_2 c_4^{\dagger} & \text{if } k_1 = k_3, k_2 \neq k_4 \\ b_1 b_3^{\dagger} & \text{if } k_1 \neq k_3, k_2 = k_4 \\ 0 & \text{otherwise.} \end{cases} \tag{8.26}$$

We define $n_{\mathbf{q}}$ and $B_{\mathbf{q}}$ as in Eqs. (8.14) and (8.15), use Eq. (8.26) and obtain

$$[n_{\mathbf{q}},\ B_{\mathbf{q}}] + B_{\mathbf{q}} + \sum_{\mathbf{k}} n_{\mathbf{k}+\frac{1}{2}\mathbf{q}} B_{\mathbf{kq}} = 0, \tag{8.27}$$

$$[n_{\mathbf{q}},\ B_{\mathbf{q}}^{\dagger}] - B_{\mathbf{q}}^{\dagger} - \sum_{\mathbf{k}} B_{\mathbf{kq}}^{\dagger} n_{\mathbf{k}+\frac{1}{2}\mathbf{q}} = 0. \tag{8.28}$$

The vacuum state, $|0\rangle$, satisfying

$$a_{\mathbf{k}} |0\rangle = b_{\mathbf{k}} |0\rangle = 0, \qquad (\text{all } \mathbf{k}) \tag{8.29}$$

is defined. One-pair states $\left|n'_{\mathbf{q}} = 1\right\rangle$ is constructed by

$$\left|n'_{\mathbf{q}} = 1\right\rangle \equiv |1\rangle = B_{\mathbf{q}}^{\dagger} |0\rangle, \qquad n_{\mathbf{q}} |1\rangle = |1\rangle. \tag{8.30}$$

The two states ($|0\rangle$, $|1\rangle$) are the only pair-number states at $\mathbf{q}$ which can be constructed without violating the restriction imposed by Eq. (8.28). In fact, applying Eq. (8.28) to $|1\rangle$ we obtain

$$\begin{aligned}([n_{\mathbf{q}},\ B_{\mathbf{q}}^{\dagger}] - B_{\mathbf{q}}^{\dagger} - \sum_{\mathbf{k}} B_{\mathbf{kq}}^{\dagger} n_{\mathbf{k}+\frac{1}{2}\mathbf{q}})\,|1\rangle &= (n_{\mathbf{q}} B_{\mathbf{q}}^{\dagger} - 2B_{\mathbf{q}}^{\dagger} - \sum_{\mathbf{k}} B_{\mathbf{kq}}^{\dagger})\,|1\rangle \\ &= (n_{\mathbf{q}} B_{\mathbf{q}}^{\dagger} - 3B_{\mathbf{q}}^{\dagger})\,|1\rangle = 0,\end{aligned}$$

or

$$n_{\mathbf{q}} B_{\mathbf{q}}^{\dagger}\,|1\rangle = 3B_{\mathbf{q}}^{\dagger}\,|1\rangle, \tag{8.31}$$

indicating that no two-pair state can be constructed in a regular manner. That is, $|2\rangle \neq B_{\mathbf{q}}^{\dagger}\,|1\rangle$. Hence fermion-boson composites move as fermions.

(D) Identical boson composite. We introduce pair operators:

$$B_{12}^{\dagger} \equiv a_{\mathbf{k}_1}^{\dagger} a_{\mathbf{k}_2}^{\dagger} \equiv a_1^{\dagger} a_2^{\dagger}, \qquad B_{12} \equiv a_2 a_1. \tag{8.32}$$

We compute commutators among B and $B^{\dagger}$ and obtain

$$[B_{12},\ B_{34}] = 0, \tag{8.33}$$

$$[B_{12},\ B_{34}^{\dagger}] = \begin{cases} 1 + n_1 + n_2 & \text{if } \mathbf{k}_1 = \mathbf{k}_3, \mathbf{k}_2 = \mathbf{k}_4, \mathbf{k}_2 \neq \mathbf{k}_3, \mathbf{k}_1 \neq \mathbf{k}_4 \\ 2 + 4n_1 & \text{if } \mathbf{k}_1 = \mathbf{k}_3, \mathbf{k}_2 = \mathbf{k}_4, \mathbf{k}_2 = \mathbf{k}_3, \mathbf{k}_1 = \mathbf{k}_4 \\ a_2 a_4^{\dagger} & \text{if } \mathbf{k}_1 = \mathbf{k}_3, \mathbf{k}_2 \neq \mathbf{k}_4, \mathbf{k}_2 \neq \mathbf{k}_3, \mathbf{k}_1 \neq \mathbf{k}_4 \\ a_2 a_4^{\dagger} + a_2 a_3^{\dagger} & \text{if } \mathbf{k}_1 = \mathbf{k}_3, \mathbf{k}_2 \neq \mathbf{k}_4, \mathbf{k}_2 \neq \mathbf{k}_3, \mathbf{k}_1 = \mathbf{k}_4 \\ a_2 a_4^{\dagger} + a_4^{\dagger} a_1 & \text{if } \mathbf{k}_1 = \mathbf{k}_3, \mathbf{k}_2 \neq \mathbf{k}_4, \mathbf{k}_2 = \mathbf{k}_3, \mathbf{k}_1 \neq \mathbf{k}_4 \\ a_3^{\dagger} a_1 & \text{if } \mathbf{k}_1 \neq \mathbf{k}_3, \mathbf{k}_2 = \mathbf{k}_4, \mathbf{k}_2 \neq \mathbf{k}_3, \mathbf{k}_1 \neq \mathbf{k}_4 \\ a_3^{\dagger} a_1 + a_2 a_3^{\dagger} & \text{if } \mathbf{k}_1 \neq \mathbf{k}_3, \mathbf{k}_2 = \mathbf{k}_4, \mathbf{k}_2 \neq \mathbf{k}_3, \mathbf{k}_1 = \mathbf{k}_4 \\ a_3^{\dagger} a_1 + a_4 a_1^{\dagger} & \text{if } \mathbf{k}_1 \neq \mathbf{k}_3, \mathbf{k}_2 = \mathbf{k}_4, \mathbf{k}_2 \neq \mathbf{k}_3, \mathbf{k}_1 = \mathbf{k}_4 \\ 0 & \text{otherwise.} \end{cases} \tag{8.34}$$

Consider a pair creation operator

$$B_{\mathbf{q}}^{\dagger} \equiv {\sum_{\mathbf{k}\neq 0}}' B_{\mathbf{kq}}^{\dagger} + B_{0\mathbf{q}}^{\dagger}. \tag{8.35}$$

Multiplying this equation from the left by $n_{\mathbf{q}}$ and from the right by $|\Phi_0\rangle$, we obtain

$$n_{\mathbf{q}} B_{\mathbf{q}}^{\dagger}\,|\Phi_0\rangle \equiv n_{\mathbf{q}}({\sum_{\mathbf{k}\neq 0}}' B_{\mathbf{kq}}^{\dagger} + B_{0\mathbf{q}}^{\dagger})\,|\Phi_0\rangle = ({\sum_{\mathbf{k}\neq 0}}' B_{\mathbf{kq}}^{\dagger} + 2B_{0\mathbf{q}}^{\dagger})\,|\Phi_0\rangle, \tag{8.36}$$

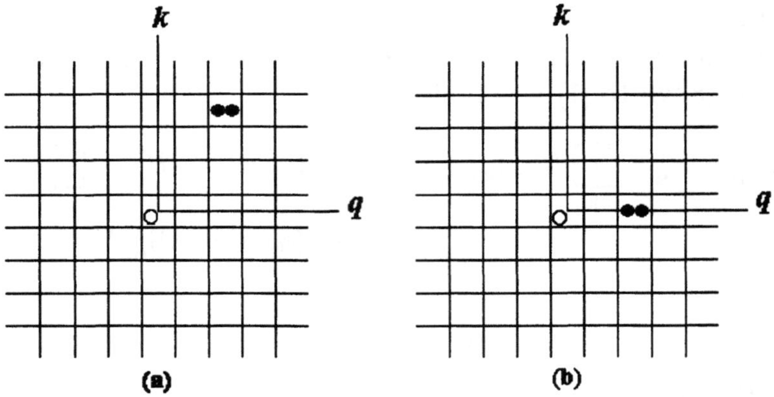

Figure 8.2: (a) State $B^\dagger_{\mathbf{kq}}\left|\Phi_0\right\rangle, k \neq 0$. (b) State $B^\dagger_{0\mathbf{q}}\left|\Phi_0\right\rangle$.

which indicates that $B^\dagger_{\mathbf{q}}\left|\Phi_0\right\rangle$ is not the eigenstate of $n_{\mathbf{q}}$. This is significant. The state corresponding to $B^\dagger_{\mathbf{kq}}\left|\Phi_0\right\rangle$, $k \neq 0$ in one dimension is shown in Fig. 8.2(a). The corresponding occupation number $n_{\mathbf{kq}} = B^\dagger_{\mathbf{kq}} B_{\mathbf{kq}}$ has the eigenvalue one since

$$\begin{aligned} n_{\mathbf{kq}} B^\dagger_{\mathbf{kq}}\left|\Phi_0\right\rangle &= a^\dagger_{\mathbf{k}+\frac{1}{2}\mathbf{q}} a^\dagger_{-\mathbf{k}+\frac{1}{2}\mathbf{q}} a_{-\mathbf{k}+\frac{1}{2}\mathbf{q}} a_{\mathbf{k}+\frac{1}{2}\mathbf{q}} a^\dagger_{\mathbf{k}+\frac{1}{2}\mathbf{q}} a^\dagger_{-\mathbf{k}+\frac{1}{2}\mathbf{q}}\left|\Phi_0\right\rangle \\ &= B^\dagger_{\mathbf{kq}}\left|\Phi_0\right\rangle, \qquad k \neq 0. \end{aligned} \tag{8.37}$$

For $\mathbf{k} = 0$ a straightforward calculation gives

$$n_{0\mathbf{q}} B^\dagger_{0\mathbf{q}}\left|\Phi_0\right\rangle = a^\dagger_{\frac{1}{2}\mathbf{q}} a^\dagger_{\frac{1}{2}\mathbf{q}} a_{\frac{1}{2}\mathbf{q}} a_{\frac{1}{2}\mathbf{q}} a^\dagger_{\frac{1}{2}\mathbf{q}} a^\dagger_{\frac{1}{2}\mathbf{q}}\left|\Phi_0\right\rangle = 2B^\dagger_{0\mathbf{q}}\left|\Phi_0\right\rangle. \tag{8.38}$$

Thus the operator $n_{0\mathbf{q}}$ has the eigenvalue 2. The state $B^\dagger_{0\mathbf{q}}\left|\Phi_0\right\rangle$ is represented by Fig. 8.2(b). These last two results generate Eq. (8.36). In the presence of the double occupancy at $k = 0$, we find no one-pair number state. This anomaly does not occur when dealing with elementary fermions since the double occupancy is excluded by Pauli's principle.

8.3 Discussion

In 1940 Pauli established the spin-statistics theorem [2]: half-integral spin elementary particles are fermions while integral spin particles are bosons.

He derived it by applying general principles of quantum theory and relativity to elementary particles. Just as elementary particles, composites are experimentally found to be indistinguishable and move either as bosons or as fermions (quantum statistical principle). This can be understood simply if the CM of a composite moves, following the same general principles, and if the spin-statistics theorem is applied. We take Democritos' atomistic view: every matter is composed of massive "atoms" (elementay particles). (Massless quantum particles such as photons, neutrinos will not be considered hereafter.) We saw in case (**D**) that no one-pair state for the identical boson pair can be constructed. Hence this composite moves neither as a boson nor as a fermion in violation of the quantum statistical principle. The arguments quoted earlier for the EOB's rule fail in this case. Electrons and nucleons have half spins while pions have zero spin. Hence the other three cases (**A**)-(**C**) are in accord with the spin-statistics theorem and also with the EOB's rule.

In our derivation we omitted consideration of spin, isospin, ... We now discuss this point. Following Dirac [1], we define the indistinguishability of a system of identical elementary particles in terms of the permutation symmetry:

$$[P,\ H] = [P,\ \xi] = 0, \qquad (\text{all } \xi \text{ and all } P) \tag{8.39}$$

$$[P,\ \rho] = 0, \tag{8.40}$$

where

$$H = H(\eta_1, \eta_2, \ldots, \eta_N) \tag{8.41}$$

is the Hamiltonian of N particle-variables η containing position, momentum and other quantum variables such as spin, isospin, ... ; ξ is a system-dynamical function such as the center-of mass and the total momentum; $P's$ are permutation operators of N particle indices. The density operator ρ is defined in the form:

$$\rho \equiv \sum_{\nu} |\nu\rangle\, P_\nu\, \langle\nu|, \qquad \sum_{\nu} P_\nu = 1, \qquad P_\nu = \text{probability}, \tag{8.42}$$

where $|\nu\rangle$ are symmetric (antisymmetric) kets for bosons (fermions). The particle state is characterized by momentum $\mathbf{k}$, spin-component σ, isospin-component τ, ... The state may equivalently be represented by the continuous position, $\mathbf{r}$, σ, τ, ... The set of momenta, $\{\mathbf{k}\}$, is infinite since the position conjugate to the momentum is a continuous variable. Dirac's relativistic wave equation [1] indicates that the electron and antiparticle (positron) has spin 1/2. Pauli's spin-statistics theorem [2] originates in the relativistic quantum motion of the particles in the ordinary three-dimensional space [2]. In contrast other sets $\{\sigma\}, \{\tau\}$, ... are all finite, and hence these variables play secondary roles in quantum statistics. This is so because the quantum statistics of the particles must be defined with the condition that there are an infinite set of particle-states. In fact, if there were only one state, neither symmetric nor antisymmetric states can be constructed. If there were only two states, no antisymmetric states for three or more particles can be constructed. Limiting the number of particles is unnatural.

We have studied the eigenvalues of the pair-number operators (n_{12}, n_{kq}, n_q), which are observables in Dirac's sense [1]. All of our results are obtained without introducing the Hamiltonian. Hence the results are likely to be valid independently of any interaction and energy (bound or unbound). This is significant, and is supported by the following arguments. We consider a system of interacting particles and write the Hamiltonian H in the form

$$H = H_0 + \lambda V, \tag{8.43}$$

where H_0 is the sum of the single-particle Hamiltonian:

$$H_0 = \sum_{j=1}^{N} h_0(\eta_j), \tag{8.44}$$

and V is an interaction Hamiltonian and λ a coupling constant. For $\lambda = 0$ quantum statistics is postulated. Consider now a continuous limit:

$$\lambda \to 1. \tag{8.45}$$

No continuous limit can change discrete (permutation in our case) symmetry. Hence the quantum statistics arising from the particle-permutation symmetry and relativistic quantum dynamics is unchanged in the limit (8.45). Such demonstration can be extended to the case of an interaction Hamiltonian with other particle-fields. All experiments appear to support our view: independence of the statistics upon interaction.

We saw in (**A**) that the CM motion of a pairon is bosonic while its motion with both $(\mathbf{k}, \mathbf{q})$ specified is fermionic. This means that the fermionic nature of the constituents (electrons) is important for the total description of a composite (pairon). This is a general character of any composite. In fact Bardeen, Cooper and Schrieffer, in their historic paper on superconductivity [8], used the fermionic property (8.3) to construct the ground-state of a BCS system, the state of the pairons bosonically condensed all at zero CM momentum. By assuming the spin-statistics theorem for composites Feynman argued that the pairons move as bosons [10], and proceeded to derive the Josephson equations [11], which will be discussed later, Chapter 12. Both fermionic and bosonic properties of the pairons were used in the total description of superconductivity [12].

Let us now consider a three-identical fermion composite. Triplet operators $(T, T^{\dagger})$ are defined by

$$T^{\dagger}_{123} \equiv c^{\dagger}_{1}c^{\dagger}_{2}c^{\dagger}_{3}, \qquad T_{123} = c_3 c_2 c_1. \tag{8.46}$$

If any two of the momenta (k_1, k_2, k_3) are the same, T's vanish due to Pauli's exclusion principle. We shall show that the CM motion of the triplet is fermionic. Decompose the triplet into a system of a two-fermion composite and a fermion. The CM motion of the pair composite is bosonic according to our study in case (A). Applying the result in case (C) to the system, we then deduce that the CM motion of the triplet is fermionic. The above line of argument can be extended to the case of an N-nucleon system. First, eliminate the multi-occupancy states. Second, split it into a system of $(N-1)$-nucleon composite and a nucleon. Third, apply the arguments in either (B) or (C), and deduce that the addition of one nucleon changes quantum statistics. Next we consider an atom composed of a nucleus and one electron. By the same argument the addition of the electron changes quantum statistics. Further addition of an electron generates the change in statistics.

In summary, the quantum statistics for the CM motion of *any* composite is determined by the total number of the constituting elementary fermions. If this number is odd (even), the composite moves as a fermion (boson). Composites may contain no massive elementary bosons. The EOB´s rule with respect to the CM motion of a composite follows directly from the commutation relations (8.3)-(8.4) and their generalizations. We stress that this rule cannot be derived from the arguments based on the symmetry of a composite wavefunction equivalent to the symmetry property of the product of the creation operators alone. The quantum statistics of the constituent particles must be treated separately. For example, the CM of hydrogen molecules $(2e, 2p)$ move as bosons. But ortho-and para-hydrogens have different internal structures and behave differently because the quantum statistics of the two constituting protons play a role [13].

Experiments show that photons are bosons. A photon in a vacuum runs with the light speed and cannot stop. Hence the photon does not have the position variable as a quantum observable. In this respect, it is essentially different from other elementary fermions such as the electron and nucleon. Pions (π), and kaons (K) are experimentally found to be massive bosons. As we saw in Section 8.2, no massive elementary bosons exist. These π and K must be regarded as composites. Fermi and Yang [14] regarded π as a composite of nucleon and antinucleon. In the standard model π is regarded as a composite of two quarks [15]. These theoretical approaches are in line with our theory.

In codensed matter physics, many elementary excitations such as phonons, magnons, plasmons, etc. appear. These particles cannot travel as fast as photons and they cannot be considered as relativistic quantum particles in principle. Hence they cannot have non-zero spins. They must therefore be bosons.

Chapter 9

Bose-Einstein Condensation

Pairons can multiply occupy the same CM momentum state. They move freely as bosons having a linear dispersion relation: $\varepsilon = cp$. The system of the pairons undergoes a B-E condensation transition of the second (third) order in 3 (2)D with the critical temperature $T_c = 1.01\,\hbar v_F \mathrm{k}_B^{-1} n^{1/3}$ $(1.24\,\hbar v_F \mathrm{k}_B^{-1} n^{1/2})$, where n is the pairon density.

9.1 Free Massless Bosons Moving in 2D

In this section we consider a system of free bosons having a linear dispersion relation: $\varepsilon = cp$, moving in 2D. The system undergoes a B-E condensation. [1] The results obtained here are useful in the discussion of high-T_c superconductors.

The total number of bosons, N, and the Bose distribution function:

$$f(\varepsilon;\beta,\mu) \equiv \frac{1}{e^{\beta(\varepsilon-\mu)}-1} \equiv f(\varepsilon) \quad (>0) \qquad (\alpha \equiv \beta\mu) \tag{9.1}$$

are related by

$$N = \sum_{\mathbf{p}} f(\varepsilon_{\mathbf{p}};\beta,\mu) = N_0 + \sum_{\substack{\mathbf{p}\\ \varepsilon_p>0}}{}' f(\varepsilon_{\mathbf{p}}), \tag{9.2}$$

where μ is the chemical potential and $\beta \equiv (\mathrm{k}_B T)^{-1}$.

$$N_0 \equiv (e^{-\beta\mu}-1)^{-1} \tag{9.3}$$

is the number of zero-momentum bosons. The prime on the summation indicates the omission of the zero-momentum state. For notational convenience we write

$$\varepsilon = cp = (2/\pi)v_F p \quad (> 0), \tag{9.4}$$

where $c \equiv (2/\pi)v_F$ is the pairon speed and v_F is the Fermi speed.

We divide Eq. (9.2) by the normalization area L^2, take the bulk limit:

$$N \to \infty, \quad L \to \infty \quad \text{while} \quad NL^{-2} \equiv n, \tag{9.5}$$

and obtain

$$n - n_0 \equiv \frac{1}{(2\pi\hbar)^2} \int d^2p \; f(\varepsilon), \tag{9.6}$$

where $n_0 \equiv N_0/L^2$ is the number density of zero-momentum bosons and n the total boson density. After performing the angular integration and changing integration variables, we obtain from Eq. (9.6) (Problem 9.1.1)

$$2\pi\hbar^2 c^2 \beta^2 (n - n_0) = \int_0^\infty dx \, \frac{x}{\lambda^{-1} e^x - 1}, \qquad [x = \beta\varepsilon] \tag{9.7}$$

$$\lambda \equiv e^{\beta\mu}. \quad (< 1) \tag{9.8}$$

The fugacity λ is less than unity for all temperatures. After expanding the integrand in (9.7) in powers of λe^{-x} (< 1), and carrying out the x-integration, we obtain

$$n_x \equiv n - n_0 = \frac{\mathrm{k}_B^2 T^2 \phi_2(\lambda)}{2\pi\hbar^2 c^2}, \tag{9.9}$$

$$\phi_m(\lambda) \equiv \sum_{k=1}^{\infty} \frac{\lambda^k}{k^m}. \qquad (0 \le \lambda \le 1) \tag{9.10}$$

We need $\phi_2(\lambda)$ here, but we introduced ϕ_m for later reference. Eq. (9.9) gives a relation among λ, n and T.

The function $\phi_2(\lambda)$ monotonically increases from zero to the maximum value $\phi_2(1) = 1.645$ as λ is raised from zero to one. In the low-temperature limit, $\lambda = 1$, $\phi_2(\lambda) = \phi_2(1) = 1.645$, and the density of excited bosons, n_x, varies like T^2 as seen from Eq. (9.9). This temperature behavior of n_x persists as long as the rhs of Eq. (9.9) is smaller than n; the *critical temperature* T_c is

$$n = \frac{\mathrm{k}_B^2 T_c^2 \phi_2(1)}{2\pi\hbar^2 c^2}$$

or

$$k_B T_c = 1.954\, \hbar c n^{1/2} \quad [= 1.24\, \hbar v_F n^{1/2}]. \tag{9.11}$$

If temperature is increased beyond T_c, the density of zero momentum bosons, n_0, becomes vanishingly small, and the fugacity λ can be determined from

$$n = \frac{k_B T^2 \phi_2(\lambda)}{2\pi\hbar^2 c^2}, \qquad T > T_c. \tag{9.12}$$

In summary the fugacity λ is equal to unity in the condensed region: $T < T_c$, and it becomes smaller than unity for $T > T_c$, where its value is determined from Eq. (9.12).

The internal energy density u, which is equal to the thermal average of the system energy per unit area, is given by

$$u = \frac{1}{(2\pi\hbar)^2} \int d^2p\, \varepsilon f(\varepsilon). \tag{9.13}$$

This can be calculated in the same way as Eq. (9.6). We obtain (Problem 9.1.2)

$$u = \frac{\phi_3(\lambda)}{\pi\hbar^2 c^2 \beta^3} = 2nk_B \frac{T^3}{T_c^2} \frac{\phi_3(\lambda)}{\phi_2(1)}. \tag{9.14}$$

The molar heat capacity at constant density, C_n, is

$$C \equiv C_n \equiv R(nk_B)^{-1} \frac{\partial u(T,n)}{\partial T}, \tag{9.15}$$

where R is the gas constant. The partial derivative $\partial u/\partial T$ may be calculated through

$$\begin{aligned} \frac{\partial u(T,n)}{\partial T} &= \frac{\partial u(T,\lambda)}{\partial T} + \frac{\partial u(T,\lambda)}{\partial \lambda} \frac{\partial \lambda(T,n)}{\partial T} \\ &= \frac{\partial u(T,\lambda)}{\partial T} - \frac{\partial u(T,\lambda)}{\partial \lambda} \frac{\partial n(T,\lambda)/\partial T}{\partial n(T,\lambda)/\partial \lambda}. \end{aligned} \tag{9.16}$$

(Problem 9.1.3). All quantities (n, u, C) can now be expressed in terms of $\phi_m(\lambda)$. After straightforward calculations, the molar heat C is (Problem 9.1.4)

$$\begin{aligned} C &= R(\pi\hbar^2 c^2 n k_B)^{-1} k_B^3 T^2 \left[3\phi_3(\lambda) - \frac{2\phi_2^2(\lambda)}{\phi_1(\lambda)} \right] \\ &= 6R \left(\frac{T}{T_c} \right)^2 \frac{\phi_3(\lambda)}{\phi_2(1)} - 4R \frac{\phi_2(\lambda)}{\phi_1(\lambda)}. \end{aligned} \tag{9.17}$$

In the condensed region $T < T_c$, $\lambda = 1$. We observe that as $\lambda \to 1$,

$$\begin{aligned} \phi_1(\lambda) &\to \sum_1^\infty k^{-1} = \infty, & \phi_2(\lambda) \to \phi_2(1) = 1.645, \\ \phi_3(\lambda) &\to \phi_3(1) = 1.202, & \phi_4(\lambda) \to \phi_4(1) = 1.082. \end{aligned} \tag{9.18}$$

Using these, we obtain from Eqs. (9.14) and (9.17)

$$u = 2nk_B \frac{\phi_3(1)}{\phi_2(1)} \frac{T^3}{T_c^2}, \tag{9.19}$$

$$C = 6R \frac{\phi_3(1)}{\phi_2(1)} \left(\frac{T}{T_c}\right)^2. \qquad (T < T_c) \tag{9.20}$$

Observe the T-quadratic dependence of C. Also note that the molar heat capacity C at T_c is given by

$$C(T_c) \equiv C_{\max} = 6R \frac{\phi_3(1)}{\phi_2(1)} = 4.38\, R. \tag{9.21}$$

For $T > T_c$, the temperature dependence of λ, given by Eq. (9.12), is quite complicated. We can numerically solve Eq. (9.12) for λ by a computer, and substitute the solution in Eq. (9.17) to obtain the temperature behavior of C. The result is shown in Fig. 9.1. Eqs. (9.17) and (9.16) allow us not only to examine the analytical behavior of C near $T = T_c$ but also to obtain C without numerically computing the derivative $\partial u(T, n)/\partial T$.

In summary the molar heat capacity C for a 2D massless bosons rises like T^2 in the condensed region, reaches $4.38\, R$ at $T = T_c$, and then decreases to the high temperature limit value $2R$. The heat capacity changes continuously at $T = T_c$, but its temperature derivative $\partial C(T, n)/\partial T$ jumps at this point. Thus the B-E condensation is a third-order phase transition. The condensation of massless bosons in 2D is noteworthy [1]. This is not a violation of Hohenberg's theorem [2] that there can be no long range order in 2D, which is derived with the assumption of an $f-$sum rule representing the mass conservation law. In fact no B-E condensation occurs in 2D for finite-mass bosons [1].

Problem 9.1.1. Verify Eq. (9.7).

Problem 9.1.2. Verify Eq. (9.14).

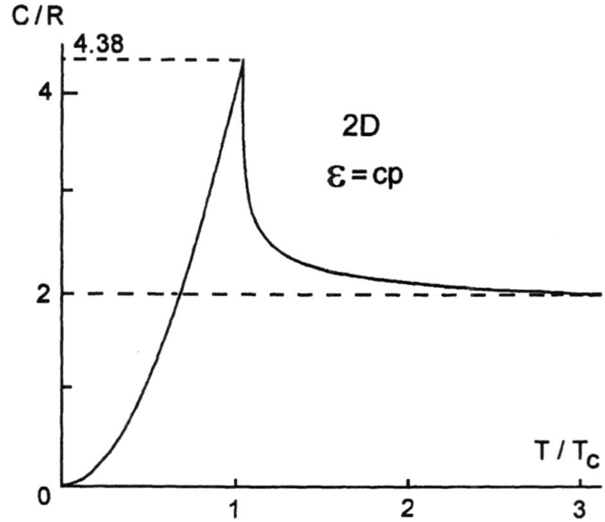

Figure 9.1: The molar heat capacity C for 2D massless bosons rises like T^2, reaches $4.38\,R$ at T_c, and then decreases to $2R$ in the high-temperature limit.

Problem 9.1.3. Prove Eq. (9.16).

Problem 9.1.4. Verify Eq. (9.17).

9.2 Free Massless Bosons in 3D

The case of free massless bosons moving in 3D can be treated similarly. We state the theories and results concisely. The results will be used to describe a B-E condensation of pairons for the 3D BCS system [3] in Section 9.3.

In the bulk limit the normalization condition is given by

$$n - n_0 \equiv \frac{1}{(2\pi\hbar)^3} \int d^3p \; f(\varepsilon; \beta, \mu), \tag{9.22}$$

which is reduced to [Problem 9.2.1]

$$n_x \equiv n - n_0 = \frac{k_B^3 T^3 \phi_3(\lambda)}{\pi^2 \hbar^3 c^3}. \tag{9.23}$$

This equation gives a relation among (n, T, λ).

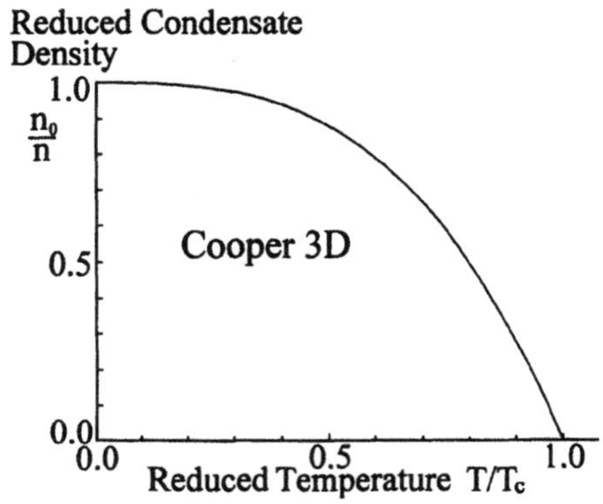

Figure 9.2: The temperature variation of the ground boson density n_0.

The function $\phi_3(\lambda)$ has the maximum value 1.202 at $\lambda = 1$, and it decreases monotonically as λ is reduced to zero. Examination of Eq. (9.23) indicates that

$$\lambda = 1 \qquad \text{for } T < T_c, \tag{9.24}$$

where the critical temperature T_c is given by

$$\mathrm{k}_B T_c = \left[\frac{\pi^2 \hbar^3 c^3 n}{\phi_3(1)}\right]^{1/3} = 2.017\, \hbar c n^{1/3}, \tag{9.25}$$

and that λ becomes less than unity for $T > T_c$, where λ is determined from

$$\pi^2 \hbar^3 c^3 n = \mathrm{k}_B^3 T^3 \phi_3(\lambda). \qquad (T > T_c) \tag{9.26}$$

Substitution of Eq. (9.24) into Eq. (9.23) indicates that the excited boson density n_x in the condensation region ($T < T_c$) rises like T^3. This means that the density of zero-momentum bosons, n_0, varies as

$$\frac{n_0}{n} = 1 - \left(\frac{T}{T_c}\right)^3 \qquad \text{for } T < T_c, \tag{9.27}$$

which is shown in Fig. 9.2. The internal energy density u is [Problem 9.2.2]

$$u = \frac{1}{(2\pi\hbar)^3} \int d^3p\, \varepsilon f(\varepsilon) = \frac{3n\mathrm{k}_B T^4 \phi_4(\lambda)}{T_c^3 \phi_3(1)}. \tag{9.28}$$

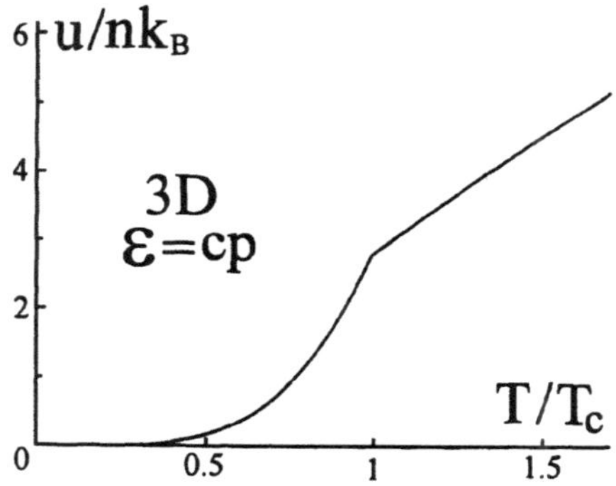

Figure 9.3: The internal energy density u for 3D massless bosons rise like T^4 below T_c, and grows less steeply after passing T_c with a discontinuous change in the slope at $T = T_c$.

This u rises like T^4 in the condensation region:

$$u \propto T^4, \qquad (T < T_c) \tag{9.29}$$

which is similar to the case of black-body radiation (Stephan-Boltzmann law). The u is continuous at T_c, but its T-derivative has discontinuity. See Fig. 9.3.

The molar heat capacity $C \equiv R(nk_B)^{-1}\partial u(T,n)/\partial T$ is [Problem 9.2.3]

$$C = 12R\left(\frac{T}{T_c}\right)^3 \frac{\phi_4(1)}{\phi_3(1)} = 10.8R\left(\frac{T}{T_c}\right)^3, \qquad \text{if } T < T_c, \tag{9.30}$$

$$C = 12R\left(\frac{T}{T_c}\right)^3 \frac{\phi_4(\lambda)}{\phi_3(1)} - 9R\frac{\phi_3(\lambda)}{\phi_2(\lambda)}, \qquad \text{if } T > T_c. \tag{9.31}$$

The temperature behavior of C is shown in Fig. 9.4. We see here that the molar heat C has a discontinuous drop ΔC at T_c equal to

$$\Delta C = \frac{9R\phi_3(1)}{\phi_2(1)} = 6.57\,R. \tag{9.32}$$

The ratio of this jump to the maximum heat capacity $C_{\max}$ is a universal constant:

$$\frac{\Delta C}{C_{\max}} = \frac{6.57\,R}{10.8\,R} = 0.608. \tag{9.33}$$

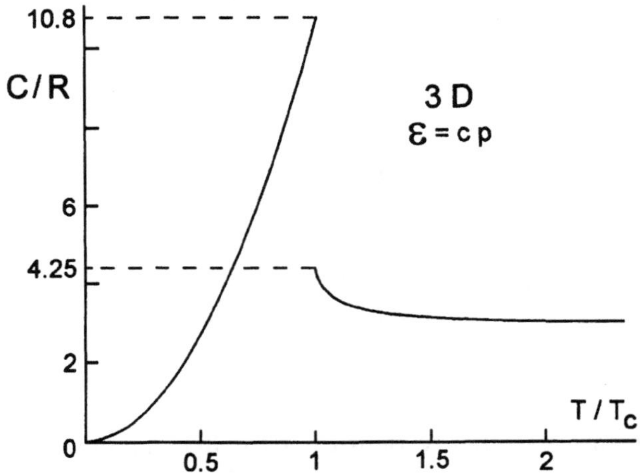

Figure 9.4: The molar heat capacity C for 3D massless bosons rises like T^3, and reaches $10.8\,R$ at the transition temperature $T_c = 2.02\,\hbar c n^{1/3}$. It then drops abruptly by $6.57\,R$ and approaches the high-temperature limit $3R$.

The heat capacity $C(T)$ just below T_c obeys a T^3-law, which is similar to Debye's T^3-law for the heat capacity of phonons at low temperatures. Thus the phase transition is of second order in contrast to the third-order phase transition obtained for the 2D bosons.

Problem 9.2.1. Verify Eq. (9.23).

Problem 9.2.2. Verify Eq. (9.28).

Problem 9.2.3. Verify Eqs. (9.30) and (9.31).

9.3 B-E Condensation of the Pairons

In the generalized BCS system there are ± pairons. The dispersion relation for each type of pairon above T_c is linear:

$$\varepsilon_j = c_j p + w_0. \tag{9.34}$$

This relation changes below T_c because of the presence of the supercondensate, see Chapter 10.

We consider the 3D BCS system in this section, and assume that

$$c_j = (1/2)v_F^{(j)}. \tag{9.35}$$

In practice "electron" and "hole" masses (m_1, m_2) may differ greatly. Then the Fermi velocities $v_F^{(1)}$ and $v_F^{(2)}$ are different, yielding different critical temperatures as seen from Eq. (9.25). We call those pairons for which T_c is the higher the *predominant pairons* (predominant for the B-E condensation).

We are primarily interested in the temperature region where a superconducting transition occurs at T_c, much lower than the Debye temperature Θ_D. We hereafter neglect the temperature dependence of the dispersion relation. Then the total number of pairons is still given by N_0:

$$\sum_j \sum_{\mathbf{q}} n_{\mathbf{q}}^{(j)} = N_0 = \hbar\omega_D \mathcal{N}(0). \tag{9.36}$$

Pairons above T_c move like free massless bosons having energy

$$\varepsilon_q \equiv w_q - w_0 = (1/2)v_F q, \tag{9.37}$$

where we shifted the energy constant, and omitted the superscripts denoting the predominant pairons. This system undergoes a B-E condensation transition of the second order at the critical temperature T_c:

$$k_B T_c = 1.01\hbar v_F n_0^{1/3}, \tag{9.38}$$

9.4 Discussion

We obtained formulas (9.11) and (9.38) for the critical temperature T_c for systems of free dominant pairons moving in 2 and 3D with the linear dispersion relation. We discuss the underlaying assumptions and important results.

9.4.1 Pairons Move as Bosons.

Pairons move as bosons. Since this is an important element in the present theory, we elaborate on its physical significance. Let us take a superconductor of macroscopic dimensions at 0 K. There exist a great number of zero-momentum pairons in the system. If the size of the system is made ten times greater, the number of zero-momentum pairons will be ten times greater. Thus the occupation number N_0 at zero momentum has no upper limit. The number of zero-momentum pairons per unit volume n_0 is finite, and it is given by $n_0 = \hbar\omega_D \mathcal{N}(0)/V$.

9.4.2 Pairons Move as Massless Particles.

Pairons move in 3D with the linear Cooper-Schrieffer relation:

$$\varepsilon_q \equiv w_q - w_0 = \frac{v_F q}{2}. \tag{9.39}$$

They move with the common speed (Problem 9.4.1)

$$c = \frac{v_F}{2}, \tag{9.40}$$

regardless of their momenta q for small q. Eq. (9.39) was obtained in Section 6.2 in the Cooper problem with the assumption of a Fermi sphere and the small-q limit. As we discussed earlier, $\pm$ pairons in a type I superconductor are generated near hyperboloidal Fermi surfaces. We must therefore recalculate the energy-momentum relation. The results are however the same, and the linear relation (9.39) still holds (see Problem 9.4.1).

9.4.3 Pairons Do Not Overlap in Space.

The pairon size can be estimated by the BCS zero-temperature coherence length

$$\xi_0 \equiv \frac{\hbar v_F}{\pi \Delta} = 0.18 \frac{\hbar v_F}{\mathrm{k}_B T_c}, \tag{9.41}$$

which was obtained in their historic paper [3] after calculating the current density and comparing with the Pippard's equation. [4] The average separation r_0 between two pairons can be defined by

$$r_0 \equiv n^{-1/3}, \tag{9.42}$$

where n is the pairon density. Using Eqs. (9.38) and (9.42), we obtain

$$r_0 = 1.01 \frac{\hbar v_F}{\mathrm{k}_B T_c} = 5.61\, \xi_0, \tag{9.43}$$

indicating that the interpairon distance is several times greater than the pairon size. Thus pairons do not overlap in space, meaning that the B-E condensation of pairons takes place *before* the picture of free pairons breaks down. This is the reason why the theory of superconductivity can be developed simply in terms of the independent pairon picture throughout.

9.4.4 Pairons Move Freely in the Crystal.

In Section 3.2 we discussed the Fermi liquid picture in which the Bloch electron moves unhindered by the static lattice potential. By extending this picture, we can say that a pairon also moves freely in a perfect crystal. Impurities can interrupt the free motion of electrons and pairons but they are omitted from our consideration.

9.4.5 Pairons Hardly Interact with Each Other.

The Bloch electron moves freely in a perfect crystal, and two Bloch electrons interact with each other through a screened Coulomb interaction. By extending this picture, two pairons interact with each other through a screened Coulomb force. But the average distance between the two pairons is on the order 10^4 Å. At this separation the screened Coulomb interaction is completely negligible.

9.4.6 B-E Condensation of Massless Bosons.

A system of massless bosons characterized by the linear dispersion relation $\varepsilon = cp$ undergoes a B-E condensation in 2 and 3D.

The 2D case was treated in Section 9.1. The critical temperature T_c is,

$$\mathrm{k}_B T_c = 1.95\, \hbar c n^{1/2}. \tag{9.44}$$

The phase transition is of the third order. The superconducting transition in high-T_c and layered organic superconductors is connected with the B-E condensation of massless bosons moving in 2D. We discuss this topic in more detail in Chapter 14.

The 3D case was treated in section 9.2. The critical temperature T_c is

$$\mathrm{k}_B T_c = 2.01\, \hbar c n^{1/3}. \tag{9.45}$$

The phase transition is of the second order. The molar heat capacity C versus temperature is shown in Fig. 9.4. The heat capacity has a jump ΔC equal to $6.57\, R$ at T_c. The maximum heat capacity at T_c, $C_{\max}$, is equal to $10.8\, R$. The ratio $\Delta C/C_{\max} = 0.608$ is a universal constant. This number 0.608 is close to $(\Delta C/C_{\max})_{BCS} = 0.588$, obtained in the finite-temperature BCS theory [3]. Table 9.1 shows the measured values of $\Delta C/C_{\max}$ for a selection of elements. The closeness of the measured values to the BCS value 0.588 has been regarded as one of the most important support of the finite-temperature BCS theory. The ratio 0.608 obtained in our theory is even closer to the measured values of 0.6 to 0.7. The main reasons for the small but non negligible discrepancy are: there are *two* types of pairons, each having the linear dispersion relation (9.39); and this relation changes substantially below T_c, which we discuss in Chapter 11.

Element	$\Delta C/C_{\max}$
Hg	0.706
In	0.630
La(hcp)	0.600
Nb	0.655
Pb	0.730
Sn	0.615
Ta	0.615
Tl	0.600
V	0.600

Table 9.1. Measured ratios $\Delta C/C_{\max}$ at the critical temperature T_c.

9.4.7 No B-E Condensation in One Dimension

In general the critical temperature T_c for free bosons moving in D dimensions can be found from

$$n = \frac{1}{(2\pi\hbar)^D} \int d^D p \, \frac{1}{exp(\varepsilon/k_B T) - 1}, \qquad \varepsilon = cp. \tag{9.46}$$

The solutions for D = 2 and 3 are given in Eqs. (9.44) and (9.45) respectively. For D = 1, Eq. (9.46) has no solution. In other words, there is no B-E condensation in 1D. (Problem 9.4.2).

To clearly see this, we take a dispersion relation:

$$\varepsilon = ap^{\alpha}, \tag{9.47}$$

where a and α are constants. If we substitute this ε in Eq. (9.46), we can find a solution if $\alpha < 1$. The index α, however, must be greater than, or equal to, unity; otherwise the boson has an infinite speed at zero momentum.

9.4.8 Critical Temperature for Predominant Pairons

In the present theory the superconducting temperature T_c is identified as the B-E condensation temperature for the predominant pairons. According to Eq. (9.38), T_c is proportional to the Fermi speed v_F. Thus if the pairon density is the same for both charge types, T_c is higher for pairons of that type for which v_F is higher. The smaller the effective mass m^*, the higher is the Fermi speed $v_F = (2\varepsilon_F/m^*)^{1/2}$. For Pb the "electron" effective mass m_1 is smaller, by a factor of about 35, than the "hole" effective mass m_2 at the necks (hyperboloidal surface of one sheet). Thus the "electron" pairon is predominant in Pb. Experimentally the predominant type may be determined by examining the sign of the Hall voltage.

9.4.9 Law of Corresponding States for T_c

The critical temperature T_c for the superconducting transition is very much lower than the Fermi temperature T_F. The ratio T_c/T_F can be computed from Eq. (9.38). After eliminating the Fermi velocity $v_F = (2\varepsilon_F/m^*)^{1/2}$ from this expression, we obtain (Problem 9.4.3)

$$\frac{T_c}{T_F} = \frac{2 \times 1.01}{(3\pi^2)^{1/3}} \frac{R_0}{r_0} = 0.653 \frac{R_0}{r_0}, \qquad R_0 \equiv n_{el}^{-1/3}, \tag{9.48}$$

where R_0 is the mean electron separation distance. Thus the ratio T_c/T_F is proportional to the ratio R_0/r_0. The interelectron distance R_0 can be calculated from data from the lattice constants. Therefore the distance r_0 can be calculated from Eq. (9.48) accurately with the knowledge of T_c, T_F, and the lattice constants. Table 9.2 shows data on R_0 and r_0 for selected elements. It is quite remarkable that the interpairon distance r_0 for elemental superconductors is much greater than the the interelectron distance R_0. Metals Al and Pb are both fcc. The significant differences in r_0 must come from the electron energy band structures, see below.

9.4.10 The Pairon Formation Factor

As discussed in Section 7.3, $\pm$ pairons are generated only near specific parts of the Fermi surface. For elemental superconductors these parts are either necks or inverted double caps, which makes the actual density n of pairons relatively very small. We discuss this point here. The maximum pairon density n is

Elements	Lattice	T_c (K)	T_F (10^4 K)	Θ_D (K)	R_0 (Å)	r_0 (10^4 Å)	α (10^{-4})
Al	*fcc*	1.14	13.49	428	1.77	13.68	0.97
Pb	*fcc*	7.193	10.87	105	1.96	1.93	11.29
Zn	*hcp*	0.875	10.90	327	1.97	16.02	0.94
Sn	*diam.*	3.722	11.64	200	2.05	4.19	4.51
In	*tetr.*	3.403	9.98	108	2.06	3.94	5.59
Nb	*bcc*	9.50	6.18	450	2.62	1.11	13.37
Hg	*rhomb.*	4.153	8.29	100	2.27	2.96	7.64
Tl	*hcp*	2.39	9.46	96	2.12	5.48	4.24
$r_0 = 0.653\, R_0 T_F/T_c$				$\alpha = 1.339\, T_c(T_F^2\Theta_D)^{-1/3}$			

Table 9.2. Interpairon distance r_0 and pairon formation factor α.

$$n = \hbar\omega_D \mathcal{N}(0)/V \quad (= r_0^{-1/3}). \tag{9.49}$$

If we assume a Cooper system with a spherical Fermi surface, we can calculate the ratio R_0^3/r_0^3 and obtain (Problem 9.4.4)

$$\frac{R_0^3}{r_0^3} = \frac{3}{4}\frac{\Theta_D}{T_F}. \tag{9.50}$$

Using this, we obtain from Eq. (9.48)

$$\frac{T_c}{T_F} = 0.653\left(\frac{3}{4}\right)^{1/3}\left(\frac{\Theta_D}{T_F}\right)^{1/3} = 0.593\left(\frac{\Theta_D}{T_F}\right)^{1/3}. \qquad \text{(experimental)} \tag{9.51}$$

For the actual elemental superconductors, the critical temperature T_c is much lower than that predicted by Eq. (9.51). To facilitate interpretation of the experimental data, we introduce a correction factor α, called a *pairon formation factor*, such that

$$\frac{T_c}{T_F} \equiv 0.593\,\alpha\left(\frac{\Theta_D}{T_F}\right)^{1/3}. \qquad \text{(experimental)} \tag{9.52}$$

Here the factor α represents the ratio of the experimentally observed (T_c/T_F) to the ideal value calculated by Eq. (9.51) with the observed Θ_D/T_c. Data for the factor α are given in Table 9.2. Since T_F appears on both sides, we may rewrite Eq. (9.52) as

$$T_c = 0.593\,\alpha\Theta_D^{1/3}T_F^{2/3}, \tag{9.53}$$

which indicates that the critical temperature T_c is high if the Fermi temperature T_F is high and if the Debye temperature Θ_D is high.

The pairon formation factor α may come from the following three main sources: First, elemental superconductors must have hyperboloidal Fermi surfaces where $\pm$ pairons are generated with the help of acoustic phonons. Since these surfaces are only small parts of the total Fermi surface, the factor α is a small fraction. Second, the $\pm$ pairons are generated in equal numbers from the physical vacuum. This means that the density of states for non predominant electrons, e.g. "holes" in Pb, is the relevant density of states that enters in Eq. (9.49). This makes α even smaller. Third, necks are more favorable than inverted double caps, since the density of states are, by geometry, higher around the neck. This feature appear to explain why fcc lead (Pb) has a higher T_c than fcc aluminum (Al). For these and possibly other reasons, the pairon density n in actual superconductors is very small. This makes the interpairon distance r_0 much greater than interelectron distance R_0 as discussed in Section 9.4.9.

In our theory, *the existence of $\pm$ pairons is the key criterion for superconductivity.* If the pairon formation α factor that depends on electron and phonon band structures, is finite, the material is a superconductor. The value of α can be discussed qualitatively in terms of the Fermi surface. If the Fermi surface is spherical or ellipsoidal in the first Brillouin zone as in Na and K, $\alpha = 0$. Then the metals remains normal down to 0 K. If a metal's Fermi surface is known to contain necks and inverted double caps as in Al, Pb, Be, W, such a metal has a finite α, and it is a superconductor. To find the pairon formation factor numerically requires a precise knowledge of the Fermi surface, which is beyond the scope of the the present-day theory.

Problem 9.4.1. Derive Eq. (9.40). Use Hamilton's equations of motion.

Problem 9.4.2. Show that there are no T_c satisfying Eq. (9.46) in 1D.

Problem 9.4.3. Derive Eq. (9.48).

Problem 9.4.4. Derive Eq. (9.50).

Problem 9.4.5. Derive Eq. (9.51).

Chapter 10

The Energy Gap Equations

Below T_c there is a supercondensate made up of ±pairons. The energy of unpaired electrons (quasi-electrons) is affected by the presence of a supercondensate: the energy of the quasi-electron changes from $\varepsilon_k^{(j)}$ to $(\varepsilon_k^{(j)2} + \Delta_j^2)^{1/2}$, which is shown from the energy "gap" equations. The density of condensed pairons, n_0, is the greatest at 0 K, and monotonically decreases to zero as temperature approaches T_c. The energy constants $\Delta_j(T)$ decrease to zero at T_c.

10.1 Introduction

In the energy-minimum principle calculation of the ground state energy, the energy gap equations appear mysteriously. The quantities in these equations refer to quasi-electrons while the Hamiltonian H_0 and the ground state Ψ contain the pairon variables only. In the present chapter we re-derive the energy gap equations, using the equation-of-motion method. We obtain a physical interpretation: an unpaired electron in the presence of the supercondensate, has the energy

$$E_p^{(i)} = (\varepsilon_p^{(i)2} + \Delta_i^2)^{1/2}.$$

Extending this theory to a finite T, we obtain the temperature-dependent energy gap equations (10.31) and the temperature-dependent quasi-electron energy.

For an elemental superconductor, the gaps for quasi-electrons of both charge types are the same: $\Delta_1 = \Delta_2 = \Delta$. The gap $\Delta(T)$, obtained as the solution of the T-dependent energy gap equation, is the greatest at 0 K and declines to zero at T_c. This is so because the unpaired electron regains the normal (Bloch) energy ε_p in the absence of the supercondensate. The maximum gap $\Delta_0 \equiv \Delta(T = 0)$ is that gap which appeared in the variational calculation of the BCS ground state energy in Chapter 7. In the weak coupling limit ($v_0 \to 0$) the maximum energy gap Δ_0 can be related to the critical temperature T_c by

$$2\Delta_0 \simeq 3.53\, k_B T_c,$$

which is the famous BCS formula.

10.2 Energies of Quasi-electrons at 0 K

Below T_c, where the supercondensate is present, quasi-electrons move differently from those above T_c. In this section we study the energies of quasi-electrons at 0 K. The ground state is described in terms of the reduced Hamiltonian:

$$\begin{aligned} H_0 &= \sum_{\mathbf{k}} 2\varepsilon_k^{(1)}\, b_{\mathbf{k}}^{(1)\dagger} b_{\mathbf{k}}^{(1)} + \sum_{\mathbf{k}} 2\varepsilon_k^{(2)}\, b_{\mathbf{k}}^{(2)\dagger} b_{\mathbf{k}}^{(2)} \\ &\quad -\sum_{\mathbf{k}'}{}' \sum_{\mathbf{k}'}{}' [v_{11} b_{\mathbf{k}}^{(1)\dagger} b_{\mathbf{k}'}^{(1)} + v_{21} b_{\mathbf{k}}^{(1)\dagger} b_{\mathbf{k}'}^{(2)\dagger} + v_{12} b_{\mathbf{k}}^{(2)} b_{\mathbf{k}'}^{(1)} + v_{22} b_{\mathbf{k}}^{(2)} b_{\mathbf{k}'}^{(2)\dagger}]. \end{aligned} \tag{10.1}$$

By using this H_0, we obtain (Problem 10.2.1)

$$[H_0,\, c_{\mathbf{p}\uparrow}^{(1)\dagger}] = \varepsilon_p^{(1)} c_{\mathbf{p}\uparrow}^{(1)\dagger} - [v_{11} \sum_{\mathbf{k}}{}' b_{\mathbf{k}}^{(1)\dagger} + v_{12} \sum_{\mathbf{k}}{}' b_{\mathbf{k}}^{(2)}] c_{-\mathbf{p}\downarrow}^{(1)}, \tag{10.2}$$

$$[H_0, c_{-\mathbf{p}\downarrow}^{(1)\dagger}] = -\varepsilon_{\mathbf{p}}^{(1)} c_{-\mathbf{p}\downarrow}^{(1)} - [v_{11} \sum_{\mathbf{k}}{}' b_{\mathbf{k}}^{(1)} + v_{12} \sum_{\mathbf{k}}{}' b_{\mathbf{k}}^{(2)\dagger}] c_{\mathbf{p}\uparrow}^{(1)\dagger}. \tag{10.3}$$

These two equations indicate that the dynamics of quasi-electrons describable in terms of c's are affected by stationary pairons described in terms of b's.

To find the energy of a quasi-electron, we follow the equation-of-motion method developed in Section 6.3. We multiply Eq. (10.2) from the right by $\psi_\nu^{(1)} \rho_0$, where $\psi_\nu^{(1)}$ is the "electron" energy-state annihilation operator and

$$\rho_0 \equiv |\Psi\rangle\langle\Psi| \equiv \prod_{\mathbf{k}}{}' (u_{\mathbf{k}}^{(1)} + v_{\mathbf{k}}^{(1)} b_{\mathbf{k}}^{(1)\dagger}) \prod_{\mathbf{k}'}{}' (u_{\mathbf{k}'}^{(2)} + v_{\mathbf{k}'}^{(2)} b_{\mathbf{k}'}^{(2)\dagger})\, |0\rangle\langle\Psi| \tag{10.4}$$

is the density operator describing the supercondensate, and take a grand ensemble trace. After using Eq. (6.41), the lhs can be written as

$$\mathrm{TR}\{[H_0,\, c^{(1)}_{\mathbf{p}\uparrow}]\, \psi^{(1)}_{\nu} \rho_0\} = \mathrm{TR}\{E^{(1)}_{\nu,\mathbf{p}}\, c^{(1)\dagger}_{\mathbf{p}\uparrow}\, \psi^{(1)}_{\nu} \rho_0\} \equiv E^{(1)}_{\mathbf{p}} \psi^{(1)*}_{\uparrow}(\mathbf{p}), \qquad (10.5)$$

where we dropped the subscript ν; the quasi-electron is characterized by momentum $\mathbf{p}$ and energy $E^{(1)}_{\mathbf{p}}$. The first term on the rhs simply yields $\varepsilon^{(1)}_{\mathbf{p}} \psi^{(1)*}_{\uparrow}(\mathbf{p})$. Consider now

$$\mathrm{TR}\{b^{(1)\dagger}_{\mathbf{k}} c^{(1)}_{-\mathbf{p}\downarrow}\, \psi^{(1)}_{\nu} \rho_0\} \equiv \mathrm{TR}\{b^{(1)\dagger}_{\mathbf{k}} c^{(1)}_{-\mathbf{p}\downarrow}\, \psi^{(1)}_{\nu}\, |\Psi\rangle\langle\Psi|\}.$$

The state $|\Psi\rangle$ is normalized to unity, and it is the only system-state at 0 K, hence

$$\mathrm{TR}\{b^{(1)\dagger}_{\mathbf{k}} c^{(1)}_{-\mathbf{p}\downarrow}\, \psi^{(1)}_{\nu} \rho_0\} = \langle\Psi|\, b^{(1)\dagger}_{\mathbf{k}} c^{(1)}_{-\mathbf{p}\downarrow}\, \psi^{(1)}_{\nu}\, |\Psi\rangle. \qquad (10.6)$$

We note that $\mathbf{k} \neq \mathbf{p}$ since the state must change after a phonon exchange. We examine the relevant matrix element and obtain (Problem 10.2.2.)

$$\langle 0|\, (u^{(1)}_{\mathbf{k}} + v^{(1)}_{\mathbf{k}} b^{(1)}_{\mathbf{k}}) b^{(1)\dagger}_{\mathbf{k}} (u^{(1)}_{\mathbf{k}} + v^{(1)}_{\mathbf{k}} b^{(1)\dagger}_{\mathbf{k}})\, |0\rangle = v^{(1)}_{\mathbf{k}} u^{(1)}_{\mathbf{k}}. \qquad (10.7)$$

We can therefore write

$$\mathrm{TR}\{b^{(1)\dagger}_{\mathbf{k}} c^{(1)}_{-\mathbf{p}\downarrow}\, \psi^{(1)}_{\nu} \rho_0\} = u^{(1)}_{\mathbf{k}} v^{(1)}_{\mathbf{k}} \psi^{(1)}_{\downarrow}(-\mathbf{p}), \qquad (10.8)$$

$$\psi^{(1)}_{\downarrow}(-\mathbf{p}) \equiv \mathrm{TR}\{c^{(1)}_{-\mathbf{p}\downarrow} \psi^{(1)}_{\nu} \rho_0\}. \qquad (10.9)$$

Collecting all contributions, we obtain

$$E^{(1)}_{\mathbf{p}} \psi^{(1)*}_{\uparrow}(\mathbf{p}) = \varepsilon^{(1)}_{p} \psi^{(1)*}_{\uparrow}(\mathbf{p}) - [v_{11} {\sum_{\mathbf{k}}}' u^{(1)}_{\mathbf{k}} v^{(1)}_{\mathbf{k}} + v_{12} {\sum_{\mathbf{k}}}' u^{(2)}_{\mathbf{k}} v^{(2)}_{\mathbf{k}}] \psi^{(1)}_{\downarrow}(-\mathbf{p}). \qquad (10.10)$$

Using Eqs. (7.24) and (7.25), we get

$$\Delta_1 \equiv v_{11} {\sum_{\mathbf{k}}}' u^{(1)}_{\mathbf{k}} v^{(1)}_{\mathbf{k}} + v_{12} {\sum_{\mathbf{k}}}' u^{(2)}_{\mathbf{k}} v^{(2)}_{\mathbf{k}}. \qquad (10.11)$$

We can therefore simplify Eq. (10.10) to

$$E^{(1)}_{\mathbf{p}} \psi^{(1)*}_{\uparrow}(\mathbf{p}) = \varepsilon^{(1)}_{\mathbf{p}} \psi^{(1)*}_{\uparrow}(\mathbf{p}) - \Delta_1 \psi^{(1)}_{\downarrow}(-\mathbf{p}). \qquad (10.12)$$

Similarly we obtain from Eq. (10.3)

$$E^{(1)}_{-\mathbf{p}} \psi^{(1)}_{\downarrow}(-\mathbf{p}) = -\varepsilon^{(1)}_{p} \psi^{(1)}_{\downarrow}(-\mathbf{p}) - \Delta_1 \psi^{(1)*}_{\uparrow}(\mathbf{p}). \qquad (10.13)$$

The energy $E_{\mathbf{p}}^{(1)}$ can be interpreted as the *positive* energy required to create an up-spin unpaired electron at $\mathbf{p}$ in the presence of the supercondensate. The energy $E_{-\mathbf{p}}^{(1)}$ can be regarded as the energy required to remove a down-spin electron from the paired state $(\mathbf{p}\uparrow, -\mathbf{p}\downarrow)$. These two energies are equal to each other:

$$E_{\mathbf{p}}^{(1)} = E_{-\mathbf{p}}^{(1)} \equiv E_p^{(1)} > 0. \tag{10.14}$$

In the stationary state Eqs. (10.12) and (10.13) must hold simultaneously, thus yielding

$$\begin{vmatrix} E_p^{(1)} - \varepsilon_p^{(1)} & \Delta_1 \\ \Delta_1 & E_p^{(1)} + \varepsilon_p^{(1)} \end{vmatrix} = 0, \tag{10.15}$$

whose solutions are $E_{\mathbf{p}}^{(1)} = \pm(\varepsilon_{\mathbf{p}}^{(1)2} + \Delta_1^2)^{1/2}$. Since $E_p^{(1)} > 0$, we obtain

$$E_{\mathbf{p}}^{(i)} = \left(\varepsilon_{\mathbf{p}}^{(i)2} + \Delta_i^2\right)^{1/2}. \tag{10.16}$$

The theory developed here can be applied to the "hole" in a parallel manner. We included this case in Eqs. (10.16). Our calculation confirms our earlier interpretation that $E_{\mathbf{p}}^{(i)}$ is the energy of the quasi-electron. In summary unpaired electrons are affected by the presence of the supercondensate, and acquire energies $E_{\mathbf{p}}^{(i)}$.

Problem 10.2.1. Derive Eqs. (10.2) and (10.3).

Problem 10.2.2. Verify Eq. (10.7).

10.3 Energy Gap Equations at 0 K

The supercondensate is made up of $\pm$ pairons, which can be described in terms of b's. We calculate $[H_0,\, b_{\mathbf{k}}^{(1)\dagger}]$ and $[H_0,\, b_{\mathbf{k}}^{(2)}]$ and obtain: (Problem 10.3.1.)

$$\begin{aligned} [H_0,\, b_{\mathbf{k}}^{(1)\dagger}] &= E_1 b_{\mathbf{k}}^{(1)\dagger} = 2\varepsilon_k^{(1)} b_{\mathbf{k}}^{(1)\dagger} \\ &\quad -[v_{11}{\sum_{\mathbf{k}'}}' b_{\mathbf{k}'}^{(1)\dagger} + v_{12}{\sum_{\mathbf{k}'}}' b_{\mathbf{k}'}^{(2)}](1 - n_{\mathbf{k}\uparrow}^{(1)} - n_{-\mathbf{k}\downarrow}^{(1)}), \end{aligned} \tag{10.17}$$

$$\begin{aligned} [H_0,\, b_{\mathbf{k}}^{(2)}] &= -E_2 b_{\mathbf{k}}^{(2)} = -2\varepsilon_k^{(2)} b_{\mathbf{k}}^{(2)} \\ &\quad +[v_{21}{\sum_{\mathbf{k}}}' b_{\mathbf{k}'}^{(1)\dagger} + v_{22}{\sum_{\mathbf{k}'}}' b_{\mathbf{k}'}^{(2)}](1 - n_{\mathbf{k}_\uparrow}^{(2)} - n_{-\mathbf{k}_\downarrow}^{(2)}). \end{aligned} \tag{10.18}$$

These equations indicate that the dynamics of the stationary pairons depends on quasi-electrons describable in terms of $n^{(j)}$.

We now multiply Eq. (10.17) from the right by $\phi_1 \rho_0$, where ϕ_1 is the −pairon energy-state annihilation operator, and take a grand ensemble trace. From the first term on the rhs, we obtain

$$2\varepsilon_k^{(1)}\,\mathrm{TR}\{b_{\mathbf{k}}^{(1)\dagger}\phi_1\rho_0\} = 2\varepsilon_k^{(1)}\,\langle\Psi|\,b_{\mathbf{k}}^{(1)\dagger}\phi_1\,|\Psi\rangle = 2\varepsilon_k^{(1)}u_{\mathbf{k}}^{(1)}v_{\mathbf{k}}^{(1)}F_1, \tag{10.19}$$

$$F_1 \equiv \langle\Psi|\,\phi_1\,|\Psi\rangle. \tag{10.20}$$

From the first and second sums, we get (Problem 10.3.2.)

$$-\sum_{\mathbf{k}'}{}'[v_{11}u_{\mathbf{k}'}^{(1)}v_{\mathbf{k}'}^{(1)} + v_{12}u_{\mathbf{k}'}^{(2)}v_{\mathbf{k}'}^{(2)}](1 - v_{\mathbf{k}}^{(1)2} - v_{-\mathbf{k}}^{(1)2})F_1. \tag{10.21}$$

Since we are looking at the Bose-condensed state, that is, the system-state characterized by zero chemical potential, the eigenvalues E_1 and E_2 are zero:

$$E_1 = E_2 = 0. \tag{10.22}$$

Collecting all contributions, we obtain

$$\{2\varepsilon_k^{(1)}u_{\mathbf{k}}^{(1)}v_{\mathbf{k}}^{(1)} - (u_{\mathbf{k}}^{(1)2} - v_{\mathbf{k}}^{(1)2})\sum_{\mathbf{k}'}{}'[v_{11}u_{\mathbf{k}'}^{(1)}v_{\mathbf{k}'}^{(1)} + v_{12}u_{\mathbf{k}'}^{(2)}v_{\mathbf{k}'}^{(2)}]\}F_1 = 0, \tag{10.23}$$

where we used $v_{-\mathbf{k}}^{(1)} = v_{\mathbf{k}}^{(1)}$. Since $F_1 \equiv \langle\Psi|\,\phi_1\,|\Psi\rangle \neq 0$, we obtain

$$2\varepsilon_k^{(1)}u_{\mathbf{k}}^{(1)}v_{\mathbf{k}}^{(1)} - (u_{\mathbf{k}}^{(1)2} - v_{\mathbf{k}}^{(1)2})\sum_{\mathbf{k}'}{}'[v_{11}u_{\mathbf{k}'}^{(1)}v_{\mathbf{k}'}^{(1)} + v_{12}u_{\mathbf{k}'}^{(2)}v_{\mathbf{k}'}^{(2)}] = 0, \tag{10.24}$$

which is just one of Eqs. (7.23), the equations equivalent to the energy gap equations (7.27).

As noted earlier in Section 8.2, the ground state Ψ of the BCS system is a superposition of many-pairon states and hence quantities like

$$\langle\Psi|\,b_{\mathbf{k}}^{(j)\dagger}\,|\Psi\rangle\,, \qquad \langle\Psi|\,b_{\mathbf{k}}^{(j)\dagger}b_{\mathbf{k}'}^{(j)\dagger}\,|\Psi\rangle\,,\ldots \tag{10.25}$$

that connect states of different pairon numbers do not vanish. In this sense the state Ψ can be defined only in the grand ensemble.

As shown in Section 7.3, the ground state is reachable from the physical vacuum by a succession of phonon exchanges. Since pair creation and pair annihilation of pairons lower the system energy, and since pairons are bosons, all pairons available in the system are condensed into the zero-momentum state. The number of condensed pairons at any one instant may fluctuate around the equilibrium value; such fluctuations are in fact more favorable.

Problem 10.3.1. Derive Eqs. (10.17) and (10.18).

Problem 10.3.2. Verify Eq. (10.21).

10.4 Temperature-Dependent Gap Equations

In the last two sections we saw that quasi-electron energies and the energy gap equations at 0 K can be derived from the equations of motion for c's and b's. We now extend our theory to a finite temperature.

First, we make an important observation. The supercondensate is composed of equal numbers of $\pm$ pairons condensed at zero momentum. Since there is no distribution, its properties cannot show any temperature variation; only its content (density) changes with temperature. Second, we reexamine Eq. (10.17). Combining this with Eq. (10.22), we obtain

$$[H_0, b_{\mathbf{k}}^{(1)\dagger}] = 2\varepsilon_k^{(1)} b_{\mathbf{k}}^{(1)\dagger} - [v_{11} \sum_{\mathbf{k}'}{}' b_{\mathbf{k}'}^{(1)\dagger} + v_{12} \sum_{\mathbf{k}'}{}' b_{\mathbf{k}'}^{(2)}](1 - n_{\mathbf{k}\uparrow}^{(1)} - n_{-\mathbf{k}\downarrow}^{(1)}) = 0. \quad (10.26)$$

From our previous study we know that both:

$$F_1 \equiv \langle \Psi | b_{\mathbf{k}}^{(1)\dagger} | \Psi \rangle = u_{\mathbf{k}}^{(1)} v_{\mathbf{k}}^{(1)} \quad \text{and} \quad \langle \Psi | b_{\mathbf{k}}^{(2)\dagger} | \Psi \rangle = u_{\mathbf{k}}^{(2)} v_{\mathbf{k}}^{(2)} \quad (10.27)$$

are finite. The b's refer to the pairons constituting the supercondensate while $n_{\mathbf{k}\uparrow}^{(1)}$ and $n_{-\mathbf{k}\downarrow}^{(1)}$ represent the occupation numbers of electrons. This means that unpaired electrons, if present, can influence the formation of the supercondensate. In fact, we see from Eq. (10.26) that if $n_{\mathbf{k}\uparrow}^{\prime(1)} = 0$, $n_{-\mathbf{k}\downarrow}^{\prime(1)} = 1$, or $n_{\mathbf{k}\uparrow}^{\prime(1)} = 1$, $n_{-\mathbf{k}\downarrow}^{\prime(1)} = 0$, terms containing v_{1j} vanish, and there is no attractive transition: that is, the supercondensate wavefunction Ψ cannot extend over the pair state $(\mathbf{k}\uparrow, -\mathbf{k}\downarrow)$. Thus as temperature is raised from 0 K, more quasi-electrons are excited, making the supercondensate formation less favorable. Unpaired electrons (fermions) have energies $E_k^{(1)}$ and the thermal average of $1 - n_{\mathbf{k}\uparrow}^{(1)} - n_{-\mathbf{k}\downarrow}^{(1)}$ is

$$\left\langle 1 - n_{\mathbf{k}\uparrow}^{(1)} - n_{-\mathbf{k}\downarrow}^{(1)} \right\rangle = 1 - 2\{\exp[\beta E_k^{(1)}] + 1\}^{-1} = \tanh[\beta E_k^{(1)}/2], \ (\geq 0). \quad (10.28)$$

In the low temperature limit, this quantity approaches unity from below

$$\tanh[\beta E_k^{(1)}/2] = \tanh[E_k^{(1)}/2k_B T] \to 1. \qquad (T \to 0) \quad (10.29)$$

We may therefore regard $\tanh[\beta E_k^{(1)}/2]$ as the probability that the pair state $(\mathbf{k}\uparrow, -\mathbf{k}\downarrow)$ is available for the supercondensate formation. As temperature is raised, this probability becomes smaller, making the supercondensate density smaller. Third, we make a hypothesis: The behavior of quasi-electrons is affected by the supercondensate only. This is a reasonable assumption at very low temperatures, where very few elementary excitations exist.

We now examine the energy gap equations (7.27) at 0 K:

$$\Delta_j = \sum_{\mathbf{k}}{}' \sum_i v_{ji} \frac{\Delta_i}{2E_k^{(i)}}. \tag{10.30}$$

Here we see that the summation with respect to $\mathbf{k}$ and i extends over all allowed pair-states. As temperature is raised, quasi-electrons are excited, making the physical vacuum less perfect. The degree of perfection at $(\mathbf{k}, i)$ will be represented by the probability $\tanh(\beta E_k^{(i)}/2)$. Hence we modify Eqs. (10.30) as follows:

$$\Delta_j = \sum_{\mathbf{k}}{}' \sum_i v_{ji} \frac{\Delta_i}{2E_k^{(i)}} \tanh(\beta E_k^{(i)}/2). \tag{10.31}$$

Equations (10.31) are called the *generalized energy-gap equations at finite temperatures.* They reduce to Eqs. (10.30) in the zero-temperature limit. In the bulk limit, the sums over $\mathbf{k}'$ can be converted into energy integrals, yielding

$$\Delta_j = \frac{1}{2} \sum_{i=1}^{2} v_{ji} \mathcal{N}_i(0) \int_0^{\hbar\omega_D} d\varepsilon \frac{\Delta_i}{(\varepsilon^2 + \Delta_i^2)^{1/2}} \tanh\left[\frac{(\varepsilon^2 + \Delta_i^2)^{1/2}}{2k_B T}\right]. \tag{10.32}$$

For elemental superconductors $v_{ij} = v_0$, and $\mathcal{N}_1(0) = \mathcal{N}_2(0) \equiv \mathcal{N}(0)$. We then see from Eqs. (10.32) that there is a common energy gap:

$$\Delta_1 = \Delta_2 \equiv \Delta, \tag{10.33}$$

which can be obtained from

$$1 = v_0 \mathcal{N}(0) \int_0^{\hbar\omega_D} d\varepsilon \frac{1}{(\varepsilon^2 + \Delta^2)^{1/2}} \tanh\left[\frac{(\varepsilon^2 + \Delta^2)^{1/2}}{2k_B T}\right]. \tag{10.34}$$

This gap Δ is temperature-dependent. The limit temperature T_c at which Δ vanishes, is given by

$$1 = v_0 \mathcal{N}(0) \int_0^{\hbar\omega_D} d\varepsilon \, \frac{1}{\varepsilon} \tanh\left[\frac{\varepsilon}{2k_B T_c}\right]. \tag{10.35}$$

This limit temperature T_c can be identified as the *critical temperature* at which the supercondensate disappears. To see this connection, let us recall the form of the quasi-particle energies: $E_k^{(j)} \equiv (\varepsilon_k^{(j)2} + \Delta_j^2)^{1/2}$. If there is no supercondensate, no gaps can appear ($\Delta_i = 0$), and $E_k^{(j)}$ is reduced to $\varepsilon_k^{(j)}$.

The gap $\Delta(T)$, which is obtained numerically from Eq. (10.34), decreases as shown in Fig. 10.1. In the weak coupling limit:

$$\exp\left[\frac{2}{v_0\mathcal{N}(0)}\right] \gg 1, \tag{10.36}$$

the critical temperature T_c can be computed from Eq. (10.35) analytically. The result can be expressed by (Problem 10.4.1.)

$$k_B T_c \simeq 1.13\, \hbar\omega_D \exp\left[\frac{1}{v_0\mathcal{N}(0)}\right]. \tag{10.37}$$

Comparing this with the weak coupling limit of Eq. (7.35), we obtain

$$2\Delta_0 \equiv 2\Delta(T=0) = 3.53\, k_B T_c, \tag{10.38}$$

which is the famous BCS formula [1], expressing a connection between the maximum energy gap Δ_0 and the critical temperature T_c.

Problem 10.4.1. Obtain the weak-coupling analytical solution (10.37) from Eq. (10.35).

10.5 Discussion

10.5.1 Ground State

The supercondensate at 0 K is composed of equal numbers of $\pm$ pairons with the total number of pairons being $N_0(0) = N_0 \equiv \hbar\omega_D\mathcal{N}(0)$. Each pairon

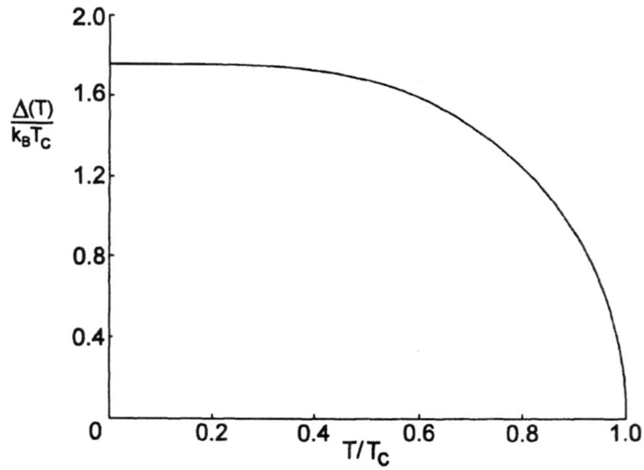

Figure 10.1: Temperature variation of energy gap $\Delta(T)$.

contributes an energy equal to the ground state energy of a Cooper pair w_0 to the system ground-state energy W:

$$W = N_0 w_0, \qquad w_0 \equiv \frac{-2\hbar\omega_D}{\exp[(2/v_0)\mathcal{N}(0)] - 1}. \tag{10.39}$$

which is in agreement with the BCS formula (7.36); this is significant. The reduced generalized BCS Hamiltonian H_0 in Eq. (10.1) satisfies the applicability condition of the equation-of-motion method (i.e. the sum of pairon energies). Thus we obtained the ground state energy W rigorously. The mathematical steps are more numerous in our approach than in the BCS-variational approach based on the minimum-energy principle. However our statistical mechanical theory has a major advantage: There is no need to guess the form of the trial wave function Ψ, which requires a great intuition. This methodological advantage of statistical mechanical theory will become clear in the finite-temperature theory.

10.5.2 Supercondensate Density

The most distinct feature of a system of bosons is the possibility of a B-E condensation. The supercondensate is identified as the condensed pairons. The property of the supercondensate cannot change with temperature because there is no distribution. But its content (density) changes with temperature. Unpaired electrons in the system hinder formation of the supercondensate, and the supercondensate density decreases as temperature is raised.

10.5.3 Energy Gap $\Delta(T)$

In the presence of the supercondensate, quasi-electrons move differently from Bloch electrons. Their energies are different:

$$E_p^{(j)} \equiv [\varepsilon_p^{(j)2} + \Delta(T)^2]^{1/2} \quad \text{(quasi-“electron”)}, \qquad \varepsilon_p^{(j)} \quad \text{(Bloch “electron”)} \tag{10.40}$$

The energy gap $\Delta(T)$ is temperature-dependent, as shown in Fig. 10.1. The gap $\Delta(T_c)$ vanishes at T_c, where there is no supercondensate.

10.5.4 Energy Gap Equations at 0 K

In Section 7.2, we saw that the energy gap Δ at 0 K mysteriously appears in the minimum-energy-principle calculation. The starting Hamiltonian H_0 and the trial state Ψ are both expressed in terms of the pairon operators b's only. Yet, the variational calculation leads to energy-gap equations (7.27), which contain the quasi-electron variables only. Using the equation-of-motion method, we discovered that the energy-eigenvalue equation for the ground pairons generates the energy gap equations.

10.5.5 Energy Gap Equations Below T_c

At a finite temperature, some quasi-electrons are excited in the system. These quasi-electrons disrupt formation of the supercondensate, making the condensed pairon density and the energy gaps Δ smaller. This in turn makes quasi-electron excitation easier. This cooperative effect is most apparent near T_c. In the original paper [1] BCS obtained the temperature-dependent energy gap equation (10.9) by the free-energy minimum principle based on the assumption that quasi-electrons are predominant elementary excitations. We have reproduced the same result (10.9) and its generalization (10.31) by the equation-of-motion method. In the process of doing so, we found that the temperature dependence of the energy gap $\Delta(T)$ arises from the temperature-dependent supercondensate density.

Chapter 11

Pairon Energy Gaps. Heat Capacity

By the phonon exchange attraction, a pair of quasi-electrons are bound to form moving pairons, which have temperature-dependent energy gaps $\varepsilon_{g,j}(T)$. These gaps are greatest at 0 K and monotonically decrease to zero at T_c. The condensed pairon density and the heat capacity are calculated under the assumption that moving pairons rather than quasi-electrons are predominant elementary excitations.

11.1 The Full Hamiltonian.

Quasi-electrons may be bound by the phonon exchange attraction to form *moving* pairons. These pairons, first considered by Fujita and Watanabe [1], move with a linear dispersion relation just as those pairons above T_c. But they have an *energy gap* $\varepsilon_g(T)$ relative to the energy level of the stationary pairons, (supercondensate). This pairon energy gap $\varepsilon_g(T)$ is the greatest at 0 K, reaching a value equal to the pairon binding energy $|w_0|$. It decreases to zero as temperature is raised to T_c. It is less than the quasi-electron gap $\Delta(T)$ in the range $(0, T_c)$.

Creation operators for "electron" (1) and "hole" (2) pairons are defined by

$$B_{12}^{(1)\dagger} \equiv B_{\mathbf{k}_1\uparrow\mathbf{k}_2\downarrow}^{(1)\dagger} \equiv c_1^{(1)\dagger}c_2^{(1)\dagger}, \qquad B_{34}^{(2)\dagger} \equiv c_4^{(2)\dagger}c_3^{(2)\dagger}. \tag{11.1}$$

The commutators among $B's$ and $B^{\dagger\prime}s$ are calculated to be,

$$[B_{12}^{(j)}, B_{34}^{(j)}] = 0, \qquad [B_{12}^{(j)}]^2 = 0. \tag{11.2}$$

$$[B_{12}^{(j)},\, B_{34}^{(j)\dagger}] = \begin{cases} 1 - n_1^{(j)} - n_2^{(j)} & \text{if } \mathbf{k}_1 = \mathbf{k}_3 \text{ and } \mathbf{k}_2 = \mathbf{k}_4 \\ c_2^{(j)} c_4^{(j)\dagger} & \text{if } \mathbf{k}_1 = \mathbf{k}_3 \text{ and } \mathbf{k}_2 \neq \mathbf{k}_4 \\ c_1^{(j)} c_3^{(j)\dagger} & \text{if } \mathbf{k}_1 \neq \mathbf{k}_3 \text{ and } \mathbf{k}_2 = \mathbf{k}_4 \\ 0 & \text{otherwise.} \end{cases} \tag{11.3}$$

Pairon operators of different types j always commute. Here

$$n_1^{(j)} \equiv c_{\mathbf{k}_1\uparrow}^{(j)\dagger} c_{\mathbf{k}_1\uparrow}^{(j)}, \qquad n_2^{(j)} \equiv c_{\mathbf{k}_2\downarrow}^{(j)\dagger} c_{\mathbf{k}_2\downarrow}^{(j)} \tag{11.4}$$

represent the number operators for "electrons" and "holes".

Let us now introduce the relative and net momenta $(\mathbf{k}, \mathbf{q})$ such that

$$\begin{aligned} \mathbf{k} &\equiv (1/2)(\mathbf{k}_1 - \mathbf{k}_2), \qquad q \equiv \mathbf{k}_1 + \mathbf{k}_2; \\ \mathbf{k}_1 &= \mathbf{k} + \mathbf{q}/2, \qquad \mathbf{k}_2 = -\mathbf{k} + \mathbf{q}/2. \end{aligned} \tag{11.5}$$

Alternatively we can represent pairon annihilation operators by

$$B_{\mathbf{kq}}^{\prime(1)} \equiv B_{\mathbf{k}_1\uparrow\mathbf{k}_2\downarrow}^{(1)} \equiv c_{-\mathbf{k}+\mathbf{q}/2\downarrow}^{(1)} c_{\mathbf{k}+\mathbf{q}/2\uparrow}^{(1)}, \qquad B_{\mathbf{kq}}^{\prime(2)} = c_{\mathbf{k}+\mathbf{q}/2\uparrow}^{(2)} c_{-\mathbf{k}+\mathbf{q}/2\downarrow}^{(2)}. \tag{11.6}$$

The prime on B will be dropped hereafter. In this k-q representation the commutation relations are re-expressed as

$$[B_{\mathbf{kq}}^{(j)},\, B_{\mathbf{k}'\mathbf{q}'}^{(i)}] = 0, \qquad [B_{\mathbf{kq}}^{(j)}]^2 = 0. \tag{11.7}$$

$$[B_{\mathbf{kq}}^{(j)}, B_{\mathbf{kq}'}^{(i)\dagger}] = \begin{cases} (1 - n_{\mathbf{k}+\mathbf{q}/2\uparrow} - n_{-\mathbf{k}+\mathbf{q}/2\downarrow})\delta_{ji} & \text{if } \mathbf{k} = \mathbf{k}' \text{ and } \mathbf{q} = \mathbf{q}' \\ c_{-\mathbf{k}+\mathbf{q}/2\downarrow}^{(j)} c_{-\mathbf{k}'+\mathbf{q}'/2\downarrow}^{(j)\dagger}\, \delta_{ji} & \text{if } \mathbf{k} + \mathbf{q}/2 = \mathbf{k}' + \mathbf{q}'/2 \\ & \text{and } -\mathbf{k} + \mathbf{q}/2 \neq -\mathbf{k}' + \mathbf{q}'/2 \\ c_{\mathbf{k}+\mathbf{q}/2\uparrow}^{(j)}\, c_{\mathbf{k}'+\mathbf{q}'/2\uparrow}^{(j)\dagger}\, \delta_{ji} & \text{if } \mathbf{k} + \mathbf{q}/2 \neq \mathbf{k}' + \mathbf{q}'/2 \\ & \text{and } -\mathbf{k} + \mathbf{q}/2 = -\mathbf{k}' + \mathbf{q}'/2 \\ 0 & \text{otherwise.} \end{cases} \tag{11.8}$$

We continue to represent the annihilation operators for ground-state pairons by $b_{\mathbf{k}}^{(j)} \equiv B_{\mathbf{k}0}^{(j)}$. Using the new notation, we can rewrite the full Hamiltonian in Eq. (5.1) as

$$\begin{aligned} H &= \sum_{\mathbf{k},s} \varepsilon_k^{(1)}\, n_{\mathbf{k},s}^{(1)} + \sum_{\mathbf{k},s} \varepsilon_k^{(2)}\, n_{\mathbf{k},s}^{(2)} \\ &\quad -\sum_{\mathbf{k}}{}' \sum_{\mathbf{q}}{}' \sum_{\mathbf{k}'}{}' [v_{11} B_{\mathbf{kq}}^{(1)\dagger} B_{\mathbf{k}'\mathbf{q}}^{(1)} + v_{12} B_{\mathbf{kq}}^{(1)\dagger} B_{\mathbf{k}'\mathbf{q}}^{(2)\dagger} + v_{21} B_{\mathbf{kq}}^{(2)} B_{\mathbf{k}'\mathbf{q}}^{(1)} + v_{22} B_{\mathbf{kq}}^{(2)} B_{\mathbf{k}'\mathbf{q}}^{(2)\dagger}] \\ &\quad -\sum_{\mathbf{k}}{}' \sum_{\mathbf{k}'}{}' [v_{11} b_{\mathbf{k}'}^{(1)\dagger} b_{\mathbf{k}}^{(1)} + v_{12} b_{\mathbf{k}'}^{(2)\dagger} b_{\mathbf{k}}^{(1)\dagger} + v_{21} b_{\mathbf{k}}^{(2)} b_{\mathbf{k}'}^{(1)} + v_{22} b_{\mathbf{k}}^{(2)} b_{\mathbf{k}'}^{(2)\dagger}]. \end{aligned} \tag{11.9}$$

Bilinear terms in b and B do not appear because of momentum conservation:

$$\mathbf{k}_1 + \mathbf{k}_2 - (\mathbf{k}_3 + \mathbf{k}_4) = \mathbf{q} - \mathbf{q}' = 0. \tag{11.10}$$

11.2 Pairon Energy Gaps

When a phonon is exchanged between two quasi-electrons, the pair formed brcomes a moving pairon. Since quasi-electrons have an energy gap Δ, moving pairons also have an energy gap ε_g relative to the energy of the ground pairons. We study the energy of moving pairons in this section. We start with the full Hamiltonian H in Eq. (11.9). Using Eqs. (11.1), and (11.7-11.8), we obtain (Problem 11.2.1.)

$$\begin{aligned}
[H,\, B^{(1)\dagger}_{\mathbf{kq}}] &= [\varepsilon^{(1)}(|\mathbf{k}+\mathbf{q}/2|) + \varepsilon^{(1)}(|-\mathbf{k}+\mathbf{q}/2|)]B^{(1)\dagger}_{\mathbf{kq}} \\
&\quad - [v_{11}\sum_{\mathbf{k}'}{}' B^{(1)\dagger}_{\mathbf{k}'\mathbf{q}} + v_{12}\sum_{\mathbf{k}'}{}' B^{(2)}_{\mathbf{k}'\mathbf{q}}](1 - n^{(1)}_{\mathbf{k}+\mathbf{q}/2\uparrow} - n^{(1)}_{-\mathbf{k}+\mathbf{q}/2\downarrow}) \\
&\quad - (c^{(1)}_{\mathbf{k}+\mathbf{q}/2\uparrow} c^{(1)}_{\mathbf{k}-\mathbf{q}/2\downarrow} + c^{(1)}_{-\mathbf{k}-\mathbf{q}/2\uparrow} c^{(1)}_{-\mathbf{k}+\mathbf{q}/2\downarrow})[v_{11}\sum_{\mathbf{k}'}{}' b^{(1)\dagger}_{\mathbf{k}'} + v_{12}\sum_{\mathbf{k}'}{}' b^{(2)}_{\mathbf{k}'}],
\end{aligned} \tag{11.11}$$

$$\begin{aligned}
[H,\, B^{(2)}_{\mathbf{kq}}] &= -[\varepsilon^{(2)}(|\mathbf{k}+\mathbf{q}/2|) + \varepsilon^{(2)}(|-\mathbf{k}+\mathbf{q}/2|)]B^{(2)}_{\mathbf{kq}} \\
&\quad - [v_{21}\sum_{\mathbf{k}'}{}' B^{(1)\dagger}_{\mathbf{k}'\mathbf{q}} + v_{22}\sum_{\mathbf{k}'}{}' B^{(2)}_{\mathbf{k}'\mathbf{q}}](1 - n^{(2)}_{\mathbf{k}+\mathbf{q}/2\uparrow} - n^{(2)}_{-\mathbf{k}+\mathbf{q}/2\downarrow}) \\
&\quad - (c^{(2)}_{-\mathbf{k}+\mathbf{q}/2\downarrow} c^{(2)\dagger}_{-\mathbf{k}-\mathbf{q}/2\uparrow} + c^{(2)\dagger}_{\mathbf{k}-\mathbf{q}/2\downarrow} c^{(2)}_{\mathbf{k}+\mathbf{q}/2\uparrow})[v_{21}\sum_{\mathbf{k}'}{}' b^{(1)\dagger}_{\mathbf{k}'} + v_{22}\sum_{\mathbf{k}'}{}' b^{(2)}_{\mathbf{k}'}].
\end{aligned} \tag{11.12}$$

These exact equations indicate that the dynamics of moving pairons depend on pairons and quasi-electrons. If we omit the interaction terms containing $B^{(1)\dagger}_{\mathbf{k}'\mathbf{q}}$ and $B^{(2)}_{\mathbf{k}'\mathbf{q}}$ in Eq. (11.11), we get

$$\begin{aligned}
[H,\, B^{(1)\dagger}_{\mathbf{kq}}] &= [\varepsilon^{(1)}(|\mathbf{k}+\mathbf{q}/2|) + \varepsilon^{(1)}(|-\mathbf{k}+\mathbf{q}/2|)]B^{(1)\dagger}_{\mathbf{kq}} \\
&\quad - (c^{(1)\dagger}_{\mathbf{k}+\mathbf{q}/2\uparrow} c^{(1)}_{\mathbf{k}-\mathbf{q}/2\downarrow} + c^{(1)}_{-\mathbf{k}-\mathbf{q}/2\uparrow} c^{(1)\dagger}_{-\mathbf{k}+\mathbf{q}/2\downarrow})[v_{11}\sum_{\mathbf{k}'}{}' b^{(1)\dagger}_{\mathbf{k}'} + v_{12}\sum_{\mathbf{k}'}{}' b^{(2)}_{\mathbf{k}'}].
\end{aligned} \tag{11.13}$$

This simplified equation is equivalent to a set of two separate equations of motion for a quasi-electron [see Eq. (10.2)]:

$$[H_0,\, c_{\mathbf{p}\uparrow}^{(1)\dagger}] = \varepsilon_p^{(1)} c_{\mathbf{p}\uparrow}^{(1)\dagger} - c_{-\mathbf{p}\downarrow}^{(1)} [v_{11} \sum_{\mathbf{k}'}{}' b_{\mathbf{k}'}^{(1)\dagger} + v_{12} \sum_{\mathbf{k}'}{}' b_{\mathbf{k}'}^{(2)}]. \tag{11.14}$$

In Sections 10.2 we analyzed Eq. (11.14) and its associate $[H_0,\, \dot{c}_{-\mathbf{p}\downarrow}^{(1)}]$, and saw that the energy of a quasi-"electron" is

$$E_p^{(j)} = [\varepsilon_p^{(j)2} + \Delta_j^2(T)]^{1/2}. \tag{11.15}$$

Using this knowledge and examining Eq. (11.13), we infer that approximate but good equations of motion for B's are represented by

$$\begin{aligned} [H,\, B_{\mathbf{kq}}^{(1)\dagger}] &= [E^{(1)}(|\mathbf{k}+\mathbf{q}/2|) + E^{(1)}(|-\mathbf{k}+\mathbf{q}/2|)] B_{\mathbf{kq}}^{(1)\dagger} \\ &\quad -[v_{11} \sum_{\mathbf{k}'}{}' B_{\mathbf{k}'\mathbf{q}}^{(1)\dagger} + v_{12} \sum_{\mathbf{k}'}{}' B_{\mathbf{k}'\mathbf{q}}^{(2)}](1 - n_{\mathbf{k}+\mathbf{q}/2\uparrow}^{(1)} - n_{-\mathbf{k}+\mathbf{q}/2\downarrow}^{(1)}), \end{aligned} \tag{11.16}$$

$$\begin{aligned} [H,\, B_{\mathbf{kq}}^{(2)}] &= -[E^{(2)}(|\mathbf{k}+\mathbf{q}/2|) + E^{(2)}(|-\mathbf{k}+\mathbf{q}/2|)] B_{\mathbf{kq}}^{(2)} \\ &\quad -[v_{21} \sum_{\mathbf{k}'}{}' B_{\mathbf{k}'\mathbf{q}}^{(1)\dagger} + v_{22} \sum_{\mathbf{k}'}{}' B_{\mathbf{k}'\mathbf{q}}^{(2)}](1 - n_{\mathbf{k}+\mathbf{q}/2\uparrow}^{(2)} - n_{-\mathbf{k}+\mathbf{q}/2\downarrow}^{(2)}). \end{aligned} \tag{11.17}$$

These two equations reduce to the correct limits Eq. (6.55) at $T = T_c$, where $\Delta_j = 0$ and $E_k^{(j)} = \varepsilon_k^{(j)}$. In the limit $T \to 0$, they generate Eqs. (11.15).

Let us now introduce the energy-(k-q) representation and rewrite Eqs. (11.16) in the bulk limit. After using the decoupling approximation and taking the low temperature limit, we obtain (Problem 11.2.2.)

$$\begin{aligned} \tilde{w}_q^{(j)} a_j(\mathbf{k},\mathbf{q}) &= [E^{(j)}(|\mathbf{k}+\mathbf{q}/2|) + E^{(j)}(|-\mathbf{k}+\mathbf{q}/2|)] a_j(\mathbf{k},\mathbf{q}) \\ &\quad -v_{jj}(2\pi\hbar)^{-3} \int' d^3k' a_j(\mathbf{k}',\mathbf{q}), \end{aligned} \tag{11.18}$$

where $\tilde{w}_q^{(j)}$ are the pairon energies and $a_j(\mathbf{k},\mathbf{q})$ the pairon wave functions. After the decoupling $\pm$ pairons can be treated separately. We drop the subscript j hereafter. Equation (11.18) is reduced to the Cooper equation

(6.57) in the zero-Δ limit. The same $\mathbf{q}$ appears in the arguments of all a's, meaning that the net momentum $\mathbf{q}$ is a constant of motion. From Eq. (11.15), we observe that

$$E(|\mathbf{k}+\mathbf{q}/2|) + E(|-\mathbf{k}+\mathbf{q}/2|) \geq 2E_k > 2\varepsilon_k. \tag{11.19}$$

Eq. (11.18) is similar to Eq. (6.11). We can use the same method to solve it to obtain

$$\widetilde{w}_q = \widetilde{w}_0 + \frac{1}{2}v_F q < 0, \tag{11.20}$$

where $\widetilde{w}_0(T)$ (< 0) is the solution of

$$1 = \mathcal{N}(0)v_0 \int_0^{\hbar\omega_D} d\varepsilon[|\widetilde{w}_0| + 2(\varepsilon^2 + \Delta^2)^{1/2}]^{-1}. \tag{11.21}$$

Note: $\widetilde{w}_0$ is temperature-dependent, since Δ is. Let us write the energy $\widetilde{w}_q$ in the form:

$$\widetilde{w}_q = w_0 + \varepsilon_g(T) + \frac{1}{2}v_F q < 0, \tag{11.22}$$

$$\varepsilon_g(T) \equiv \widetilde{w}_0(T) - w_0 > 0, \tag{11.23}$$

where ε_g is *the energy gap between excited and ground-state pairons.* (This gap is shown in energy diagram in Fig. 1.10). The gap ε_g is nonnegative, as seen by comparing Eqs. (11.21) and (6.12) and noting inequality (11.19). The gap $\varepsilon_g(T)$ depends on the temperature T, and it can be numerically computed from Eq. (11.21), which is shown in Fig. 11.1.

The temperature behavior of $\varepsilon_g(T)$ and $\Delta(T)$ are similar. Compare Figs. 10.1 and 11.1. Both energy gaps are greatest at 0 K and monotonically decrease to zero as temperature is raised to T_c. Further note that the pairon energy gap $\varepsilon_g(T)$ is smaller than the quasi-electron energy gap $\Delta(T)$ over the entire temperature range:

$$\varepsilon_g(T) < \Delta(T), \qquad (0 < T < T_c), \tag{11.24}$$

an inequality which can be verified from Eq. (11.18) (Problem 11.2.3).

Problem 11.2.1. Derive Eqs. (11.11) and (11.12).

Problem 11.2.2. Derive Eq. (11.18).

Problem 11.2.3. Show inequality Eq. (11.24) from Eqs.(11.19) and (11.20).

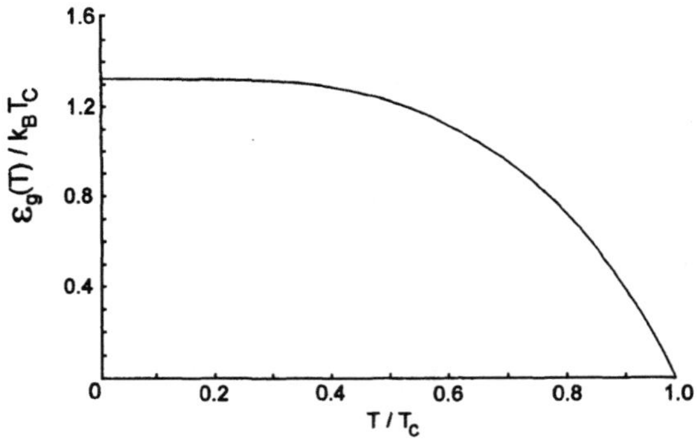

Figure 11.1: Variation of the pairon energy gap $\varepsilon_g(T)$ with temperature.

11.3 Density of Condensed Pairons

Below T_c there are a great number of condensed pairons. We discuss the temperature dependence of the condensed pairon density in this section.

At 0 K all pairons are condensed at zero momentum. At a finite temperature there are moving pairons and quasi-electrons. Pairons by construction are in negative-energy states, while quasi-electrons have positive kinetic energies. Thus by the Boltzmann principle, pairons are more numerous. In the following we assume that there are only excited pairons. There are $\pm$ pairons. The predominant pairons are those composed of "electrons" or "holes", whichever has smaller effective mass. These pairons dominate the B-E condensation and again the supercurrent flow. The supercondensate is composed of equal numbers of $\pm$ pairons. The B-E condensation of pairons can be discussed by looking at the predominant pairons only. The number density of ground predominant pairons at 0 K, $n_0(0)$, is given by

$$n_0(0) = (1/2)\hbar\omega_D \mathcal{N}_1(0)V^{-1} \equiv n, \qquad T = 0. \tag{11.25}$$

As temperature is raised, the number density of the ground pairons $n_0(T)$ decreases and vanishes at the critical temperature T_c. The sum of the density of ground pairons, $n_0(T)$, and the density of moving pairons, $n_x(T)$, is equal to n:

$$n_0(T) + n_x(T) = n. \tag{11.26}$$

Our task then is to find $n_x(T)$. [There is no way of finding $n_0(T)$ directly in the theory of B-E condensation.]

Let us recall that pairons do exist above T_c, and they have a linear dispersion relation

$$w_q = w_0 + \frac{1}{2}v_F q < 0, \qquad w_0 \equiv \frac{-2\hbar\omega_D}{\exp[2/v_0 N(0)] - 1}. \qquad (T > T_c) \quad (11.27)$$

Below T_c, moving pairons have the T-dependent relation:

$$\tilde{w} = \tilde{w}_0(T) + \frac{1}{2}v_F q < 0, \qquad (T < T_c) \quad (11.28)$$

where $\tilde{w}_0(T)$ is the solution of

$$1 = \mathcal{N}(0)v_0 \int_0^{\hbar\omega_D} d\varepsilon\, [|\,\tilde{w}_0\,| + 2(\varepsilon^2 + \Delta^2)^{1/2}]^{-1}. \quad (11.29)$$

The quasi-electron energy gap $\Delta(T)$ is given by Eq. (10.34) and the critical temperature T_c by Eq. (10.35). Note: $\Delta = 0$ at T_c. Then from Eq. (10.30), we obtain $\tilde{w}_0(T_c) = w_0$. Thus we obtain

$$\tilde{w}_q - w_0 = \tilde{w}_0 - w_0 + \frac{1}{2}v_F q \equiv \varepsilon_g + \frac{1}{2}v_F q > 0. \quad (11.30)$$

Moving pairons in equilibrium are populated by the Bose distribution function:

$$f(\varepsilon; \beta, \mu) \equiv \frac{1}{e^{\beta(\varepsilon-\mu)} - 1} \equiv f(\varepsilon). \qquad (\varepsilon = \tilde{w}_q - w_0) \quad (11.31)$$

In the condensation region ($T < T_c$), the chemical potential μ vanishes:

$$\mu = 0, \qquad T < T_c. \quad (11.32)$$

Thus the number density of excited pairons, $n_X(T)$, below T_c, is given by

$$n_X(T) = (2\pi\hbar)^{-3} \int d^3q [\exp\{\beta[\varepsilon_g(T) + v_F q/2]\} - 1]^{-1}. \quad (11.33)$$

Let us assume a Fermi sphere. We then obtain (Problem 11.3.1.)

$$n_X(T) = \frac{4(k_B T)^3}{\pi^2\hbar^3 v_F^3} \int_0^\infty \frac{x^2 dx}{\exp[\beta\varepsilon_g(T) + x] - 1}. \qquad \left(x \equiv \frac{1}{2}\beta v_F q\right) \quad (11.34)$$

Using the Bose-Einstein integral

$$\phi_\nu(\lambda) \equiv \frac{1}{\Gamma(\nu)} \int_0^\infty dx \frac{x^{\nu-1}}{\lambda^{-1}e^x - 1} \equiv \sum_{j=1}^\infty \frac{\lambda^\nu}{j^\nu}, \tag{11.35}$$

where $\Gamma(\nu)$ is the gamma function, we can reexpress Eq. (11.34) as

$$n_X(T) = \pi^{-2}[2k_BT/\hbar v_F]^3 \phi_3[\exp(-\varepsilon_g/k_BT)]. \tag{11.36}$$

At T_c, where $\widetilde{w}_0(T_c) = w_0$, $\varepsilon_g = 0$,

$$n_X(T_c) = \pi^{-2} \frac{2k_BT_c}{\hbar v_F} \phi_3(1) = n. \tag{11.37}$$

Therefore the desired $n_0(T)$ is given by

$$\begin{aligned} n_0(T) &= n - n_X(T) \\ &= \pi^{-2} \frac{2k_BT_c}{\hbar v_F} \phi_3(1) \{1 - (T/T_c)^3 \frac{\phi_3[\exp(-\varepsilon_g/k_BT)]}{\phi_3(1)}\}. \end{aligned} \tag{11.38}$$

After introducing reduced variables:

$$T/T_c \equiv x, \qquad \varepsilon_g/k_BT_c \equiv y, \tag{11.39}$$

we can then reexpress Eq. (11.38) as

$$\frac{n_0(T)}{n} = 1 - x^3 \frac{\phi_3[\exp(-y/x)]}{\phi_3(1)}. \tag{11.40}$$

The temperature dependence of $n_0(T)/n$ is shown in Fig. 11.2.

If there were no energy gap in the pairon energy $[\varepsilon_g(T) = 0]$, Eq. (11.38) simplifies to

$$\frac{n_0(T)}{n} = 1 - x^3 = 1 - (T/T_c)^3, \qquad [\varepsilon_g(T) = 0] \tag{11.41}$$

which is a known result for a system of free bosons moving with a linear dispersion relation $\varepsilon = cp$, see Eq. (9.27). Note: The effect of a nonzero energy gap $\varepsilon_g(T)$ is quite appreciable. The number density of moving pairons, $n_X(T)$, drops off exponentially a little below T_c, as seen from Eq. (11.38). The existence of the condensate is the most distinct feature of a system of bosons. Elementary excitations in the presence of a supercondensate contribute to the heat capacity of the system, which will be discussed in the following section.

Problem 11.3.1. Verify Eqs. (11.34) and (11.36).

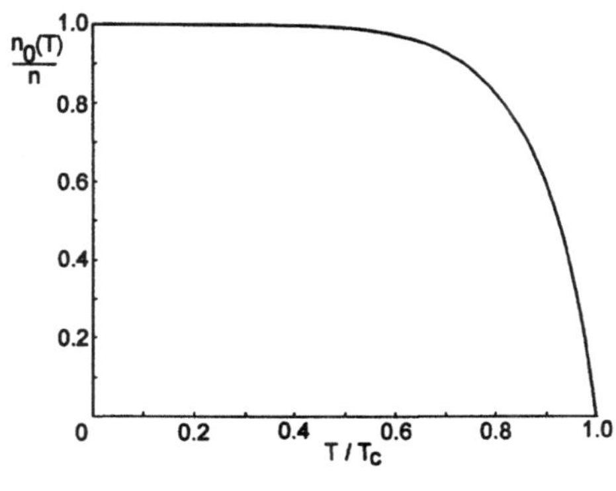

Figure 11.2: The reduced supercondensate density $n_0(T)/n$ monotonically declines with increasing temperature.

11.4 Heat Capacity

In this section we calculate the heat capacity of the generalized BCS system under the assumption that there are only predominant pairons (no quasi-electrons). For the Cooper system the dispersion relation for pairons is

$$w_q - w_0 = \frac{1}{2} v_F q. \tag{11.42}$$

For our generalized BCS system, the dispersion relation above T_c has the same form (11.42). But below T_c, the relation is altered to

$$\tilde{w}_q - w_0 = \varepsilon_g(T) + \frac{1}{2} v_F q. \tag{11.43}$$

The internal energy density $u(T)$ can be calculated as follows: (Problem 11.4.1)

$$\begin{aligned} u(T) &= \frac{1}{(2\pi\hbar)^3} \int d^3q (\frac{1}{2} v_F q) \frac{1}{\exp\{[\varepsilon_g(T) + v_F q/2]/k_B T\} - 1} \\ &= \frac{24(k_B T)^4}{\pi^2 \hbar^3 v_F^3} \frac{1}{\Gamma(4)} \int_0^\infty dx \frac{x^3}{e^{\beta\varepsilon_g} e^x - 1} = 3nk_B T_c (T/T_c)^4 \frac{\phi_4[\exp(-\varepsilon_g/k_B T)]}{\phi_3(1)}. \end{aligned} \tag{11.44}$$

Comparing Eq. (11.44) with Eq. (9.28) and noting that $\varepsilon_g(T_c) = 0$, we observe that the internal energy is continuous at T_c. Note: $u(T)$ is a sole function of T. Differentiating $u(T)$ with respect to T, we obtain (Problem 11.4.2.)

$$\begin{aligned} C(T) &= \frac{du(T)}{dT} = 12nk_B(T/T_c)^3\frac{\phi_4[\exp(-\varepsilon_g/k_BT)]}{\phi_3(1)} \\ &-3n(T/T_c)^2\frac{\phi_3[\exp(-\varepsilon_g/k_BT)]}{\phi_3(1)}\left\{\frac{T}{T_c}\frac{d\varepsilon_g(T)}{dT}-\frac{\varepsilon_g(T)}{T_c}\right\}, \quad (T<T_c). \end{aligned} \tag{11.45}$$

Above T_c heat capacity C is given by

$$C(T) = 12nk_B(T/T_c)^3\frac{\phi_4(\lambda)}{\phi_3(1)} - 9nk_B\frac{\phi_3(\lambda)}{\phi_2(\lambda)}, \tag{11.46}$$

where the fugacity λ is implicitly given by

$$(1/8)\pi^2\hbar^3v_F^3n = k_B^3T^3\phi_3(\lambda), \qquad (T>T_c). \tag{11.47}$$

Comparing Eqs. (11.45) and (11.46), we see that the heat capacity C has a *discontinuity* at T_c. The magnitude of the jump ΔC at T_c is (Problem 11.4.3.)

$$\Delta C = 3nk_B\left[\frac{3\phi_3(1)}{\phi_2(1)} - \frac{1}{k_B}\frac{d\varepsilon_g}{dT}\right]_{T=T_c}. \tag{11.48}$$

Figure 11.3. shows a plot of reduced heat capacity C/nk_B as a function of T/T_c. The effect of the pairon-energy gap ε_g on the heat capacity C appears prominently. From Eq. (11.45), the behavior of $C(T)$ far below T_c may be represented analytically by $C(T) \sim (T/T_c)^3\exp(-\varepsilon_g/k_BT)$ because $\phi_4(x)$ is a smooth function of x.

There is also a small but non-negligible correction on the maximum heat capacity $C_{\max}$ [see Eq. (11.47)] because

$$\frac{d\varepsilon_g}{dT} < 0. \tag{11.49}$$

The jump ΔC over the maximum $C_{\max}$ is given by

$$\Delta C/C_{\max} = \left[\frac{3\phi_3(1)}{\phi_2(1)} - \frac{d\varepsilon_g}{d(k_BT_c)}\right] \div \left[\frac{4\phi_4(1)}{\phi_3(1)} - \frac{d\varepsilon_g}{d(k_BT_c)}\right]. \tag{11.50}$$

If we neglect $d\varepsilon_g(T)/dT$, we have

$$\Delta C/C_{\max} = \frac{3}{4}\left[\frac{\phi_3(1)^2}{\phi_2(1)\phi_4(1)}\right] = 0.608, \tag{11.51}$$

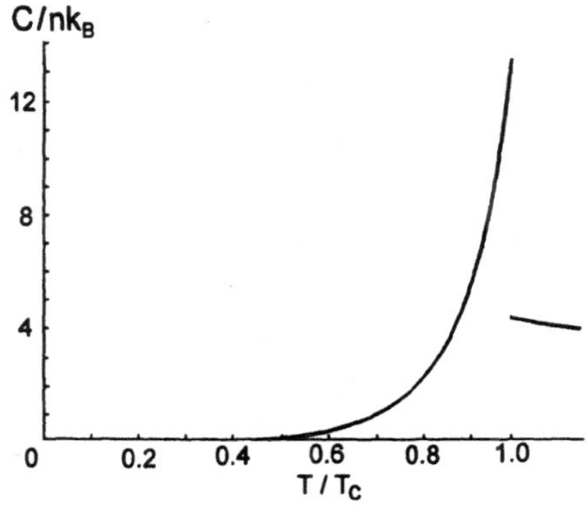

Figure 11.3: The (reduced) heat capacity C/nk_B versus temperature.

which is quite close to the value obtained by the finite-temperature BCS theory [2],

$$(\Delta C/C_{\max})_{BCS} = 0.588. \tag{11.52}$$

The numerical calculation from Eqs. (11.46)-(11.48) and (11.50) yields the following ratio

$$\Delta C/C_{\max} = 0.702. \tag{11.53}$$

The experimentally observed ratios of the heat capacity jump ΔC to the maximum $C_{\max}$ at T_c. The ratios are in the range $0.6 - 0.7$. Thus our theoretical results (11.51) and (11.53) are in very good agreement with experiments. Further discussion on our results are given in the following section.

Problem 11.4.1. Verify Eq. (11.44).

Problem 11.4.2. Verify Eq. (11.45).

Problem 11.4.3. Verify Eq. (11.48).

11.5 Discussion

In the present chapter we studied the thermodynamic properties of the BCS superconductor below T_c. The equation-of-motion method is repeatedly used to study the energy spectra of quasi-electrons, ground and moving pairons.

Quasi-electrons and pairons are found to have gaps $(\Delta_j, \varepsilon_{gj})$ in their spectra. These gaps depend on temperature. Under the assumption that excited pairons are the only elementary excitations in the superconductor, we computed the density of condensed pairons, $n_0(T)$, which is shown in Fig. 11.2. Under the same assumption, we calculated the heat capacity $C(T)$, which is shown in Fig. 11.3. The overall agreement with experimental data is very good. Our description of the superconducting state is quite different from the original BCS description [2], in particular regarding the choice of elementary excitations: moving pairons versus quasi-electrons.

11.5.1 Pairon Energy Gap ε_g

The phonon exchange attraction acts at all time and at any temperature. Thus two quasi-electrons may be bound to form moving pairons. Since quasi-electrons have an energy gap $\Delta(T)$, excited pairons also have an energy gap $\varepsilon_g(T)$. The dispersion relation for the moving pairon is

$$\tilde{w}_q = w_0 + \varepsilon_g(T) + \frac{1}{2} v_F q < 0. \qquad (w_0 < 0) \tag{11.54}$$

The temperature behavior of the gap $\varepsilon_g(T)$ is shown in Fig. 11.1.

11.5.2 Elementary Excitations in a Superconductor

BCS calculated many thermodynamic properties, including the critical temperature T_c, the energy gap $\Delta(T)$ and the heat capacity under the assumption that quasi-electrons are elementary excitations. This assumption was necessary if the free-energy minimum principle is used to determine the thermal distribution of the quasi-electrons. In our theory all thermodynamic properties are computed, using the grand canonical ensemble. The question of whether moving pairons or quasi-electrons are dominant is answered logically by examining energies of the quasi-particles. Excited pairons having negative energies are more numerous than quasi-electrons, and the former dominate the low-temperature behavior of the superconductor.

11.5.3 The Ground Pairon Density n_0 at 0 K

The energy gap $\varepsilon_g(0)$ at 0 K is equal to the binding energy $|w_0|$ of a pairon. We can obtain the values of $|w_0|$ through the quantum tunneling and/or photo-absorption experiments. Since the total binding energy W is given by

$$\frac{W}{V} = n_0 \left| w_0 \right| = \frac{B_c^2}{2\mu_0}, \tag{11.55}$$

we can estimate the pairon density n_0. For Al (Pb), $B_c = 105$ (803) G and $\varepsilon_g = 0.34$ (2.73) meV. The values of n_0 calculated from Eq. (11.55) are 0.73 (5.29) $\times\ 10^{18}$ cm^{-3}. Thus the estimated pairon density is about 10^{-4} of the conduction electron density. This means that the great majority of electrons do not participate in the pairon formation. Only those electrons near the hyperboloidal Fermi surface (necks and/or double-caps) can form pairons.

11.5.4 Heat Capacity

Heat capacity is one of the most revealing properties of any many-particle system. Precise heat capacity measurements often contribute to the basic understanding of the underlying physics. We saw earlier in Chapter 8 that the superconducting transition is a B-E condensation of free pairons moving with the linear dispersion relation. The condensation transition is of the second order. Below T_c the presence of the supercondensate generates a pairon energy gap. In the strong predominant pairon limit, we calculated the heat capacity $C(T)$, shown in Fig. 11.3. The effect of ε_g on $C(T)$ is apparent for all temperatures below T_c. Far below T_c, $C(T)$ falls almost exponentially as $\exp[-\varepsilon_g(T)/k_B T]$. Since $d\varepsilon_g/dT < 0$ near T_c, the maximum $C_{\max}$ is greater for the generalized BCS system than for the Cooper system. Thus, the ratio $\Delta C/C_{\max}$ is greater for the generalized BCS system (0.702) than for the Cooper system (0.608). Experimental values of $\Delta C/C_{\max}$ for elemental superconductors are in the $(0.6 - 0.7)$ range. The agreement is better than the BCS prediction of 0.588. Our numerical calculations are performed with the assumption of strong predominance. In real metals the ratio of absolute effective masses for "electrons" and "holes", $|m_1/m_2|$, may be far from unity. It varies from material to material. Thus if we include the finiteness of $|m_1/m_2|$ in the calculations, we may account for the material-dependence of $\Delta C/C_{\max}$. The BCS theory is based on the assumption that quasi-electrons are the only elementary excitations, while our calculations are carried out on the assumption that excited pairons are predominant. It is quite remarkable that the calculated $C(T)$ is not dissimilar in spite of the fact that Bose (Fermi) distribution functions are used for moving pairons (quasi-electrons).

Chapter 12

Quantum Tunneling

The I-V curves for the quantum tunneling are interpreted based on the pairon transport model for S-I-S and S_1-I-S_2 sandwiches. The data analysis yields a direct measurement of the pairon energy gap $\varepsilon_g(T)$ as a function of temperature T. The negative resistance ($dI/dV < 0$) in the I-V curve for the S_1-I-S_2 system arises from the bosonic nature of the moving pairon.

12.1 Introduction

In the last chapter we saw that there are two energy gaps (Δ, ε_g), Δ in the quasi-electron and ε_g in the pairon energy spectrum. Although both (Δ, ε_g) are called energy gaps, they are different in character. The pairon energy gap ε_g is the energy separation between the stationary and moving pairons as shown in Fig. 12.1(a). This gap is similar to the ionization gap between the discrete ground state energy level and the continuous ionized states energy band in an atom. In contrast, the quasi-electron energy gap Δ appears in the energy of a quasi-electron:

$$E_k = (\varepsilon_k^2 + \Delta^2)^{1/2}, \tag{12.1}$$

where ε_k is the energy of the Bloch electron. Eq. (12.1) indicates that the quasi-electron in the superconducting state has a higher energy than the electron in the normal state. Looking at the energy spectrum as a whole, we note that the quasi-electrons have the minimum excitation energy Δ relative to the Fermi energy as shown in Fig. 12.1(b). The point is that the gap Δ defined in terms of the energy of the quasi-electron is compared with the

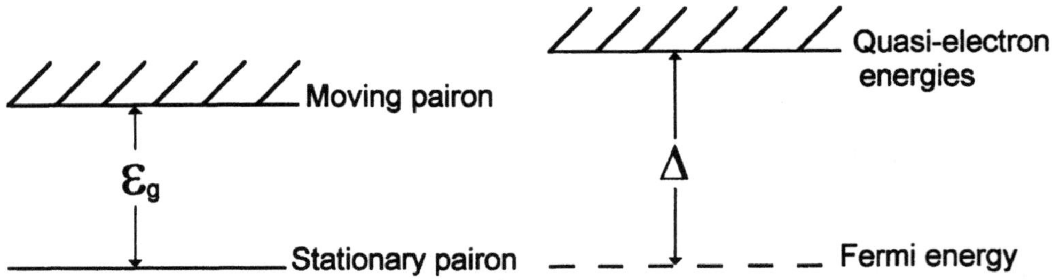

Figure 12.1: (a) The moving and stationary pairons have an energy gap ε_g. (b) Quasi-electron energies have a minimum separation Δ relative to the Fermi energy (dashed line).

energy of the electron in the normal state. As we saw in subsection 7.3.9, the Fermi surface is blurred in the superconducting state, and no sharp Fermi energy exists, which is indicated by the dash line. Hence the photo-absorption experiments, if performed, likely detect the pairon energy gap ε_g and not the electron energy gap Δ. The quantum tunneling techniques can also detect only the pairon-energy gaps ε_g.

12.2 Quantum Tunneling in S-I-S Systems

We discuss a quantum tunneling in Al-Al_2O_3-Al. A typical experimental setup is schematically shown in Fig. 12.2. Here, S_1 and S_2 are superconductors and I is an oxide (insulator thin film) of width $\sim$ 20 Å. The system S_1-I-S_2 is connected with a variable resistor and a battery. In the present section we consider the case where the same superconductors are used for both S_1 and S_2. Such a system is called an S-I-S system.

The operating principles are as follows. If the bias voltage is low, charged particles may quantum-tunnel through the oxide; the resulting current is low. When the voltage is raised high enough, the supercondensate may release moving pairons, and the resulting pairons may tunnel and generate a sudden increase in the current. By measuring the *threshold voltage* V_t, we obtain information on the energy gaps. The experimental I(current)-V(voltage) curves for Al-Al_2O_3-Al sandwich obtained by Giaever and Megerle [1] are reproduced in Fig. 12.3. The main features of the I-V curves (see the two lowest lines at T = 1.10, 1.08 K, T_c =1.14 K) are:

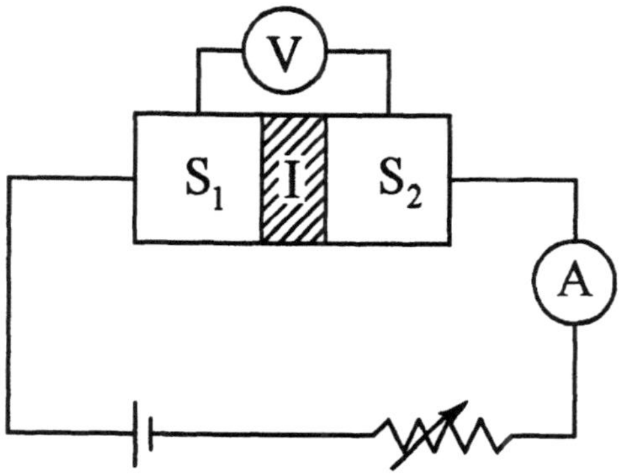

Figure 12.2: A schematic sketch of quantum tunneling circuit.

a) An (anti)symmetry with respect to the zero voltage (not shown).

b) A nearly flat small current below some threshold voltage V_t.

c) A sudden current increase near the threshold is temperature-dependent, and the slope dI/dV becomes steeper as temperature is lowered.

d) All I-V curves are below the straight line representing Ohm's law behavior.

e) There is no sudden change in the I-V curve at T_c (1.14 K).

We shall analyze the experimental data based on the *pairon transport model.* Giaever *et al.* interpreted the data based on the electron transport model [1, 2]. Earlier we gave our view that the pairon gap ε_g is the real energy gap similar to the ionization energy gap and can be probed by optical experiments. We give one more reason why we choose the pairon transport model. Tunneling experiments are done in a steady-state condition, in which all currents in the superconductors S_j are supercurrents; that is, the charges are transported by the condensed pairons. Pairons are in negative-energy states, while quasi-electrons have positive energies. By the Boltzmann principle, these pairons are much more numerous than quasi-electrons at the lowest temperatures. Besides, our pairon transport model is a natural one since we deal with pairons throughout in the S-I-S sandwich. In Section 15.1 we shall see that the Josephson tunneling, which occurs with no bias, can also be discussed in terms of the pairon motion.

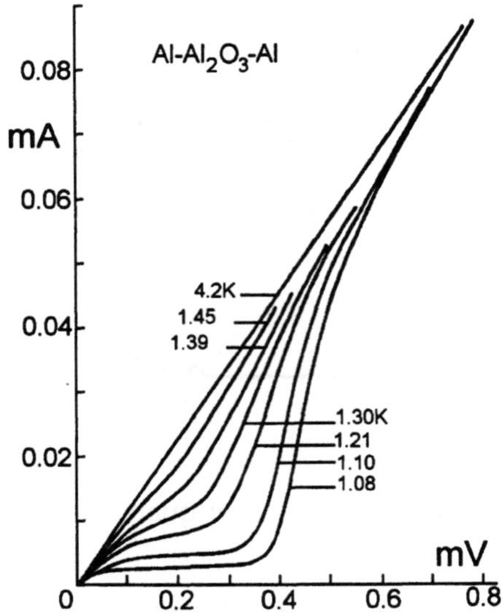

Figure 12.3: Current-voltage characteristics of an $Al\text{-}Al_2O_3\text{-}Al$ "sandwich" at various temperatures, after Giaever and Megerle [1]. $T_c = 1.14$ K.

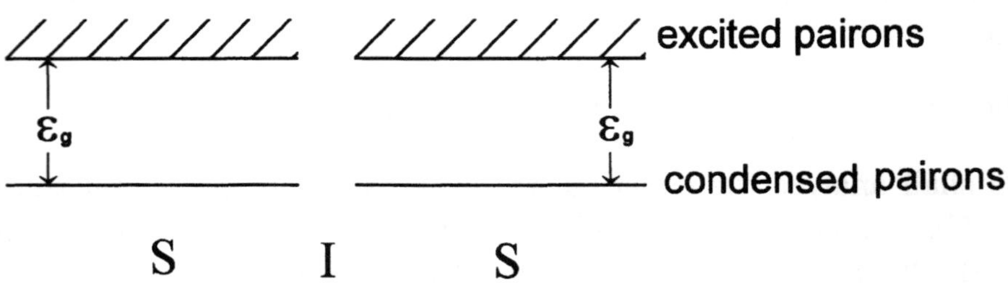

Figure 12.4: Pairon energy diagram for S-I-S.

Consider first the low-temperature limit. If the voltage is raised high enough, part of the supercondensate near the oxide may dissociate and release moving pairons, and some of these pairons gain enough energy to tunnel through the oxide. The threshold voltage V_t times the pairon charge (magnitude) $2e$ should equal twice the binding energy $|w_0|$:

$$2eV_t = 2\,|w_0| \qquad \text{or} \qquad eV_t = |w_0|\,. \tag{12.2}$$

(The factor 2 in front of $|w_0|$ will be explained later). Such behavior is in accordance with experiments [3], where an extremely sharp slope is observed.

The behavior shown in Fig. 12.3 is now interpreted with the aid of the diagrams in Fig. 12.4, where the energy spectra for pairons in S_1 and S_2 are shown. Ground pairon energy levels are chosen to have the same height, ensuring that with no bias voltage, no supercurrent flows in S-I-S. Since we have the same superconductors ($S_1 = S_2$), we have one energy gap ε_g. The energy spectra in S_1 and S_2 are symmetric with respect to I (oxide). Then *the* I-V *curve is antisymmetric* as stated in **a**). [In contrast, if different conductors are substituted for (S_1, S_2), the I-V curve is asymmetric. We discuss this case in Section 12.3].

To see this antisymmetry, first consider the case when a small voltage is applied. Those quasi-particles carrying the positive (negative) charges in the oxide tend to move right (left). Second, if we reverse the bias, then the opposite tendency holds. If the energy spectra on both sides are the same, the I-V curves must be antisymmetric. This case can be regarded as a *generalized Ohm's law*:

$$I = \frac{V}{R(V)}, \qquad \text{(small } V) \tag{12.3}$$

where R is a resistance, a material constant independent of the polarity of the voltage; R may depend on V, which is the said generalization.

Let us now derive a basic formula for the quantum tunneling. Assume that a +pairon having charge $q = 2e$ tunnels rightward from S_1 to S_2. The pairon in S_1 must arrive at the interface (S_1, I) with the positive (rightward) velocity. Inside S_1 there are excited pairons (bosons) moving independently in all directions and populated with the Planck distribution function:

$$f(\varepsilon) = \frac{1}{e^{\beta\varepsilon} - 1}, \qquad \varepsilon = cp, \qquad c \equiv v_F/2. \tag{12.4}$$

The pairon chemical potential vanishes below T_c. This condition is similar to the case of black-body radiation (inside an oven maintained at a temperature T), where photons move and are isotropically distributed.

We ask how many pairons arrive at the small interface element ΔA per unit time in a particular direction (θ, ϕ). The number density of pairons moving in the right direction within the solid angle:

$$d\Omega = \sin\theta \, d\theta \, d\phi, \tag{12.5}$$

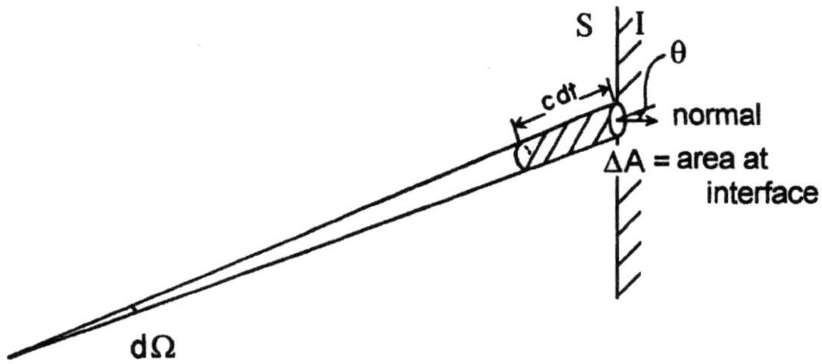

Figure 12.5: A pairon proceeding in the solid angle $d\Omega$ and located within the quasi-cylinder reaches the interface area ΔA in the time dt.

see Fig. 12.5, is given by

$$\frac{1}{(2\pi\hbar)^3} f(\varepsilon)\, p^2\, dp\, d\Omega. \tag{12.6}$$

Noting that all pairons move with the same speed c, we obtain

$$\text{pairon flux} = cf(\varepsilon)(2\pi\hbar)^{-3} p^2\, dp\, d\Omega = \frac{1}{2} v_F f(\varepsilon)(2\pi\hbar)^{-3} p^2\, dp\, d\Omega. \tag{12.7}$$

Notice that this expression is the same as that for the photon flux escaping from a small hole in the oven wall in a specified direction (θ, ϕ). Upon arriving at the interface, some pairons may be kicked in rightward through the oxide, gain energy equal to $2eV$, and reach S_2. The quantum tunneling here is similar to the elastic potential scattering. Only the final state must have an energy higher by $2eV$ than the initial state so that

$$\varepsilon_f = \varepsilon_i + 2eV, \tag{12.8}$$

which takes care of the energy gain during its passage through the oxide. We may then describe the quantum tunneling in terms of the quantum transition rate

$$R = \frac{2\pi}{\hbar} \left| \langle \mathbf{p}_f \,|v|\, \mathbf{p}_i \rangle \right|^2 \delta(\varepsilon_f - \varepsilon_i), \tag{12.9}$$

where v is a tunneling perturbation (energy). The precise nature of the perturbation does not matter; the perturbation's only role is to initiate a quantum tunneling transition from $\mathbf{p}_i$ to $\mathbf{p}_f$. Such perturbation may come from lattice defects, interface irregularities, and others. We assume that the matrix element, $|\langle \mathbf{p}_f | v | \mathbf{p}_i \rangle|^2$ is a constant:

$$|\langle \mathbf{p}_f | v | \mathbf{p}_i \rangle|^2 = M^2, \tag{12.10}$$

which is reasonable because of the incidental nature of the perturbation. Equation (12.9) may be used if and only if the energy of the final state ε_f is in an energy continuum. This may be satisfied by the excited pairon states in the final-state, depending on ε_i, ε_f, and $2eV$. If not, quantum tunneling cannot occur. We multiply Eqs. (12.7) and (12.9) together and integrate with respect to $d\Omega$ over the right-half, and further integrate with respect to dp. Since the +pairon has charge $2e$, we multiply the result by $2e$ to obtain an expression for the current density J as

$$J = \frac{4\pi e M^2}{\hbar} \left(\frac{1}{2} v_F\right) \int_{p_x > 0} f(\varepsilon) \frac{p^2 dp}{(2\pi\hbar)^3} d\Omega \, \delta(\varepsilon_f - \varepsilon - 2eV). \tag{12.11}$$

The Ω-integration equals one-half of the integration over all directions. Integration with respect to the initial-state $\mathbf{p}_i$ may be replaced by integration with the final state $\mathbf{p}_f$. If we use the identity:

$$\frac{1}{(2\pi\hbar)^3} \int d^3 p_f \, \delta(\varepsilon_f - \varepsilon_i) = \mathcal{N}_f(\varepsilon_i), \tag{12.12}$$

where $\mathcal{N}_f$ is the density of states at the final state, we can reexpress (12.11) as

$$J = \frac{2\pi}{\hbar} C M^2 \left(\frac{1}{2} e v_F\right) f_i(\varepsilon) [1 + f_f(\varepsilon + 2eV)] \mathcal{N}_f(\varepsilon + 2eV). \tag{12.13}$$

Here we added two correction factors: C (a constant) and $[1 + f_f(\varepsilon + 2eV)]$. The arguments, $\varepsilon + 2eV$ in $\mathcal{N}_f$ and f_f, represent energies measured relative to the energy level fixed in the superconductor S_i. Fermi's golden rule for a quantum transition rate, is given by

$$R(\mathbf{p}, \mathbf{p}_0) = (2\pi/\hbar) \, |\langle \mathbf{p} | v | \mathbf{p}_0 \rangle|^2 \, \mathcal{N}_f(\varepsilon). \tag{12.14}$$

Thus, Eq. (12.13) can be interpreted simply in terms of this rate R. Bardeen pointed out this important fact [4] right after Giaever's experiments [2]. The appearance of the Planck distribution function $f_i(\varepsilon)$ is significant. Since this factor arises from the initial-state pairon flux, we attach a subscript i. In the derivation we tacitly assumed that all pairons arriving at the interface (S_i, I) can tunnel to S_f and that tunneling can occur independently of the incident angle relative to the positive x-direction. Both assumptions lead to an overestimate. To compensate for this, we included the correction factor

$$C\ (<1). \tag{12.15}$$

In consideration of the boson-nature of the pairons, we also inserted the *quantum statistical factor*

$$1 + f_f(\varepsilon). \tag{12.16}$$

Let us summarize the results of our theory of quantum tunneling.

1. The dominant charge carriers are moving pairons.

2. In the rightward bias ($V_1 > V_2$), $+(-)$ pairons move preferentially right (left) through the oxide.

3. The bias voltage $V \equiv V_1 - V_2$ allows moving pairons to gain or lose an energy equal to $2eV$ in passing the oxide.

4. Quantum tunneling occurs at $\varepsilon_f = \varepsilon_i \pm 2eV$ and it does so if and only if the final state is in a continuous pairon energy band.

5. Moving pairons (bosons) are distributed according to the Planck distribution law, which makes the tunneling current temperature-dependent.

6. Some pairons may separate from the supercondensate and directly tunnel through the oxide.

7. Some condensed pairon may be excited and tunnel through the oxide, which requires an energy equal to twice the energy gaps ε_g or greater.

Statements 1-6 are self-explanatory. Statement 7 arises as follows: the minimum energy required to raise one pairon from the ground state to the excited state in S_i is equal to the energy gap ε_g. But to keep the supercondensate neutral, another pairon of the opposite charge must be taken away, which requires an extra energy equal to ε_g (or greater). Thus, the minimum energy required to move one pairon from the condensate to an excited state and keep the supercondensate intact is $2\varepsilon_g$. In the steady-state experimental

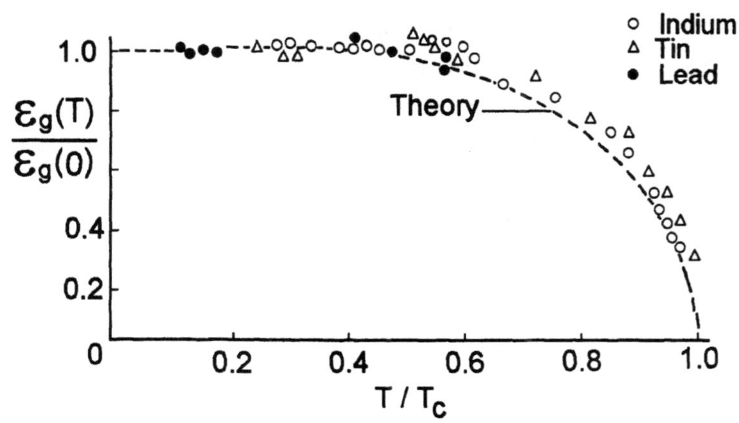

Figure 12.6: Variation of the measured energy gap $\varepsilon_g(T)$ with temperature.

condition, the initial and final states must be maintained and the supercondensate be repaired with the aid of external forces (bias voltages). The oxide is used to generate a bias. If the oxide layer is too thick, the tunneling currents are too small to measure.

We now analyze the S-I-S tunneling as follows. For a small bias V below the threshold bias V_t such that

$$V < V_t \equiv \varepsilon_g(T)/e. \tag{12.17}$$

The excited pairons already present in S_1 and S_2 may tunnel through the oxide. The current I is small since the number density of excited pairons is small. It should reach a plateau, where all excited pairons, whose total number is fixed at a given temperature, contribute; this explains feature **b)**.

Above the threshold V_t, some of the condensed pairons in the supercondensate may evaporate so that the resulting excited pairons tunnel through the oxide. The tunneling current is much greater, since the supercondensate is involved. Giaever *et al.* [1, 3] observed that the threshold voltage V_t depends on the temperature T and could determine the energy gap $\varepsilon_g(T)$ as a function of T. Fig. 12.6 represents part of their results.

Above T_c there are no energy gaps, and so electrons may be excited more easily. There are moving pairons still above T_c. Since the number densities of "electrons" and "holes" outweigh the density of pairons far above T_c, the I-V curve should eventually approach the straight line, which explains feature **d)**.

Feature **e)** supports our hypothesis that current passing through the oxide is carried by pairons. If quasi-electrons that become normal electrons above T_c were involved in the charge transport, the current I would have followed the ideal (Ohm's law) straight line immediately above T_c, in disagreement with the experimental observation.

Quantum tunneling is a relatively easy experiment to perform. It is most remarkable that the energy gap, which is one of the major superconducting features, can be obtained directly with no calculation.

The energy gap $\varepsilon_g(T)$ can also be measured in photo-absorption experiments. These experiments [5] were done before the tunneling experiments, and they give similar curves but with greater errors, particularly near T_c. We briefly discuss this case. First, consider the case of $T = 0$. As we saw in Section 7.2 the binding energy is equal to $|w_0|$. The threshold photon energy $\Delta\varepsilon$ above which a sudden increase in absorption occurs, is twice the binding energy $|w_0|$:

$$\Delta\varepsilon = 2w_0 \; [= 2\varepsilon_g(0)]. \qquad (0 \text{ K}) \tag{12.18}$$

The photon is electrically neutral and hence cannot change the charge state of the system. Thus the factor 2 in Eq. (12.18) arises from the dissociation of *two* pairons of different charges $\pm 2e$. This is similar to the electron-positron pair creation by a γ-ray, where the threshold energy is $2mc^2$ (c = light speed). The threshold energy $\Delta\varepsilon$ was found to be temperature-dependent. This can be interpreted simply by assuming that threshold energy corresponds to twice the pairon energy gap:

$$\Delta\varepsilon = 2\varepsilon_g(T). \tag{12.19}$$

12.3 Quantum Tunneling in S_1-I-S_2

Giaever and his group [1-3] carried out various tunneling experiments. The case in which two *different* superconductors are chosen for (S_1, S_2), is quite revealing. We discuss this case here. Fig. 12.7 shows the I-V curves of an Al-Al_2O_3-Pb sandwich at various temperatures, reproduced from Ref. [1]. The main features are:

a′) The curves are generally similar to those for S-I-S, shown in Fig. 12.3. They exhibit the same qualitative behaviors [features **b-e**].

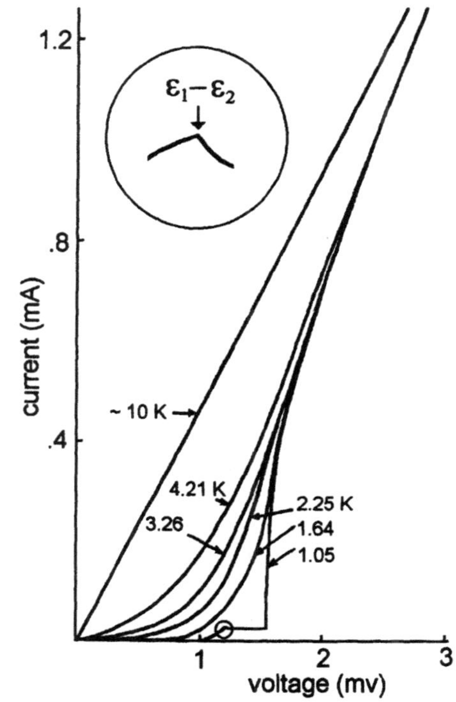

Figure 12.7: The I-V curves of Al-Al_2O_3-Pb sandwich at various temperatures. after Giaever and Megerle.

b′) There is a distinct maximum at

$$2eV_a = \varepsilon_{g1} - \varepsilon_{g2} \equiv \varepsilon_1 - \varepsilon_2, \tag{12.20}$$

where ε_i are the energy gaps for superconductors i.

c′) Above this maximum there is a *negative resistance* region ($dI/dV < 0$). (see the inset in Fig. 12.7.)

d′) Below T_c for both conductors, there is a major increase in current at

$$2eV_b = \varepsilon_{g1} + \varepsilon_2 \equiv \varepsilon_1 + \varepsilon_2. \tag{12.21}$$

The I-V curves will now be interpreted with the aid of diagrams in Fig. 12.8. There are two energy gaps (ε_1, ε_2). We assume that $\varepsilon_1 > \varepsilon_2$. No right-left symmetry exists with respect to the oxide, meaning no antisymmetry in the I-V curve, see below.

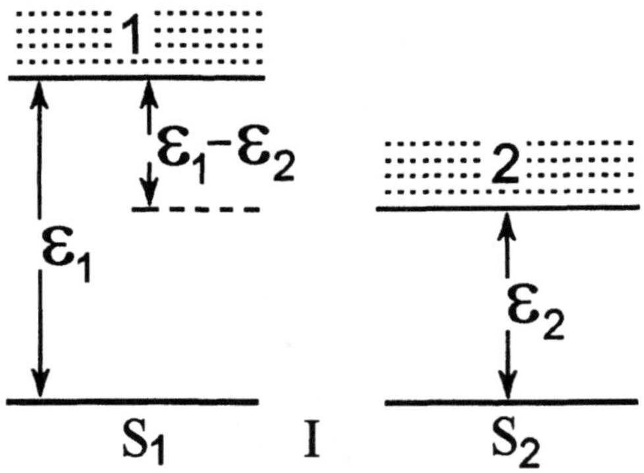

Figure 12.8: Pairon energy diagrams for S_1-I-S_2.

Consider first the case of no bias. Excited pairons in the oxide, may move right or left with no preference. Therefore there is no net current. Suppose we apply a small voltage. The preferred direction for the +pairon having $2e$ is right while that for the −pairon (having $-2e$) is left. From the diagram in Fig. 12.8, we see that the +pairons from S_1 can preferentially move (tunnel) to S_2, which contributes a small electric current running right. The −pairons from S_2 can preferentially move to S_1, which also contributes a tiny electric current running right. This contribution is tiny, since the excited pairons in S_2 must have energies greater by at least ε_1 ($> \varepsilon_2$); by the Boltzmann factor argument, the number of excited pairons must be extremely small.

We now increase the bias and examine the behavior of the net current. Earlier we enumerated a set of conditions for quantum tunneling. Below the threshold bias V_a defined by $\varepsilon_1 - \varepsilon_2 = 2eV_a$, the main contribution is due to the +pairons tunneling from S_1 to S_2. Above the threshold V_a, −pairons may preferentially tunnel from S_2 to S_1 with the aid of the bias voltage $V\,(> V_a)$. Their contribution should be much greater than that coming from those +pairons because the number of excited pairons in S_2 is much greater than that of excited pairons in S_1 since $\varepsilon_1 > \varepsilon_2$. This is the main cause of the distinct current maximum in feature **b′**). The total number of excited pairons in S_2 is fixed, and therefore there should be a second plateau.

Why does a negative resistance region occur, where $dI/dV < 0$? We may explain this most unusual feature as follows. First we note that formula (12.13) applies at a specific energy ε corresponding to the initial momentum $\mathbf{p}$. Integration with respect to $d\varepsilon$ yields the observed current. As the bias voltage V reaches and passes V_a, the extra current starts to appear and increase because more pairons can participate in the new process. The current density of the participating pairons is proportional to

$$\int_0^\infty d\varepsilon\, \mathcal{N}(\varepsilon) f_2(\varepsilon)[1 + f_1(\varepsilon + 2eV)]$$

$$= A \int_0^\infty d\varepsilon\, \varepsilon^2 f_2(\varepsilon)[1 + f_1(\varepsilon + 2eV)]. \qquad (i = 2,\ f = 1) \qquad (12.22)$$

For great values of V (not exceeding V_b), the Planck distribution $f_1(\varepsilon + 2eV)$ vanishes, and the above integral in Eq. (12.22) equals the total number density of the excited pairons n_2 in S_2:

$$A \int_0^\infty d\varepsilon\, \varepsilon^2 f_2(\varepsilon) = n_2. \qquad (12.23)$$

For an intermediate range between V_a and V_b but near V_a, the Planck distribution function $f_f(\varepsilon + 2eV)$ does not vanish, and therefore it should generate a maximum, as observed in the experiment. We emphasize that this maximum arises because of the boson-nature of pairons. If we assume the electron (fermion) transport model, expression (12.13) for the quantum-tunneling current must be modified with the quantum statistical factors:

$$f_F^{(i)}(\varepsilon)[1 - f_F^{(f)}(\varepsilon + eV)]. \qquad (12.24)$$

where f_F represents the Fermi distribution function. The modified formula cannot generate a maximum because of the negative sign.

If the bias voltage is raised further and passes the threshold voltage V_b defined by

$$\varepsilon_1 - \varepsilon_2 + 2\varepsilon_2 = \varepsilon_1 + \varepsilon_2 = 2eV_b, \qquad (12.25)$$

the new process 7 becomes active. Since this process originates in the supercondensate, the resulting current is much greater, which explains the observed feature **d′**).

The major current increases in S-I-S and S_1-I-S_2 systems both originate from the supercondensate. The threshold voltages given in terms of the energy gaps ε_j are different. Current increases occur at 2ε for S-I-S and at

$\varepsilon_1 + \varepsilon_2$ for S_1-I-S_2. Using several superconductors for S_j, one can make a great number of S_1-I-S_2 systems. Giaever and his group [1, 3] and others [6] have studied these systems. They found that the I-V curves have the similar features. Extensive studies confirm the main finding that the energy gap $\varepsilon_g(T)$ can be obtained directly from the quantum tunneling experiments. Taylor, Burstein and others [6] reported that in such systems as Sn-I-Tl, there are excess currents starting at ε_{Tl} and ε_{Sn} in addition to the principal exponential growth at $\varepsilon_{\mathrm{Tl}} + \varepsilon_{\mathrm{Sn}}$. These excess currents can be accounted for by applying the rule 6 to the non-predominant pairons, which have been neglected in our discussion.

As noted earlier the right-left symmetry is broken for the S_1-I-S_2 system. Hence the I-V curve is asymmetric. This can be shown as follows. Consider two cases

$$(\mathrm{A}) : \text{rightward bias } V_1 > V_2, \qquad (\mathrm{B}) : \text{leftward bias } V_1 < V_2. \tag{12.26}$$

In (A) $+(-)$ pairons move through the oxide preferentially right (left), while $+(-)$ pairons move preferentially left (right) in (B). We assume the same voltage difference $|V_1 - V_2|$ for both cases and compute the total currents. For (A) the total current I_A is the sum of the current I^+ arising preferentially from the $+$ pairons originating in S_1 and the current I^- arising from the $-$ pairons originating in S_2. Both currents effectively transfer the positive charge right (in the positive direction): $I^{+(1)}$, $I^{-(2)} > 0$. For (B) there are two similar contributions, which are both negative. In summary

$$(\mathrm{A}) : I_A = I^{+(1)} + I^{-(2)} > 0, \qquad I^{+(1)}, I^{-(2)} > 0, \tag{12.27}$$

$$(\mathrm{B}) : I_B = I^{+(2)} + I^{-(1)} < 0, \qquad I^{+(2)}, I^{-(1)} < 0. \tag{12.28}$$

If the same superconductors are used for the two sides ($S_1 = S_2$), then from the generalized Ohm's law we have

$$I^{+(1)} = -I^{+(2)}, \qquad I^{-(2)} = -I^{-(1)}, \tag{12.29}$$

$$I_A = -I_B \qquad \text{if} \qquad S_1 = S_2, \tag{12.30}$$

indicating that the I-V curve is antisymmetric. If different superconductors are used, Eqs. (12.29) do not hold. Then in general

$$|I_A| \neq |I_B| \qquad \text{if} \qquad S_1 \neq S_2. \quad \text{(q.e.d.)} \tag{12.31}$$

The preceding proof may be extended to any type of charge carrier including the electrons.

Let us now go back and discuss the I-V curves for the S_1-I-S_2 systems. For elemental superconductors, both $\pm$ pairons have the same energy gaps. Therefore, the magnitudes of threshold voltages are independent of the bias direction. But by inequality (12.31) the magnitudes of tunneling currents (at the same voltage) are unequal. For a high-T_c cuprate superconductor, there are two energy gaps $(\varepsilon_1, \varepsilon_2)$ for $\pm$pairons. The I-V curves for S_1-I-S_2 systems are asymmetric and more complicated (see Section 17.5).

12.4 Discussion

The quantum tunneling experiment allows a direct determination of the pairon energy gap $\varepsilon_g(T)$ as a function of temperature T. This T-dependence originates in the presence of the supercondensate. (We shall see later in Chapter 14 that $\varepsilon_g(T)$ is proportional to the density of condensed pairons.) The I-V curve for S_1-I-S_2 system has a cusp at the gap difference: $2eV_a = \varepsilon_1 - \varepsilon_2$. Using this distinctive property, we may determine the energy gap ε_g more precisely. The observed negative resistance region just above the cusp can be explained based only on the bosonic pairon transport model but not on the fermionic electron transport model.

Chapter 13

Flux Quantization

The moving supercondensate, which is made up of $\pm$ pairons all condensed at a finite momentum $\mathbf{p}$, generates a supercurrent. Flux quantization is the first quantization effect manifested on a macroscopic scale. The phase of a macro-wavefunction can change due to the pairon motion *and* the magnetic field, leading to London's equation. The penetration depth λ based on the pairon flow model is given by $\lambda = (c/e)(p/4\pi k_0 n_0 \mid v_F^{(2)} + v_F^{(1)} \mid)^{1/2}$. The quasi-wavefunction $\Psi_\sigma(\mathbf{r})$ representing the supercurrent can be expressed in terms of the pairon density operator n as $\Psi_\sigma(\mathbf{r}) \equiv \langle \mathbf{r} \, |n| \, \sigma \rangle$, where σ denotes the condensed pairon state. It changes in time, following a Schrödinger-like equation of motion.

13.1 Ring Supercurrent

The most striking superconducting phenomenon is a never-decaying ring supercurrent [1]. Why is the supercurrent not hindered by impurities which must exist in any superconductor? We discuss this basic question and flux quantization in this section. Let us take a ring-shaped superconductor at 0 K. The *ground state* for a pairon (or any quantum particle) in the absence of electromagnetic fields can be represented by a real wavefunction $\psi_0(\mathbf{r})$ having no nodes and vanishing at the ring boundary:

$$\psi_0(\mathbf{r}) = \begin{cases} \text{nearly constant} & \text{inside the body} \\ 0 & \text{at the boundary.} \end{cases} \tag{13.1}$$

Such a wavefunction corresponds to the zero-momentum state, and can generate no current. Pairons move as bosons. They neither overlap in space nor interact with each other. At 0 K a collection of free pairons therefore occupy the same zero-momentum state ψ_0. The many-pairon ground-state wavefunction $\Psi_0(\mathbf{r})$, which is proportional to $\psi_0(\mathbf{r})$, represents the supercondensate discussed earlier in Chapter 7. The supercondensate is composed of equal numbers of $\pm$ pairons with the total number being

$$N_0 = \hbar\omega_D \mathcal{N}(0). \tag{13.2}$$

We now consider current-carrying single-particle and many-particle states. There are many nonzero momentum states whose energies are very close to the ground-state energy 0. These states can be represented by the wavefunctions $\{\psi_n\}$ having a finite number of nodes, n, along the ring. The state ψ_n is represented by

$$\psi_n(\mathbf{r}) = u \exp(iq_n x/\hbar), \tag{13.3}$$

$$q_n \equiv 2\pi\hbar n/L, \qquad n = \pm 1, \pm 2, \ldots \tag{13.4}$$

where x is the coordinate along the ring circumference of length L; the factor u is real and nearly constant inside, but vanishes at the boundary. When a macroscopic ring is considered, the wavefunction ψ_n represents a state having linear momentum q_n along the ring. For small n, the value of $q_n = 2\hbar n/L$ is very small, since L is a macroscopic length. The associated energy eigenvalue is also very small.

Suppose the system of free pairons of both charge types occupy the same state ψ_n The many-pairon system-state Ψ_n so specified can carry a macroscopic current along the ring. In fact a pairon has charge $\pm 2e$ depending on charge type. There are equal numbers of $\pm$ pairons, and their speeds $c_j \equiv v_F^{(j)}/2$ are different. Hence the total electric current density j, calculated by the rule: (charge)×(number density)×(velocity),

$$j = (-2e)\frac{n_0}{2}(c_1 - c_2) = \frac{1}{2}en_0(v_F^{(2)} - v_F^{(1)}) \tag{13.5}$$

does not vanish. A schematic drawing of a supercurrent in 2D is shown in Fig. 13.1 (a). For comparison the normal current due to a random electron motion is shown in Fig. 13.1 (b). Notice the great difference between the two.

In their classic paper [2] BCS assumed that there were "electrons" and "holes" in a model superconductor, but they also assumed a spherical Fermi surface. (The contradictory nature of these two assumptions was mentioned earlier in Section 7.2). In this model "electrons" and "holes" have the same

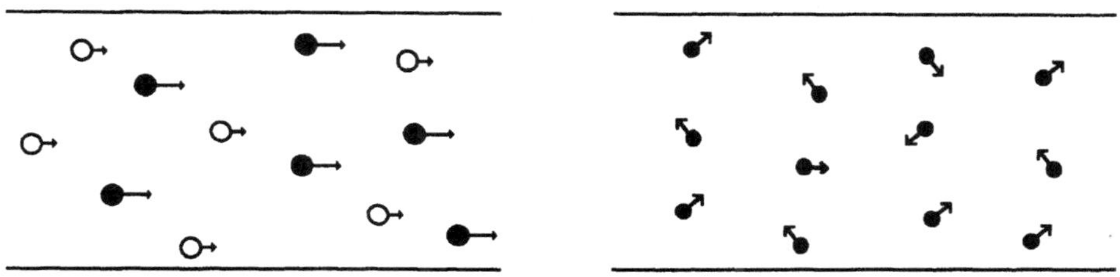

Figure 13.1: (a) Supercurrent; (b) normal current.

effective mass (magnitude) m^* and the same Fermi velocity $v_F^{(2)} = v_F^{(1)} = (2\varepsilon_F/m^*)^{1/2}$. Then the supercurrent vanishes according to Eq. (13.5). Thus a finite supercurrent cannot be treated based on the original BCS Hamiltonian.

The supercurrent arises from a *many-boson state of motion*. The many-particle state is not destroyed by impurities, phonons, etc. This is somewhat similar to the situation in which a flowing river (large object) cannot be stopped by a small stick (small object). In more rigorous terms, the change in the many-pairon-state can occur only if a transition from one system-state to another, involving a great number of pairons. The supercurrent is very similar to a laser. This analogy will be further expounded later in Chapter 15, where we discuss Josephson effects. Earlier we saw that because of charge and momentum conservation, the phonon exchange simultaneously pair-creates $\pm$ pairons of the same momentum $\mathbf{q}$ from the physical vacuum. This means that the condensation of pairons can occur at *any* momentum state $\{\psi_n\}$. In the absence of electromagnetic fields, the zero-momentum state having the minimum energy is the equilibrium state. In the presence of a magnetic field, the minimum-energy state is not necessarily the zero-momentum state but it can be a finite momentum state, see below.

The supercurrent, generated by a *neutral* supercondensate in motion, is very stable against an applied voltage since no Lorentz electric force can act on it.

Let us now consider the effect of a magnetic field. In flux quantization experiments [3] a minute flux is trapped in the ring, and this flux is maintained by the ring supercurrent (see Fig. 1.5). According to Onsager's hypothesis [4] the flux generated by a circulating electron carrying charge $-e$ is quantized in units of $\Phi_{el} \equiv h/e = 2\pi\hbar/e$. Experiments in superconductors [3], (data are summarized in Fig. 1.6) show that the trapped flux Φ is quantized as

$$\Phi = n\Phi_0, \qquad (n = 0, 1, \ldots) \qquad \Phi_0 \equiv \Phi_{pairon} \equiv \frac{h}{2e} = \frac{\pi\hbar}{e}. \tag{13.6}$$

From this Onsager concluded [5] that the particle circulating on the ring has a charge (magnitude) $2e$, in accord with the BCS picture of the super-condensate composed of pairons of charge (magnitude) $2e$. Flux quantization experiments were reported in 1961 by two teams, Deaver-Fairbank and Doll-Näbauer [3]. Their experiments are regarded as the most important confirmation of the BCS theory. They also show Onsager's great intuition about flux quantization. The integers n appearing in (13.4) and (13.6) are the same, which can be seen by applying the Bohr-Sommerfeld quantization rule:

$$\oint p dx = 2\pi\hbar(n + \gamma) \tag{13.7}$$

to the circulating pairons. (Problem 13.1.1) The phase (number) γ is zero for the present ring (periodic) boundary condition. A further discussion of flux quantization is given in Section 13.3. The supercurrent is generated by $\pm$ pairons condensed at a single momentum q_n and moving with different speeds $c_j \equiv v_F^{(j)}/2$. This picture explains why the supercurrent is unstable against a magnetic field $\mathbf{B}$. Because of the Lorentz-magnetic force:

$$\pm 2e\, \mathbf{c}_j \times \mathbf{B} = \pm e\, \mathbf{v}_F^{(j)} \times \mathbf{B}, \tag{13.8}$$

the B-field tends to separate $\pm$ pairons from each other. From this we see that the (thermodynamic) *critical field* B_c should be higher for low-v_F materials than for high-v_F materials, in agreement with experimental evidence. For example high-T_c superconductors have high B_c, reflecting the fact that they have low pairon speeds ($c_j \equiv v_F^{(j)}/2 \sim 10^5$ ms^{-1}). Since the supercurrent itself induces a magnetic field, there is a limit on the magnitude of the supercurrent, called a *critical current.*

Problem 13.1.1. Apply Eq. (13.7) to the ring supercurrent and show that $\gamma = 0$. Note: The phase γ does not depend on the quantum number n, suggesting a general applicability to the Bohr-Sommerfeld quantization rule with a high quantum number.

13.2 Phase of the Quasi-Wavefunction

The supercurrent at a small section along the ring is represented by

$$\Psi_p(x) = A\exp(ipx/\hbar), \tag{13.9}$$

where A is a constant amplitude. We put $q_n \equiv p$; the pairon momentum is denoted by the more conventional symbol p. The quasi-wavefunction Ψ_p in Eq. (13.9) represents a system-state of pairons all condensed at $\mathbf{p}$ and the wavefunction ψ in Eq. (13.3) the single-pairon state. Ψ_p and ψ_p are the same function except for the normalization constant (A). In this chapter we are mainly interested in the supercondensate quasi-wavefunction. We simply call Ψ the wavefunction hereafter (omitting quasi). In a SQUID shown in Fig. 1.8. two supercurrents macroscopically separated ($\sim$ 1 mm) can interfere just as two laser beams coming from the same source. In wave optics two waves are said to be *coherent* if they can interfere. Using this terminology, two supercurrents are coherent within the *coherence range* of 1 mm. The coherence of the wave traveling through a region means that if we know the phase and amplitude at any space-time point, we can calculate the same at any other point from a knowledge of the k-vector ($\mathbf{k}$) and angular frequency (ω). In the present section we discuss the phase of a general wavefunction and obtain an expression for the phase difference at two space-time points in the superconductor.

First consider a monochromatic plane wave running in the x-direction

$$\Psi = A\,e^{i2\pi(x/\lambda - t/T)} = A\,e^{i(kx-\omega t)} = A\,e^{i(px-Et)/\hbar}, \tag{13.10}$$

where the conventional notations: $2\pi/\lambda \equiv k$, $2\pi/T \equiv \omega$, $p \equiv \hbar k$, $E \equiv \hbar\omega$ are used. We now take two points $(\mathbf{r}_1, t_1, \mathbf{r}_2, t_2)$. The phase difference $(\Delta\phi)_{12}$ between them,

$$(\Delta\phi)_{12} = k(x_1 - x_2) - \omega(t_1 - t_2), \tag{13.11}$$

depends on the time difference $t_1 - t_2$ only. In the steady-state condition, $\omega(t_1 - t_2)$ is a constant, which will be omitted hereafter. The positional phase difference is

$$(\Delta\phi)_{\mathbf{r}_1,\mathbf{r}_2} \equiv \phi(\mathbf{r}_1) - \phi(\mathbf{r}_2) = k(x_1 - x_2). \tag{13.12}$$

The phase-difference $\Delta\phi$ for a plane wave proceeding in a $\mathbf{k}$ direction is given by

$$(\Delta\phi)_{\mathbf{r}_1,\mathbf{r}_2} = \int_{\mathbf{r}_2 \to \mathbf{r}_1} \mathbf{k}\cdot d\mathbf{r}, \tag{13.13}$$

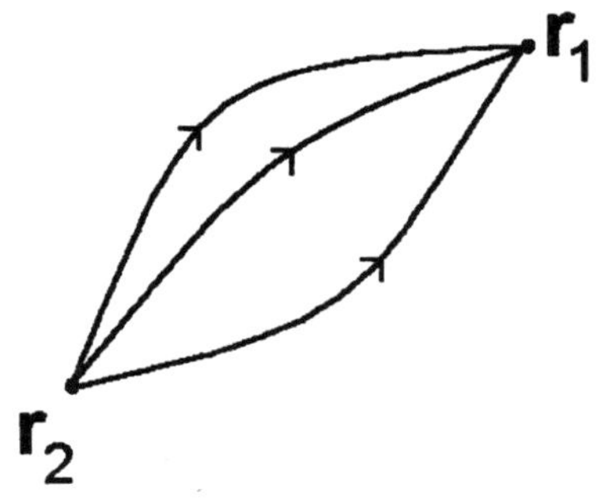

Figure 13.2: Directed paths from $\mathbf{r}_2$ to $\mathbf{r}_1$.

where the integration is along a directed straight line path from $\mathbf{r}_2$ to $\mathbf{r}_1$. When the plane wave extends over the whole space, the line integral $\int_{\mathbf{r}_2 \to \mathbf{r}_1} \mathbf{k} \cdot d\mathbf{r}$, along *any* curved path joining the points $(\mathbf{r}_1, \mathbf{r}_2)$, see Fig. 13.2, has the same value (Problem 13.2.1):

$$(\Delta\phi)_{\mathbf{r}_1,\mathbf{r}_2} \equiv \int\limits_{\mathbf{r}_2 \to \mathbf{r}_1} \mathbf{k} \cdot d\mathbf{r} = \int_{\mathbf{r}_2}^{\mathbf{r}_1} \mathbf{k} \cdot d\mathbf{r}. \tag{13.14}$$

The line integral now depends on the end points only, and it will be denoted by writing out the limits explicitly as indicated in the last member of Eqs. (13.14). The same property can equivalently be expressed by (Problem 13.2.2)

$$\oint\limits_{C} \mathbf{k} \cdot d\mathbf{r} = 0, \tag{13.15}$$

where the integration is carried out along *any* closed directed path C.

We consider a ring supercurrent as shown in Fig. 13.3 (a). For a small section, the enlarged section containing point A, see (b), the supercurrent can be represented by a plane wave having the momentum $\mathbf{p} = \hbar\mathbf{k}$. The phase difference $(\Delta\phi)_{\mathbf{r}_1,\mathbf{r}_2}$, where $(\mathbf{r}_1, \mathbf{r}_2)$ are any two points in the section, can be represented by Eqs. (13.14). If we choose a closed path ABA, the line integral vanishes:

$$(\Delta\phi)_{ABC} = \oint\limits_{ABC} \mathbf{k} \cdot d\mathbf{r} = 0. \tag{13.16}$$

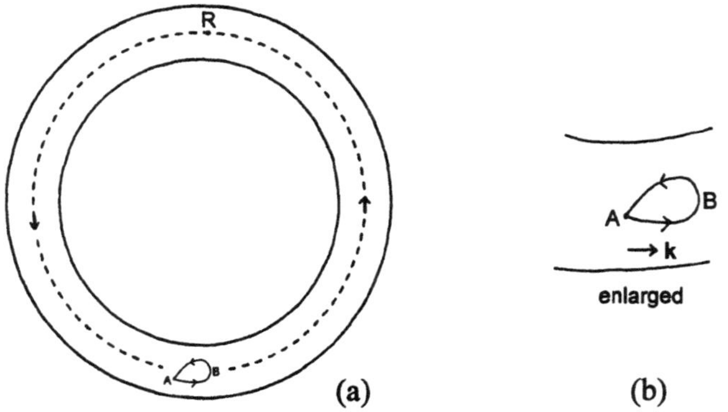

Figure 13.3: (a) A ring supercurrent; (b) an enlarged section.

Let us now calculate the line integral along the ring circumference ARA indicated by the dotted line in (a). Note: The vector **k** changes its direction along the ring; for each small section, we may use Eq. (13.14). Summing over all sections, we obtain

$$(\Delta\phi)_{ARA} = \oint_{ARA} \mathbf{k}\cdot d\mathbf{r} = kL, \qquad (13.17)$$

where L is the ring length. Thus the line integral along the closed path ARA does not vanish. In fact if we choose a closed path C that circles the cavity counterclockwise N_1 times and clockwise N_2 times, the integral along the closed path is

$$(\Delta\phi)_C = \oint_C \mathbf{k}\cdot d\mathbf{r} = (N_1 - N_2)kL. \qquad (13.18)$$

Cases ABA (ARA), represented by Eq. (13.16) [(13.17)] can be obtained from this general formula by setting $N_1 = N_2 = 0$ ($N_1 = 1, N_2 = 0$). Since the momentum $\mathbf{p} \equiv \hbar\mathbf{k}$ is quantized such that $p_n = 2n\hbar/L$, Eq. (13.18) can be reexpressed as

$$(\Delta\phi)_C = \oint_C \mathbf{k}\cdot d\mathbf{r} = 2\pi(N_1 - N_2)n. \qquad (13.19)$$

Problem 13.2.1. (a) Calculate the line integral in Eq. (13.14) by assuming a straight path from $\mathbf{r}_2$ to $\mathbf{r}_1$, and verify the equivalence of Eqs. (13.14) and (13.15). (b) Assume that the path is composed of two straight paths. Verify the equivalence. (c) Treat a general curved path.

Problem 13.2.1. (a) Assume Eq. (13.14) and verify Eq. (13.15). (b) Prove the inverse: assume Eq. (13.15) and verify Eq. (13.14).

13.3 London's Equation. Penetration Depth

In 1935 the brothers F. and H. London published classic papers [6] on the electrodynamics of a superconductor. By using London's equation (13.30) together with Maxwell's equation (13.37), they demonstrated that an applied magnetic field $\mathbf{B}$ does not drop to zero abruptly inside the superconductor but penetrates to a certain depth. In this section we derive London's equation and its generalization. We also discuss flux quantization once more.

In Hamiltonian mechanics the effect of electromagnetic fields $(\mathbf{E}, \mathbf{B})$ is included by replacing the Hamiltonian without the fields, $H(\mathbf{r}, \mathbf{p})$, with

$$H' = H(\mathbf{r}, \mathbf{p} - q\mathbf{A}) + q\Phi, \qquad (q = charge) \tag{13.20}$$

where $(\mathbf{A}, \Phi)$ are the vector and scalar potentials generating the electromagnetic fields:

$$\mathbf{B} = \nabla \times \mathbf{A}, \qquad \mathbf{E} = -\nabla\Phi - \frac{\partial \mathbf{A}}{\partial t}, \tag{13.21}$$

and then use Hamilton's equations of motion. In quantum mechanics we may use the same Hamiltonian H' (the prime dropped hereafter) as a linear operator and generate Schrödinger's equation of motion. Detailed calculations show (Problem 13.3.1.) that the phase difference $(\Delta\phi)_{\mathbf{r}_1,\mathbf{r}_2,\mathbf{B}}$ changes in the presence of the magnetic field such that

$$(\Delta\phi)_{\mathbf{r}_1,\mathbf{r}_2,\mathbf{B}} = \frac{1}{\hbar}\int_{\mathbf{r}_2}^{\mathbf{r}_1} \mathbf{p} \cdot d\mathbf{r} + \frac{q}{\hbar}\int_{\mathbf{r}_2}^{\mathbf{r}_1} \mathbf{A} \cdot d\mathbf{r} \equiv (\Delta\phi)^{(motion)}_{\mathbf{r}_1,\mathbf{r}_2} + (\Delta\phi)^{(\mathbf{B})}_{\mathbf{r}_1,\mathbf{r}_2}. \tag{13.22}$$

We call the first term on the rhs the phase difference due to particle motion and the second term, the phase difference due to the magnetic field. [The motion of charged particles generates an electric current, so $(\Delta\phi)^{(motion)}_{\mathbf{r}_1,\mathbf{r}_2}$ may also be referred to as the phase difference due to the current.]

We now make a historical digression. Following London-London [6], let us assume that the supercurrent is generated by hypothetical superelectrons.

A superelectron has mass m and charge $-e$. Its momentum $\mathbf{p}$ is related to its velocity $\mathbf{v}$ by

$$\mathbf{v} = \mathbf{p}/m. \tag{13.23}$$

The supercurrent density $\mathbf{j}_s$ is then

$$\mathbf{j}_s = -en_s\mathbf{v}. \tag{13.24}$$

Using the last two equations we calculate the motional phase difference:

$$(\Delta\phi)^{(motion)}_{\mathbf{r}_1,\mathbf{r}_2} \equiv \frac{1}{\hbar}\int_{\mathbf{r}_2}^{\mathbf{r}_1} \mathbf{p}\cdot d\mathbf{r} = -\int_{\mathbf{r}_2}^{\mathbf{r}_1} \left[\frac{m}{e\hbar n_s}\right] \mathbf{j}_s \cdot d\mathbf{r}. \tag{13.25}$$

Let us consider an infinite homogeneous medium, for which the line-integral of the phase along any closed path vanishes [see (13.15)]

$$\oint \left[\mathbf{j}_s + \frac{e^2 n_s}{m}\mathbf{A}\right] \cdot d\mathbf{r} = 0. \tag{13.26}$$

Employing Stoke's theorem (Problem 13.3.2.)

$$\oint d\mathbf{r}\cdot\mathbf{C} = \iint d\mathbf{S}\cdot\nabla\times\mathbf{C}, \tag{13.27}$$

where the rhs is a surface integral with differential element $d\mathbf{S}$ pointing in a direction according to the right-hand screw rule, we obtain from Eq. (13.26)

$$\mathbf{j}_s + e^2 m^{-1} n_s \mathbf{A} + \nabla\chi = 0, \tag{13.28}$$

where $\chi(\mathbf{r})$ is an arbitrary scalar field. Now the connection between the magnetic field $\mathbf{B}$ and the vector potential $\mathbf{A}$, as represented by $\mathbf{B} = \nabla\times\mathbf{A}$, has a certain arbitrariness; we may add the gradient of the scalar field χ to the original vector field $\mathbf{A}$ since

$$\nabla\times(\mathbf{A} + \nabla\chi) = \nabla\times\mathbf{A}. \tag{13.29}$$

Using this *gauge-choice property* we may re-write Eq. (13.28) as

$$\mathbf{j}_s = -e^2 m^{-1} n_s \mathbf{A} \equiv -\,\Lambda_{el}\,\mathbf{A}, \qquad \Lambda_{el} \equiv e^2 m^{-1} n_s. \tag{13.30}$$

This is known as *London's equation.* The negative sign indicates that the supercurrent is diamagnetic. The physical significance of Eq. (13.30) will be discussed later.

Let us go back to the condensed pairon picture of the supercurrent. First consider a +pairon having charge $+2e$. Since the pairon has the linear energy-momentum relation

$$\varepsilon_2 = w_0 + (1/2)v^{(2)}p \equiv w_0 + c_2 p, \tag{13.31}$$

the velocity $\mathbf{v}_2$ has magnitude $c_2 \equiv v_F^{(2)}/2$ and direction $\hat{\mathbf{p}}$ along the momentum $\mathbf{p}$:

$$\mathbf{v}_2 = c_2 \hat{\mathbf{p}}. \tag{13.32}$$

Thus if $p \neq 0$, there is a supercurrent density equal to

$$\mathbf{j}_s^{(2)} = (2e)(n_0/2)c_2\hat{\mathbf{p}} = en_0\mathbf{v}_2. \tag{13.33}$$

Similarly, −pairons contribute

$$\mathbf{j}_s^{(1)} = -en_0\mathbf{v}_1 \qquad \mathbf{v}_1 = c_1\mathbf{p}. \tag{13.34}$$

Using the last three relations we repeat the calculations and obtain

$$\mathbf{j}_s = -\Lambda_{pairon}\,\mathbf{A}, \qquad \Lambda_{pairon} \equiv 2e^2 n_0\,(c_2 + c_1)\,p^{-1} \equiv \Lambda, \tag{13.35}$$

which we call a *generalized London's equation* or simply *London's equation.* This differs from the original London equation (13.30) merely by a constant factor.

Let us now discuss a few physical consequences derivable from London's equation (13.35). Taking the curl of this equation we obtain

$$\nabla \times \mathbf{j}_s = -\Lambda\,\nabla \times \mathbf{A} = -\Lambda \mathbf{B}. \tag{13.36}$$

Using this and one of Maxwell's equations

$$\nabla \times \mathbf{H} = \mathbf{j}_s, \qquad [\frac{\partial \mathbf{D}}{\partial t} = 0, \quad \mathbf{j}_n = 0] \tag{13.37}$$

[where we neglected the time-derivative of the dielectric displacement $\mathbf{D}$ and the normal current $\mathbf{j}_n$], we obtain (Problem 13.3.3.)

$$\lambda^2\,\nabla^2\mathbf{B} = \mathbf{B}, \qquad \lambda \equiv (c_0/e)\{p/[8\pi k_0 n_0\,(|c_2 + c_1|)]\}^{1/2}. \quad (\varepsilon_0\mu_0 \equiv c_0^{-2}) \tag{13.38}$$

Here λ is called a *generalized London penetration depth.*

We consider the boundary of a semi-infinite slab (superconductor). When an external magnetic field **B** is applied parallel to the boundary, the **B**-field computed from Eq. (13.38) can be shown to fall off exponentially (Problem 13.3.4):

$$B(x) = B(0)\exp(-x/\lambda). \tag{13.39}$$

(This solution is shown in Fig. 1.13.) Thus the interior of the superconductor, far from the surface, will show the Meissner state: $B = 0$. Experimentally, the penetration depth at lowest temperatures is on the order of 500 Å.

We emphasize that London's equation (13.35) holds for the supercurrent $\mathbf{j}_s$ only. Since this is a strange equation, we rederive it by a standard calculation. The effective Hamiltonian H for a + pairon moving with speed $c_2 \equiv v_F^{(2)}/2$ is

$$H = c_2\,|\mathbf{p}|. \tag{13.40}$$

In the presence of a magnetic field $\mathbf{B} = \nabla \times \mathbf{A}$, this Hamiltonian is modified to

$$H' = c_2\,|\mathbf{p} - q\mathbf{A}|. \tag{13.41}$$

Assume now that the momentum $\mathbf{p}$ points in the x-direction. The velocity v_x is

$$v_x \equiv \partial H'/\partial p_x = c_2 p^{-1}(p_x - qA_x). \tag{13.42}$$

We calculate the quantum mechanical average of v_x, multiply the result by the charge $2e$ and the +pairon density, $n_0/2$, and obtain

$$(2e)(n_0/2)\,\langle v_x\rangle = en_0c_2\,\langle p_x/\,|\mathbf{p}|\rangle - 2e^2n_0c_2p^{-1}A_x. \tag{13.43}$$

Adding the contribution of $-$pairons, we obtain the total current density j_s:

$$j_s = en_0(c_2 - c_1)\,\langle p_x/\,|\mathbf{p}|\rangle - 2e^2n_0(c_2 + c_1)p^{-1}A_x. \tag{13.44}$$

Comparing this with London's equation (13.35) we see that the supercurrent arises from the magnetic field term for the pairon velocity **v**.

As another important application of Eq. (13.22) let us consider a supercurrent ring. By choosing a closed path around the central line of the ring and integrating $(\Delta\phi)_\mathbf{B}$ along the path ARA, we obtain from Eqs. (13.18) and (13.22)

$$(\Delta\phi)^{motion}_{ARA} + 2e\hbar^{-1}\oint_{ARA} \mathbf{A}\cdot d\mathbf{r} = kL = 2\pi n. \tag{13.45}$$

The closed path integral can be evaluated by using Stoke's theorem as

$$\oint_{ARA} \mathbf{A} \cdot d\mathbf{r} = \iint d\mathbf{S} \cdot \nabla \times \mathbf{A} = \iint d\mathbf{S} \cdot \mathbf{B} = BS \equiv \Phi, \tag{13.46}$$

where S is the area enclosed by the path ARA, and $\Phi \equiv BS$ the magnetic flux enclosed. The phase difference due to the pairon motion $(\Delta)^{motion}_{ARA}$ is zero since there is no supercurrent along the central part of the ring. We obtain from Eqs. (13.45) and (13.46)

$$\Phi = n\frac{\pi\hbar}{e} \equiv n\,\Phi_0, \tag{13.47}$$

reconfirming the flux quantization discussed [see Eq. (13.6)]

Let us now choose a second closed path around the ring cavity but one where the supercurrent does not vanish. For example choose a path within a penetration depth around the inner side of the ring. For a small section where the current runs in the x-direction (see Fig. 13.6), the current due to a +pairon is

$$(2e)c_2\,\langle p_x/p\rangle - 4e^2 c_2 p^{-1} A_x, \tag{13.48}$$

where p_x/p is the component of the unit vector $\hat{\mathbf{p}}$ pointing along the momentum $\mathbf{p}$. If we sum $\hat{\mathbf{p}}$ over the entire circular path, the net result is Eq. (13.19)

$$\oint_{ARA} \hat{\mathbf{p}} \cdot d\mathbf{r} = 2\pi\hbar L. \tag{13.49}$$

This example shows that the motional contribution to the supercurrent, part of Eq. (13.43): $en_0c_2\,\langle p_x/p\rangle$, does not vanish. Thus in this case the magnetic flux by itself is not quantized, but the lhs of Eq. (13.45), called the *fluxoid*, is quantized.

Problem 13.3.1. Derive Eq. (13.22).

Problem 13.3.2. Prove Stoke's theorem Eq. (13.27) for a small rectangle and then for a general case.

Problem 13.3.3. Derive Eq. (13.38).

Problem 13.3.4. Solve Eq. (13.38) and obtain Eq. (13.39).

13.4 Quasi-Wavefunction and Its Evolution

The quasi-wavefunction $\Psi_\nu(\mathbf{r})$ for a quasiparticle in the state ν is defined by

$$\Psi_\nu(\mathbf{r}) \equiv \mathrm{TR}\{\psi_\nu^\dagger \phi(\mathbf{r})\rho\}, \tag{13.50}$$

or

$$\Psi_\nu(\mathbf{r}) = \langle \mathbf{r} \,| n |\, \nu \rangle. \tag{13.51}$$

Here n is the pairon density operator; the corresponding density matrix elements are represented by

$$\langle \mu \,| n |\, \nu \rangle \equiv \mathrm{TR}\{\psi_\nu^\dagger \psi_\mu \rho\}. \tag{13.52}$$

The density operator n, like the system-density operator ρ, can be expanded in the form:

$$n = \sum_\mu |\mu\rangle \, P_\mu \, \langle \mu|, \qquad P_\mu \geq 0, \tag{13.53}$$

where $\{P_\mu\}$ denote the relative probabilities that particle-states $\{\mu\}$ are occupied. It is customary in quantum many-body theory to adopt the following normalization condition:

$$\sum_\mu P_\mu = \langle \hat{N} \rangle, \tag{13.54}$$

where $\hat{N}$ is the total number operator. Using this, we obtain

$$\mathrm{tr}\{n\} \equiv \sum_\nu \langle \nu \,| n |\, \nu \rangle = \sum_\nu \int d^3r \, \langle \nu |\, \mathbf{r} \rangle \, \langle \mathbf{r} \,| n |\, \nu \rangle = \int d^3r \, \psi_\nu^*(\mathbf{r}) \Psi_\nu(\mathbf{r}) = \langle \hat{N} \rangle, \tag{13.55}$$

where

$$\psi_\nu^*(\mathbf{r}) \equiv \langle \nu |\, \mathbf{r} \rangle, \qquad [\psi_\nu(\mathbf{r}) = \langle \mathbf{r} |\, \nu \rangle] \tag{13.56}$$

is the wavefunction for a single pairon. If an observable X for the system is the sum of single-particle observables ξ:

$$X = \sum\nolimits_j \xi^{(j)}, \tag{13.57}$$

then the grand-ensemble average $\langle X \rangle$ can be calculated from

$$\langle X \rangle \equiv \mathrm{TR}\{X\rho\} = \mathrm{tr}\{\xi n\}, \tag{13.58}$$

where the lhs means the many-particle average and the rhs the single-particle average. Using Eq. (13.51) we can reexpress $\mathrm{tr}\{\xi n\}$ as:

$$\mathrm{tr}\{\xi n\} = \int d^3\mathbf{r} \sum_\nu \langle \nu | \, \mathbf{r}\rangle \, \langle \mathbf{r} \, |\xi n| \, \nu\rangle = \sum_\nu \int d^3 r \, \psi_\nu^*(\mathbf{r}) \, \xi(\mathbf{r}, -i\hbar\nabla)\Psi_\nu(\mathbf{r}), \tag{13.59}$$

where we assumed that ξ is a function of position $\mathbf{r}$ and momentum $\mathbf{p}$: $\xi = \xi(\mathbf{r}, \mathbf{p})$. Thus the average $\langle X\rangle$ for the many-particle system can be calculated in terms of the quasi-wavefunction Ψ.

The system-density operator $\rho(t)$ changes, following the quantum Liouville equation:

$$i\hbar\frac{\partial \rho}{\partial t} = [H, \, \rho]. \tag{13.60}$$

Using this, we study the time evolution of the quasi-wavefunction $\Psi_\nu(\mathbf{r}, t)$.

First we consider the supercondensate at 0 K. The supercondensate at rest can be constructed using the reduced Hamiltonian H_0:

$$\begin{aligned} H_0 \;\; &= \;\; \sum_{\mathbf{k}}{}' 2\varepsilon_k^{(1)} b_{\mathbf{k}}^{(1)\dagger} b_{\mathbf{k}}^{(1)} + \sum_{\mathbf{k}}{}' 2\varepsilon_k^{(2)} b_{\mathbf{k}}^{(2)\dagger} b_{\mathbf{k}}^{(2)} \\ &\quad - \sum_{\mathbf{k}}{}' \sum_{k'}{}' [v_{11} b_{\mathbf{k}}^{(1)\dagger} b_{\mathbf{k}'}^{(1)} + v_{12} b_{\mathbf{k}}^{(1)\dagger} b_{\mathbf{k}'}^{(2)\dagger} + v_{21} b_{\mathbf{k}}^{(2)} b_{\mathbf{k}'}^{(1)} + v_{22} b_{\mathbf{k}}^{(2)} b_{\mathbf{k}'}^{(2)\dagger}]. \end{aligned} \tag{13.61}$$

Note: the electron kinetic energies are expressed in terms of the ground pairon operators b's. Let us recall that the supercondensate is generated from the physical vacuum by a succession of pair-creation, pair-annihilation and pair-transition via phonon exchanges. In this condition, "electrons" (and "holes") involved are confined to a shell of energy-width $\hbar\omega_D$ about the Fermi surface, and up- and down-spin electrons are always paired ($\mathbf{k}\uparrow, -\mathbf{k}\downarrow$) to form ground pairons. This stationary supercondensate cannot generate a supercurrent.

The moving supercondensate can be generated from the physical vacuum via phonon exchanges. The relevant reduced Hamiltonian $H_{\mathbf{q}}$ is

$$\begin{aligned} H_{\mathbf{q}} \;\; &\equiv \;\; \sum_{\mathbf{k}}{}' \sum_{j} [\varepsilon^{(j)}(|\mathbf{k}+\mathbf{q}/2|) + \varepsilon^{(j)}(|-\mathbf{k}+\mathbf{q}/2|)] B_{\mathbf{kq}}^{(j)\dagger} B_{\mathbf{kq}}^{(j)} \\ &\quad - \sum_{\mathbf{k}}{}' \sum_{\mathbf{k}'}{}' [v_{11} B_{\mathbf{kq}}^{(1)\dagger} B_{\mathbf{k'q}}^{(1)} + v_{12} B_{\mathbf{kq}}^{(1)\dagger} B_{\mathbf{k'q}}^{(2)\dagger} + v_{21} B_{\mathbf{kq}}^{(2)} B_{\mathbf{k'q}}^{(1)} + v_{22} B_{\mathbf{kq}}^{(2)} B_{\mathbf{k'q}}^{(2)\dagger}]. \end{aligned} \tag{13.62}$$

which reduces to Eq. (13.61) when $\mathbf{q} = 0$. Supercondensation can occur at any momentum $\mathbf{q}$. The quasi-wavefunction $\Psi_{\mathbf{q}}$ representing the moving supercondensate is

$$\Psi_{\mathbf{q}} = A\exp[i(\mathbf{q}\cdot\mathbf{r} - \omega_j t)], \tag{13.63}$$

where the angular frequency ω_j is given by

$$\omega_j = qv_F^{(j)}/2, \tag{13.64}$$

a relation arising from the fact that pairons have energies

$$\varepsilon^{(j)} = w_0 + qv_F^{(j)}/2 \tag{13.65}$$

The Hamiltonian $H_{\mathbf{q}}$ in Eq. (13.62) is a sum of single-pairon energies. Hence we can describe the system in terms of one-pairon density operator n. This operator $n(t)$ changes in time, following the one-body quantum Liouville equation:

$$i\hbar\frac{\partial n}{\partial t} = [h,\, n]. \tag{13.66}$$

Let us now take a mixed representation of this equation. Introducing the quasi-wavefunction for the moving supercondensate,

$$\langle \mathbf{r}\, | n |\, \sigma\rangle \equiv \Psi_\sigma(\mathbf{r}), \tag{13.67}$$

we obtain

$$i\hbar\frac{\partial}{\partial t}\Psi_\sigma(\mathbf{r},t) = h(\mathbf{r}, -i\hbar\nabla, t)\Psi_\sigma(\mathbf{r},t), \tag{13.68}$$

which is formally identical with the Schrödinger equation for a quantum particle.

Since London's macrowavefunction were introduced by the intuition of the great men for a stationary state problem, it is not immediately clear how to describe its time-evolution. Our quantum statistical calculations show that the quasi-wavefunction satisfy the familiar Schrödinger equation of motion. This is a significant result. Important applications of Eq. (13.68) will be discussed in Chapter 15.

13.5 Discussion

Supercurrents exhibit many unusual behaviors. We enumerate their important features in the following subsections.

13.5.1 Supercurrents

The supercurrent is generated by a moving supercondensate composed of equal numbers of $\pm$ pairons condensed at a finite momentum $\mathbf{p}$. Since $\pm$ pairons, having charges $\pm 2e$ and momentum $\mathbf{p}$, move with different speeds $c_i = v_F^{(i)}/2$, the net electric current density:

$$j = en_0(v_F^{(2)} - v_F^{(1)})/2$$

does not vanish. Supercurrents totally dominate normal currents condensed pairons have lower energies by at least the gap ε_g than noncondensed pairons, and hence they are more numerous.

13.5.2 Supercurrent is Not Hindered by Impurities

The macroscopic supercurrent generated by a moving supercondensate is not hindered by impurities that are microscopic by comparison. The fact that no microscopic perturbation causes a resistance (energy loss), is due to the quantum statistical nature of the supercondensate; the change in the many-pairon-state can occurs only if a transition from one (supercondensate) state to another is induced. Large lattice imperfections and constrictions can however affect the supercurrent significantly.

13.5.3 The Supercurrent Cannot Gain Energy from a DC Voltage

The supercondensate is electrically neutral, and hence it is stable against Lorentz-electric force. The supercurrent cannot gain energy from the applied voltage.

13.5.4 Critical Fields and Critical Currents

In the superconducting state, $\pm$ pairons move in the same direction (because they have the same momentum) with different speeds. If a magnetic field $\mathbf{B}$ is applied, the Lorentz-magnetic force tends to separate $\pm$ pairons. Hence there must be a critical (magnetic) field B_c. The magnetic force is proportional to pairon speeds $v_F^{(i)}/2$. Thus the superconductors having lower Fermi velocities, like high-T_c cuprates, are much more stable than elemental superconductors, and they have higher critical fields. Since the supercurrents themselves generate magnetic fields, there are critical currents J_c.

13.5.5 Supercurrent Ring and Flux Quantization

A persistent supercurrent ring exhibits striking superconducting properties: resistanceless surface supercurrent, Meissner state, and flux quantization.

The superconducting state Ψ_n is represented by a momentum-state wavefunction: $\Psi_n(x) = A\exp(-i\hbar^{-1}p_n x)$, $p_n \equiv 2\pi\hbar n/L$. Since the ring circumference L is macroscopic, the quantized momentum p_n is vanishingly small under normal experimental condition ($n \sim 1 - 100$). The actual value of n in the flux quantization experiments is determined by the Onsager's rule: $\Phi = n\Phi_0 \equiv n\pi\hbar/e$. Each condensed pairon has an extremely small momentum p_n and therefore an extremely small energy $\varepsilon = v_F^{(i)} p_n/2$.

13.5.6 Meissner Effect and Surface Supercurrent

A macroscopic superconductor expels an applied magnetic field **B** from its interior; this is the Meissner effect. Closer examination reveals that the B-field within the superconductor vanishes excluding the surface layer. In fact a finite magnetic field penetrates the body within a thin layer of the order of 500 Å, and in this layer diamagnetic surface supercurrents flow such that the B-field in the main body vanishes, as shown in Fig. 1.17. Why does such condition exist? The stored magnetic field energy for the system in the Meissner state is equal to $VB^2/2\mu_0$. This may be pictured as magnetic pressure acting near the surface and pointing inward. This is balanced by a *Meissner pressure* pointing outwards that tries to keep the state superconductive. This pressure is caused by the Gibbs free energy difference between super and normal states, $G_N - G_S$. Thus if

$$G_N - G_S > VB^2/2\mu_0, \tag{13.69}$$

the Meissner state is maintained in the interior with a steep but continuously changing B-field near the surface. If the B-field is raised beyond the critical field B_c, the inequality (13.69) no longer holds, and the normal state will return. (The kinetic energy of the surface supercurrent is neglected here.

13.5.7 London's Equation

The steady-state supercurrent in a small section can be represented by a plane wavefunction: $\Psi_{\mathbf{p}}(\mathbf{r}) = A\exp(i\hbar^{-1}\mathbf{p}\cdot\mathbf{r}) = A\exp(i\hbar^{-1}px)$. The momentum $\mathbf{p}$ appears in the phase. The Londons assumed, based on the Hamiltonian mechanical consideration of a system of superelectrons, that the phase difference at two points $(\mathbf{r}_1, \mathbf{r}_2)$ in the presence of a magnetic field B has a field component, and obtained London's equation: $\mathbf{j}_s = -e^2m^{-1}n_s\mathbf{A}$.

We assumed that the supercurrent is generated by the $\pm$ pairons condensed at a momentum p, and used the standard Hamiltonian mechanics to obtain

$$\mathbf{j}_s = -2e^2n_0(c_2 + c_1)p^{-1}\mathbf{A}. \tag{13.70}$$

The revised London equation has a proportionality factor different from that of the original London equation.

13.5.8 Penetration Depth

The existence of a penetration depth λ was predicted by the London brothers [6] and it was later confirmed by experiments. This is regarded as an important historical step in superconductivity theory. The qualitative agreement between theory and experiment established a tradition that electromagnetism as represented by Maxwell's equations can, and must, be applied to describe the superconducting state.

The Londons used their equation and Maxwell's equations to obtain an expression for the penetration depth:

$$\lambda_{London} = \frac{c}{e}\left[\frac{m}{4\pi k_0 n_s}\right]^{1/2}.$$

If we adopt the pairon flow model, we obtain instead

$$\lambda = \frac{c}{e}\left[\frac{p}{8\pi k_0 \left|c_2 + c_1\right|}\right]^{1/2}.$$

Note: there is no mass in this expression, since pairons move as massless particles. The $n_0^{-1/2}$-dependence is noteworthy. The supercondensate density n_0 approaches zero as temperature is raised to T_c. The penetration depth λ therefore increases indefinitely as $T \to T_c$.

Chapter 14

Ginzburg-Landau Theory

The pairon field operator $\psi^\dagger(\mathbf{r},t)$ changes, following Heisenberg's equation of motion. If the Hamiltonian H contains a pairon kinetic energy h_0, a condensation energy $\alpha(<0)$ and a repulsive point-like interpairon interaction $\beta\delta(\mathbf{r}_1-\mathbf{r}_2)$, $\beta(>0)$, the evolution equation for ψ is non-linear, from which we derive the Ginzburg-Landau (GL) equation:

$$h_0(\mathbf{r},-i\hbar\nabla)\Psi'_\sigma(\mathbf{r})+\alpha\Psi'_\sigma(\mathbf{r})+\beta\left|\Psi'_\sigma(\mathbf{r})\right|^2\Psi'_\sigma(\mathbf{r})=0$$

for the GL wavefunction $\Psi'_\sigma(\mathbf{r})\equiv\langle\mathbf{r}\left|n^{1/2}\right|\sigma\rangle$, where σ denotes the state of the condensed pairons, and n the pairon density operator. The GL equation with $\alpha=-\varepsilon_g(T)$ holds for all temperatures (T) below T_c, where $\varepsilon_g(T)$ is the T-dependent pairon energy gap. Its equilibrium solution yields that the condensed pairon density $n_0(T)=\left|\Psi_\sigma(\mathbf{r})\right|^2$ is proportional to $\varepsilon_g(T)$. The T-dependence of the expansion parameters near T_c conjectured by GL: $\alpha=-b(T_c-T)$, $\beta=$ constant are confirmed. A new formula for the penetration depth: $\lambda=(c/e)[p\varepsilon_0/n_0(v_F^{(1)}+v_F^{(2)})]^{1/2}$ is obtained, where $c=$ light speed, $v_F^{(i)}=(2\varepsilon_F/m_i)^{1/2}$ are the Fermi velocities for "electrons" (1) and "holes" (2), and p is the condensed pairon momentum.

14.1 Introduction

In 1950 Ginzburg and Landau (GL) [1] proposed a revolutionary idea that below T_c a superconductor has a *complex order parameter* (GL *wavefunction*) Ψ' just as a ferromagnet possesses a real order parameter (spontaneous magnetization). Based on Landau's theory of second-order phase transition

[2], GL expanded the free energy density $f(\mathbf{r})$ of a superconductor in powers of small $|\Psi(\mathbf{r})|$ and $|\nabla\Psi(\mathbf{r})|$:

$$f(\mathbf{r}) = f_0 + \alpha\left|\Psi'(\mathbf{r})\right|^2 + \frac{1}{2}\beta\left|\Psi'(\mathbf{r})\right|^4 + \frac{\hbar^2}{2m_0}\left|\boldsymbol{\nabla}\Psi'(\mathbf{r})\right|^2, \tag{14.1}$$

where f_0, α and β are constants, and m_0 is the superelectron mass. To include the effect of a magnetic field $\mathbf{B}$, they used a quantum replacement:

$$\boldsymbol{\nabla} \to \boldsymbol{\nabla} - (iq/\hbar)\mathbf{A}, \qquad q = charge, \tag{14.2}$$

where $\mathbf{A}$ is a vector potential generating $\mathbf{B} = \boldsymbol{\nabla}\times\mathbf{A}$, and added a magnetic energy term $B^2/2\mu_0$. The integral of the so modified $f(\mathbf{r})$ over the sample volume V gives the Helmholtz free energy F. After minimizing F with variations in Ψ'^* and $\mathbf{A}_j$, GL obtained

$$\frac{1}{2m}\left|-i\hbar\boldsymbol{\nabla} - q\mathbf{A}\right|^2\Psi'(\mathbf{r}) + \alpha\Psi'(\mathbf{r}) + \beta\left|\Psi'(\mathbf{r})\right|^2\Psi'(\mathbf{r}) = 0, \tag{14.3}$$

$$\mathbf{j} = -\frac{iq\hbar}{2m}(\Psi'^*\boldsymbol{\nabla}\Psi' - \Psi'\boldsymbol{\nabla}\Psi'^*) - \frac{q^2}{2m}\Psi'^*\Psi'\mathbf{A}. \tag{14.4}$$

With the density condition:

$$\Psi'^*(\mathbf{r})\Psi'(\mathbf{r}) = n_s(\mathbf{r}) = \text{superelectron density}, \tag{14.5}$$

Eq. (14.4) for the current density $\mathbf{j}$ in the homogeneous limit ($\boldsymbol{\nabla}\Psi' = 0$) reproduces London's equation [3]. Eq. (14.3) is the celebrated Ginzburg-Landau equation, which is quantum mechanical and nonlinear. Since the smallness of $|\Psi'|^2$ is assumed, the GL equation were thought to hold only near T_c, $T_c - T \ll T_c$. Below T_c there is a supercondensate whose motion generates a supercurrent and whose presence generates gaps in the elementary excitation energy spectra. The GL wavefunction Ψ' represents the quantum state of this supercondensate. The usefulness of the GL theory has been well known [4]. The most remarkable results are GL's introduction of the concept of a coherence length [1] and Abrikosov's prediction of a vortex structure in a type II superconductor [5], which was later confirmed by experiments [6]. In their original work GL adopted a superelectron model. The later quantum flux experiments [7, 8] however show that the charge carriers in the supercurrent are pairons [9] having charge (magnitude) $2e$, confirming the basic physical picture of the BCS theory [10]. Here we take the view that Ψ' represents the state of the condensed pairons (rather than the superelectrons). In this chapter we microscopically derive [11] the GL equation (14.3) with a revised density condition:

$$|\Psi'(\mathbf{r})|^2 = \text{condensed pairon density} = n_0(\mathbf{r}). \tag{14.6}$$

We obtain physical interpretation of the expansion parameters (α, β), and discuss their temperature dependence. We further show that the revised GL equations are valid for all temperatures below T_c. The equilibrium solution of this equation with no fields yields a remarkable result that the condensed pairon density $n_0(T)$ is proportional to the pairon energy gap $\varepsilon_g(T)$, observed in the quantum tunneling experiments [12]. We propose a new expression (14.39) for the pairon kinetic energy. Using this we obtain a revised expression (14.41) for the penetration depth.

14.2 Derivation of the G-L Equation

In later twenties Dirac [13] and others [14] showed that quantum field operators, $\psi(\mathbf{r},t)$ and $\psi^\dagger(\mathbf{r},t)$, for bosons (fermions) satisfying the Bose (Fermi) commutation (anticommutation) rules:

$$\begin{aligned} [\psi(\mathbf{r},t),\, \psi^\dagger(\mathbf{r}',t)]_\mp &\equiv \psi(\mathbf{r},t)\psi^\dagger(\mathbf{r}',t) \mp \psi^\dagger(\mathbf{r}',t)\psi(\mathbf{r},t) = \delta^{(3)}(\mathbf{r}-\mathbf{r}'), \\ [\psi(\mathbf{r},t),\, \psi(\mathbf{r}',t)]_\mp &= 0, \end{aligned} \tag{14.7}$$

evolve in time, following Heisenberg's equation of motion:

$$-i\hbar\frac{\partial\psi(\mathbf{r},t)}{\partial t} = [H,\, \psi(\mathbf{r},t)]_-, \tag{14.8}$$

where H represents the Hamiltonian of a system under consideration. In this section the commutators (anticommutators) are represented by $[\mathrm{A},\mathrm{B}]_-$ ($[\mathrm{A},\mathrm{B}]_+$) rather than by $[\mathrm{A},\mathrm{B}]$ ($\{\mathrm{A},\mathrm{B}\}$). If an interparticle interaction Hamiltonian

$$\begin{aligned} H_I &= \frac{1}{2}\sum_{i\neq j}\sum v(|\mathbf{r}_i-\mathbf{r}_j|) \\ &= \frac{1}{2}\int d^3r_1\int d^3r_2\, v(|\mathbf{r}_1-\mathbf{r}_2|)\psi^\dagger(\mathbf{r}_1,t)\psi^\dagger(\mathbf{r}_2,t)\psi(\mathbf{r}_1,t)\psi(\mathbf{r}_2,t) \end{aligned} \tag{14.9}$$

is assumed, the commutator $[H_I,\ \psi(\mathbf{r},t)]_-$ generates (Problem 14.2.1)

$$-\int d^3r_1\, v(|\mathbf{r}-\mathbf{r}_1|)\psi^\dagger(\mathbf{r}_1,t)\psi(\mathbf{r}_1,t)\psi(\mathbf{r},t), \tag{14.10}$$

indicating that the evolution equation for the quantum field ψ is intrinsically nonlinear in the presence of the interparticle interaction [14]. Two pairons, having like charge, repel each other. We represent this by a repulsive point-like potential

$$v(\mathbf{r}) = \beta\delta^{(3)}(\mathbf{r}), \qquad \beta > 0. \tag{14.11}$$

We then obtain

$$[H_I,\ \psi(\mathbf{r},t)]_- = -\beta n(\mathbf{r},t)\,\psi(\mathbf{r},t) \equiv -\beta\psi^\dagger(\mathbf{r},t)\psi(\mathbf{r},t)\,\psi(\mathbf{r},t), \tag{14.12}$$

which *vanishes for fermions.* Hereafter we shall consider bosons only.

If we assume a kinetic energy $h_0(\mathbf{r},\mathbf{p})$ and a constant condensation energy $\alpha(<0)$, the following equation is obtained from Eq. (14.8):

$$i\hbar\frac{\partial\psi(\mathbf{r},t)}{\partial t} = h_0(\mathbf{r},-i\hbar\frac{\partial}{\partial\mathbf{r}})\psi(\mathbf{r},t) + \alpha\psi(\mathbf{r},t) + \beta n(\mathbf{r},t)\,\psi(\mathbf{r},t). \tag{14.13}$$

Let us consider a persistent ring supercurrent. The wavefunction at the point $\mathbf{r}$ in a small section along the ring is characterized by a discrete momentum $(p_\nu,0,0)$ with

$$p_\nu = \frac{2\pi\hbar}{L}\nu, \qquad (\nu = 0,\pm1,\pm2,...) \tag{14.14}$$

where L is the ring circumference. We note that the absolute value $|\nu|$ characterizes the number of flux quanta $\Phi_0 \equiv \pi\hbar/e$ enclosed by the ring [8]. Further note that the momentum eigenstate $(p_\nu,0,0) \equiv \mathbf{p}$ defines the state σ of the supercondensate.

Let us define a quasiwavefunction $\Psi_\sigma(\mathbf{r})$ by

$$\Psi_\sigma(\mathbf{r}) \equiv \mathrm{TR}\{\psi(\mathbf{r})\,\rho\,\phi_\sigma^\dagger\} \equiv \langle\mathbf{r}\,|n|\,\sigma\rangle, \tag{14.15}$$

where $\phi_\sigma^\dagger$ is the condensed-pairon-state (σ) creation operator, ρ the system density operator, and the symbol TR denotes a grand ensemble trace; the pairon density operator n is defined through the position density matrix elements:

$$\mathrm{TR}\{\psi(\mathbf{r})\,\rho\,\psi^\dagger(\mathbf{r}')\} = \langle\mathbf{r}\,|n|\,\mathbf{r}'\rangle, \tag{14.16}$$

normalized such that

$$\frac{1}{V}\int d^3r \left\langle \mathbf{r}\left|n\right|\mathbf{r}\right\rangle \equiv \frac{1}{V}\mathrm{tr}\{n\} = \text{average pairon density}, \tag{14.17}$$

where the symbol tr means a one-body trace. The GL wavefunction $\Psi'_\sigma(\mathbf{r})$ can be related with the pairon density operator n by

$$\Psi'_\sigma(\mathbf{r}) = \left\langle \mathbf{r}\left|n^{1/2}\right|\sigma\right\rangle, \qquad (n^{1/2})^2 = n. \tag{14.18}$$

The two wavefunctions, $\Psi'_\sigma(\mathbf{r})$ and $\Psi_\sigma(\mathbf{r})$, have different amplitudes; $|\Psi'_\sigma(\mathbf{r})|^2 = \left\langle \mathbf{r}\left|n^{1/2}\right|\sigma\right\rangle\left\langle \sigma\left|n^{1/2}\right|\mathbf{r}\right\rangle \equiv n_\sigma(\mathbf{r})$ is the condensed pairon density while $\Psi_\sigma(\mathbf{r})$ itself is proportional to $n_\sigma(\mathbf{r})$.

We now multiply Eq. (14.13) by $\rho\phi_\sigma^\dagger$ from the right and take a grand ensemble trace. Writing the results in terms of $\Psi_\sigma(\mathbf{r})$ and using a factorization approximation, we obtain

$$i\hbar\frac{\partial\Psi_\sigma(\mathbf{r})}{\partial t} = h_0(\mathbf{r}, -i\hbar\nabla)\Psi_\sigma(\mathbf{r}) + \alpha\Psi_\sigma(\mathbf{r}) + \beta\left|\Psi_\sigma(\mathbf{r})\right|^2\Psi_\sigma(\mathbf{r}). \tag{14.19}$$

For the steady state the time derivative vanishes, yielding

$$h_0\Psi_\sigma(\mathbf{r}) + \alpha\Psi_\sigma(\mathbf{r}) + \beta\left|\Psi_\sigma(\mathbf{r})\right|^2\Psi_\sigma(\mathbf{r}) = 0, \tag{14.20}$$

which is precisely the GL equation (14.3).

In our derivation we assumed that pairons move as bosons, which is essential, see the sentence following Eq. (14.12). Bosonic pairons can multiply occupy the net momentum state $\mathbf{p}$ while fermionic superelectron cannot. The correct density condition (14.6) instead of (14.5) must therefore be used.

14.3 Condensation Energy

Our microscopic derivation allows us to interpret the expansion parameters (α, β) in the original GL theory as follows: α represents the pairon condensation energy, and β the repulsive interaction strength. The latter, from its mechanical origin, is temperature-independent:

$$\beta = \text{constant} > 0, \qquad T_c > T. \tag{14.21}$$

BCS showed [10] that the ground state energy W for the BCS system is

$$W = \hbar\omega_D\mathcal{N}(0)w_0, \qquad w_0 \equiv -2\hbar\omega_D\{\exp[2/(v_0\mathcal{N}(0))] - 1\}^{-1}, \tag{14.22}$$

where $\mathcal{N}(0)$ is the density of states per spin at the Fermi energy and w_0 the pairon ground-state energy. Hence we can choose

$$\alpha = w_0 < 0, \qquad T = 0. \tag{14.23}$$

In the original work [1] GL considered a superconductor immediately below T_c, where $|\Psi'_\sigma|^2$ is small. Gorkov [15] used Green's functions to interrelate the GL and the BCS theory near T_c. Werthamer and Tewardt [16] extended the Ginzburg-Landau-Gorkov theory to all temperatures below T_c, and arrived at more complicated equations. Here, we derived the original GL equation by examining the superconductor at 0 K from the condensed pairons point of view. The transport property of a superconductor below T_c is dominated by the condensed pairons. Since there is no distribution, the qualitative property of the condensed pairons cannot change with temperature. The pairon size (the minimum of the coherence length derivable directly from the GL equation) naturally exists. In summary there is only one supercondensate whose behavior is similar at all temperatures below T_c; only the density of condensed pairons can change. Thus there will be a quantum nonlinear equation (14.20) for $\psi_\sigma(\mathbf{r})$ valid everywhere below T_c.

The pairon energy spectrum below T_c has a discrete ground-state energy, which is separated from the energy continuum of moving pairons [17]. This separation $\varepsilon_g(T)$, called the pairon energy gap, is T-dependent. This energy gap, as in the well-known case of the atomic energy spectra, can be detected in photo-absorption [18] and quantum tunneling experiments [12]. Inspection of the pairon energy spectrum with a gap suggests that

$$\alpha = -\varepsilon_g(T) < 0, \qquad T_c > T. \tag{14.24}$$

Solving Eq. (14.20) with $h_0\Psi_\sigma = 0$ (no currents, no fields), we obtain

$$n_0(T) = |\Psi_\sigma|^2 = \beta^{-1}\varepsilon_g(T), \tag{14.25}$$

indicating that the *condensate density* $n_0(T)$ *is proportional to the energy gap* $\varepsilon_g(T)$.

We now consider an ellipsoidal macroscopic sample of a type I superconductor below T_c subject to a weak magnetic field $\mathbf{H}_a$ applied along its major axis. Because of the Meissner effect, the magnetic fluxes are expelled from the main body, and the magnetic energy is higher by $(1/2)\mu_0 H_a^2 V$ in the super state than in the normal state. If the field is sufficiently raised, the sample reverts to the normal state at a critical field H_c, which can be computed in terms of the free-energy expression (14.1) with the magnetic field included. We obtain after using (14.6) and (14.22)

$$H_c \simeq (\mu_0 \beta)^{-1/2} \varepsilon_g(T) \propto n_0(T), \tag{14.26}$$

indicating that the measurements of H_c give the T-dependence of $n_0(T)$ approximately. The field-induced transition corresponds to the evaporation of condensed pairons, and not to their break-up into electrons. Moving pairons by construction have negative energies while quasi-electrons have positive energies. Thus, the moving pairons are more numerous at the lowest temperatures, and they are dominant elementary excitations. Since the contribution of the moving pairons was neglected in the above calculation, Eq. (14.24) is approximate, see below.

We stress that the pairon energy gap ε_g is distinct from the quasi-electron energy gap Δ, which is the solution of [Eq. (3.27), Ref. 9]

$$1 = v_0 \mathcal{N}(0) \int_0^{\hbar\omega_D} d\varepsilon \frac{1}{(\varepsilon^2 + \Delta^2)^{1/2}} \tanh \left[\frac{(\varepsilon^2 + \Delta^2)^{1/2}}{2k_B T} \right]. \tag{14.27}$$

In the presence of a supercondensate the energy-momentum relation for an unpaired (quasi) electron changes:

$$\varepsilon_k \equiv \left| k^2/(2m) - \varepsilon_F \right| \rightarrow E_k \equiv (\varepsilon_k^2 + \Delta^2)^{1/2}. \tag{14.28}$$

Since the density of condensed pairons changes with T, the gap Δ is T-dependent and is determined from Eq. (14.27) (originated in the BCS energy gap equation). Two unpaired electrons can be bound by the phonon-exchange attraction to form a moving pairon whose energy $\tilde{w}_q$ is given by

$$\tilde{w}_q = \tilde{w}_0 + \frac{1}{2} v_F q < 0, \tag{14.29}$$

$$1 = v_0 \mathcal{N}(0) \int_0^{\hbar\omega_D} d\varepsilon \, [|\tilde{w}_0| + 2(\varepsilon^2 + \Delta^2)^{1/2}]^{-1}. \tag{14.30}$$

Note that $\tilde{w}_0$ is T-dependent since Δ is. At T_c, $\Delta = 0$ and the lower band edge $\tilde{w}_0$ is equal to the pairon ground-state energy w_0. If $T < T_c$, $\Delta > 0$. We may then write

$$\tilde{w}_q = w_0 + \varepsilon_g(T) + \frac{1}{2} v_F q, \qquad \varepsilon_g(T) \equiv \tilde{w}_0 - w_0 \geq 0, \tag{14.31}$$

and call $\varepsilon_g(T)$ the pairon energy gap. The two gaps (Δ, ε_g) have similar T-behavior; they are zero at T_c and they both grow monotonically as temperature is lowered. The rhs of Eq. (14.27) is a function of $(T,\ \Delta^2)$; T_c is a regular point such that a small variation $\delta T \equiv T_c - T$ generates a small variation in Δ^2. Hence we obtain

$$\Delta(T) \cong a(T_c - T)^{1/2}, \qquad T_c - T \ll T_c, \qquad a = \text{constant}, \tag{14.32}$$

showing that Δ falls off steeply near T_c. Using similar arguments we get from Eqs. (14.30)-(14.32)

$$\varepsilon_g(T) \cong b(T_c - T), \qquad T_c - T \ll T_c, \qquad b = \text{constant}, \tag{14.33}$$

As noted earlier, moving pairons have finite (zero) energy gaps in the super (normal) state, which makes Eq. (14.26) approximate. But the gaps disappear at T_c, and hence the linear-in-$(T_c - T)$ behavior should hold for the critical field H_c:

$$H_c = c(T_c - T), \qquad T_c - T \ll T_c \qquad c = \text{constant}, \tag{14.34}$$

which is supported by experimental data. Tunneling and photo absorption data appear to support the linear law in Eq. (14.33) less strongly. But, in transport and optical experiments the signal become weaker (since n_0 gets smaller) and the relative errors get greater (since ε_g is very small) as temperature approaches T_c. Further careful experimental studies and analyses are called for here.

In the original GL theory [1], the following signs and T-dependence of the expansion parameters (α, β) near T_c were assumed and tested:

$$\alpha \cong -b(T_c - T) < 0, \qquad \beta = \text{constant} > 0, \tag{14.35}$$

all of which are reestablished by our microscopic calculations.

In summary we reached a significant conclusion that the GL equation is valid for all temperatures below T_c. Our interpretation of $(\alpha,\ \beta)$ involves assumptions. The usefulness of such an equation can only be judged by working out its solutions and comparing with experiments. As noted earlier, the most important results in the GL theory include GL's introduction of a coherence length [1] and Abrikosov's prediction of a vortex structure [5], both concepts holding not only near T_c but for all temperatures below T_c. Also the upper critical field $H_{c2}(T)$ for all temperatures is known to be described in terms of the GL equation [4].

14.4 Penetration Depth

As an application of our theory we consider a cylindrical superconductor trapping $|\nu|$ flux quanta and maintaining a persistent supercurrent near its inner side, see Fig. 14.1. The wavefunction near the inner surface may be represented by

$$\Psi_\sigma(\mathbf{r}) = \text{constant} \times \exp(i\mathbf{p}\cdot\mathbf{r}/\hbar), \qquad \sigma \equiv \mathbf{p} \equiv (p_\nu, 0, 0), \qquad p_\nu = 2\pi\hbar\nu/L. \tag{14.36}$$

This Ψ_σ clearly satisfies Eq. (14.20) with the kinetic energy $\varepsilon_0(p)$ and $\alpha + \beta\,|\Psi_\sigma|^2 = -\varepsilon_g + \beta n_0$. This is the lowest-energy state of the system with a magnetic flux trapped at any temperature below T_c. By losing the magnetic flux the system may go down to the true equilibrium state with no current. Since L is a macroscopic length, $p_\nu = 2\pi\hbar\nu/L$ and its associated energy $\varepsilon_0(p)$ are both very small. This example also supports our conclusion that the GL equation is valid everywhere below T_c. Since Ψ_σ representing a stationary state is characterized by momentum $(p_\nu, 0, 0) \equiv \mathbf{p}$, the GL equation is valid independently of materials, meaning that the law of corresponding states works well for all superconductors.

GL adopted that superelectron model in which the kinetic energy in the presence of a magnetic field $\mathbf{B}$ is given by

$$h_0 = \frac{1}{2m}[-i\hbar\nabla + e\mathbf{A}(\mathbf{r})]^2. \tag{14.37}$$

We now seek an appropriate expression for h_0. At 0 K all pairons are zero-momentum pairons, which do not generate a supercurrent. Cooper studied the energy w_p of a moving pairon [9], and obtained [unpublished but recorded in Schrieffer's book, Eq. (2-15) of Ref. [19]],

$$w_p = w_0 + \frac{1}{2}v_F p, \qquad v_F \equiv (2\varepsilon_F/m^*)^{1/2}, \tag{14.38}$$

where w_0 (< 0) is the pairon ground state energy, see Eqs. (14.22). Note that the energy-momentum relation is linear; the pairons move with the common speed $v_F/2$.

We now propose a new kinetic energy term:

$$h_0 = \frac{1}{2}v_F^{(1)}\,|-i\hbar\nabla^{(1)} + 2e\mathbf{A}(\mathbf{r})\,| + \frac{1}{2}v_F^{(2)}\,|-i\hbar\nabla^{(2)} - 2e\mathbf{A}(\mathbf{r})\,|\,. \tag{14.39}$$

If we use this h_0 , and assume that the pairons are condensed at $\mathbf{p}$, we obtain

$$\mathbf{j} = -\frac{e}{2}(v_F^{(1)} - v_F^{(2)})\,|\Psi_\sigma(\mathbf{r})|^2\,\hat{\mathbf{p}} - e^2(v_F^{(1)} + v_F^{(2)})p_\nu^{-1}\,|\Psi_\sigma(\mathbf{r})|^2\,\mathbf{A}, \qquad (14.40)$$

where $\hat{\mathbf{p}}$ is the unit vector pointing in the direction of $\mathbf{p}$. Note that the motional component (first term) reproduces Eq. (13.5). Omitting this component, we obtain the corrected London equation (13.35). By using Maxwell's equations $\nabla^2\mathbf{A} = -\mu_0\mathbf{j}$, we obtain a new expression for the penetration depth:

$$\lambda \equiv \frac{c}{e}\{\varepsilon_0 p_\nu/[n_0 \mid v_F^{(2)} + v_F^{(1)} \mid]\}^{1/2} \qquad \text{(new)} \qquad (14.41)$$

in contrast with London's result [3]:

$$\lambda \equiv \frac{c}{e}\left(\frac{\varepsilon_0 m}{n_s}\right)^{1/2}. \qquad \text{(London)} \qquad (14.42)$$

The $n_0^{-1/2}$-dependence in Eq. (14.41) is noteworthy; the penetrations depth λ increases to ∞ as $T \to T_c$ in agreement with experiment. We also note that our expression (14.40) contains no undetermined parameter like the superelectron mass. Thus it can be used to determine $(p_\nu, n_0, v_F^{(1)} + v_F^{(2)})$.

Finally, Eq.(14.19) represents an evolution equation for the G-L wavefunction, which is significant. Since the GL wavefunction was intuitively introduced for a stationary state problem, it is not immediately clear how to describe its time evolution. In the absence of interparticle interactions, Eq. (14.19) is reduced to a standard Schrödinger equation of motion.

14.5 Discussion

We derived the GL equation from first principles. In the derivation we found that the particles which are described by the GL wavefunction $\Psi'(\mathbf{r})$ must be bosons. We took the view that $\Psi'(\mathbf{r})$ represents the bosonically condensed pairons. This explains the quantum nature of the wavefunction. In fact $\Psi'_\sigma(\mathbf{r}) = \langle \mathbf{r}\,|n^{1/2}|\,\sigma\rangle$ is a mixed representation of the pairon square-root density operator $n^{1/2}$ in terms of the position $\mathbf{r}$ and the momentum state σ. The new density condition is given by $\Psi'^{*}_\sigma(\mathbf{r})\Psi'_\sigma(\mathbf{r}) = n_\sigma(\mathbf{r})$ = condensed pairon density. The nonlinearity of the GL equation arises from the point-like repulsive interpairon interaction. In 1950 when Ginzburg and Landau published their work, the Cooper pair (pairon) was not known. They simply assumed the superelectron model.

The expansion parameters (α, β) in the GL theory are identified as the negative of the pairon binding energy and the repulsive interpairon interaction strength. This eventually leads to a remarkable result that the temperature-dependent condensed pairon density $n_0(T)$ is proportional to the pairon energy gap $\varepsilon_g(T)$.

Chapter 15

Josephson Effects

Josephson effects are quantum statistical effects manifested on a macroscopic scale. A close analogy emerges between a supercurrent and a laser. Supercurrents, not lasers, respond to electromagnetic fields, however. Basic equations for the current passing a Josephson junction are derived. They are used to discuss SQUID and AC Josephson effects. Analyses of Shapiro steps in the V-I diagram show that the quasi-wavefunction Ψ_σ rather than the G-L wavefunction represents correct pairon dynamics.

15.1 Josephson Tunneling and Interference

In 1962 Josephson predicted a supercurrent tunneling through a small barrier with no energy loss [1]. Shortly thereafter Anderson and Rowell [2] demonstrated this experimentally. Consider the circuit shown in Fig. 12.1. The circuit contains two superconductors S_1 and S_2 with a Josephson junction consisting of a very thin oxide film of thickness $\sim$ 10 Å, see Fig. 15.1. The two superconductors are made of the same material. The I-V curves observed are shown in Fig. 15.2. Finite current I_0 appears (a) even at $V = 0$, and its magnitude is of the order of mA; it is very sensitive to the presence of a magnetic field. When a weak field $B = \mu_0 H$ ($\sim$ 0.4 Gauss) is applied, the current I_0 drops significantly as shown in (b). When voltage ($\sim$ mV) is raised high enough, the normal tunneling current appears, whose behavior is similar to that of Giaever tunneling in the S-I-S system. (see Fig. 12.2). To see the physical significance of Josephson tunneling, let us consider the same system above T_c. The two superconductors above T_c show potential drops, and the oxide layer generates a large potential drop. Below T_c the two superconductors having no resistance show no potential drops. Moreover, the oxide layer exhibits no potential drop! This is an example of quantum tunneling, which we learn in elementary quantum theory. The quasi-wavefunctions $\Psi_j(\mathbf{r})$ in the superconductors S_j do not vanish abruptly at the S_j-I interfaces. If the oxide layer is small (10 Å), the two wavefunctions Ψ_1 and Ψ_2 may be regarded as a *single wavefunction* extending over both regions. Then pairons can tunnel through the oxide layer with no energy loss.

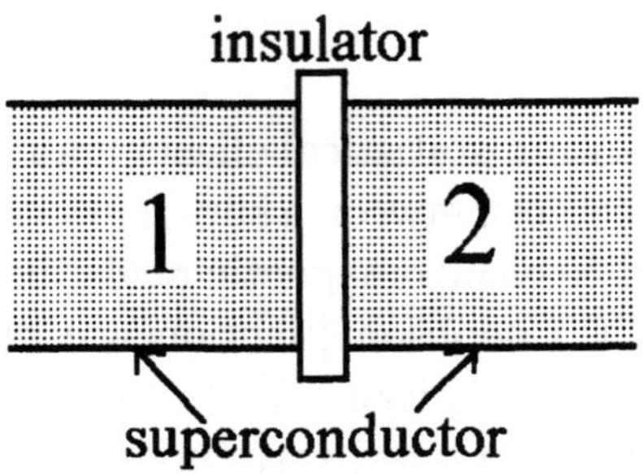

Figure 15.1: Insulator sandwiched by two superconductors.

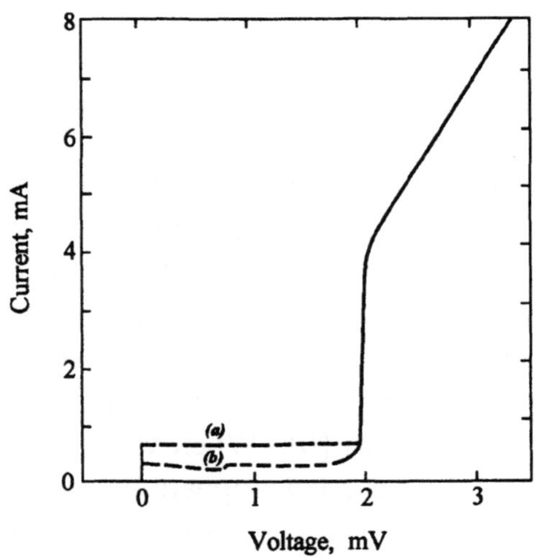

Figure 15.2: The I-V curves indicating a Josephson tunneling current. (a) $B = 0$, (b) $B = 0.4$ Gauss.

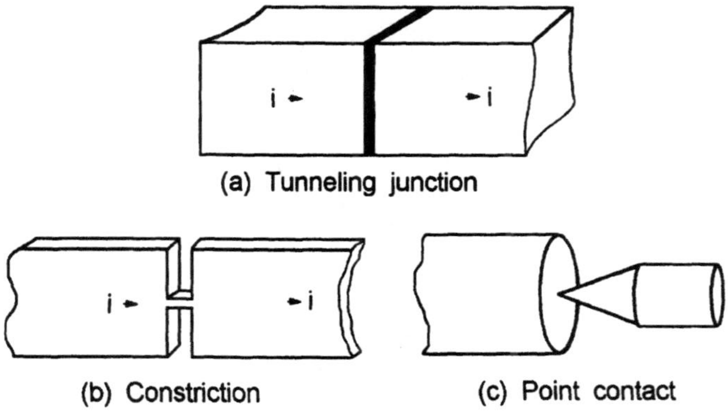

Figure 15.3: Three types of weak links.

The oxide layer that allows supercurrent tunneling, shown in Fig. 15.3 (a), is called a *tunneling junction.* Similar effects can be produced by *constriction* (b) and *point contact* (c). Any of the three is called a *weak link* or a *Josephson junction.*

We now take a ring-shaped superconductor with two Josephson junctions as shown in Fig. 1.8. Below T_c the current may split in two branches and rejoin. If a very weak magnetic field is applied normal to the ring and is varied, the current I has an oscillatory component, as shown in Fig. 1.9 [3]. The oscillatory part can roughly be represented by

$$I = I_{\max} \cos(\pi\Phi/\Phi_0), \qquad \Phi_0 \equiv \pi\hbar/e, \tag{15.1}$$

where Φ is the magnetic flux enclosed by the ring:

$$\Phi = BA, \qquad A = \pi r^2 \quad (r \sim 1 \text{ mm}); \tag{15.2}$$

and $I_{\max}$ is a constant. This is a *supercurrent interference.* The two supercurrents separated by 1 mm can interfere just as two laser beams from the same source.

This interference may be explained as follows. We divide the steady supercurrents in two, as shown in Fig. 15.4, where (a) represents the current in the absence of B, and (b) the diamagnetic current going through two junctions. The diamagnetic current is similar to that appearing in the flux quantization experiment discussed in Section 13.1. Junctions allow the magnetic flux Φ to change continuously, which generates a continuous current.

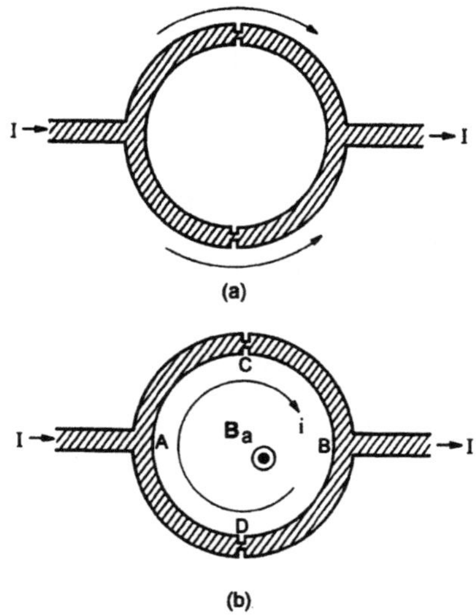

Figure 15.4: (a) the supercurrent at $B = 0$, (b) the diamagnetic supercurrent is generated and shows an interference pattern as the B field is increased.

Detailed calculations, (see Section 15.2) show that the period of oscillation is $\Phi/2\Phi_0$, as indicated in Eq. (15.1). To appreciate the physical significance, consider the same circuit above T_c. Application of a B-field generates a diamagnetic normal current around the ring, which dies out due to resistance, and cannot contribute to the steady current.

We close this section by pointing out a close analogy between supercurrent and laser. Both can be described by the wavefunction $U\exp(i\mathbf{k}\cdot\mathbf{r})$, $U =$ constant, representing a system of massless bosons all occupying the same momentum state $\hbar\mathbf{k}$. Such a monochromatic massless boson flux has a *self-focusing power* (capability). A flux of photons in a laser is slowed down by atomic electrons in a glass plate, but it can refocus by itself into the original state due to the photon's boson nature. Similarly the pairon flux (supercurrent) becomes monochromatic after passing a Josephson junction. Thus both laser and supercurrent can interfere at a *macroscopic* distance. The self-focusing power comes from the quantum statistical factor:

$$N_{\mathbf{p}} + 1, \tag{15.3}$$

with $N_{\mathbf{p}}$ denoting the number of pairons associated with the condensation process in which a pairon joins the group numbering $N_{\mathbf{p}}$ in the state $\mathbf{p}$.

The importance of the quantum statistical factor has been well established. Ueling and Uhlenbeck [4] derived the Ueling-Uhlenbeck collision terms by quantum-correcting the Boltzmann collision terms. Tomonaga [5] solved the corrected Boltzmann equation and obtained a T^2-law behavior of the viscosity coefficient for a highly degenerate fermion gas. Earlier in Section 12.3 we discussed the negative resistance region in the I-V curve for the S_1-I-S_2 system, arising from the quantum statistical factor. The number $N_p + 1$ is an enormous enhancement factor. But part of the enhancement is compensated by the factor N_p associated with the decondensation process in which a pairon leaves the group. Feynman argued for such a *boson enhancement effect* in his provocative discussion of the supercurrent [6]. He discussed this effect in terms of the probability amplitude, and therefore the factor $N_p^{1/2}$ appears in his argument. The macroscopic interference may be observed for massless bosons (photons, pairons) only. It is interesting to note that such a self-focusing power is not known for a fermion (electron, nucleon, neutrino) flux. Quantum diffraction is observed for both fermions and bosons.

Supercurrents and lasers are however different. Pairons carry charge, but photons do not. Hence only pairons can contribute to the charge transport; moreover supercurrents generate magnetic fields and react against electromagnetic fields. We also note that pairons can stop, while phonons cannot and they always move.

15.2 Equations Governing a Josephson Current

In this section we derive basic equations governing the supercurrent passing through a Josephson junction. Our derivation essentially follows Feynman's in his lecture notes on the supercurrent [6, 7]. Our theory however is based on the independent pairon model, while Feynman assumed the superelectron model.

Consider an insulator of width Δx sandwiched between two identical superconductors as shown in Fig. 15.1. If the width Δx is large, the two superconductors do not affect each other, and the quantum equations of motion in each superconductor are uncoupled:

$$i\hbar\frac{d}{dt}\Psi_1(t) = E_1\Psi_1, \qquad i\hbar\frac{d}{dt}\Psi_2(t) = E_2\Psi_2, \tag{15.4}$$

where E_j are the energies of the supercondensate with E_1 differing from E_2 if there is a voltage across the insulator:

$$E_2 - E_1 = qV \, (= 2eV). \tag{15.5}$$

Now if the width Δx is very small ($\sim$ 10 Å), then Josephson tunneling can occur, and the two wavefunctions are correlated. We represent this by

$$i\hbar\frac{d}{dt}\Psi_1(t) = E_1\Psi_1 + K\Psi_2, \qquad i\hbar\frac{d}{dt}\Psi_2(t) = E_2\Psi_2 + K\Psi_1, \tag{15.6}$$

where K is a real coupling constant (energy). If there is no bias voltage, we have a single energy $E_1 + K$ for the combined system. Thus in this case, the reality of the energy constant K is justified. The wavelength $\lambda = 2\pi/k$ of the pairon is very much greater than the junction width Δx:

$$\lambda \gg \Delta x. \tag{15.7}$$

Then the phase of the quasi-wavefunction:

$$\phi(x) = kx \equiv 2\pi x/\lambda \tag{15.8}$$

changes little when x is measured in units (lengths) Δx. We may simply assume that each superconductor has a position-independent phase θ_j.

Let us assume a wavefunction of the form:

$$\Psi_j(t) = n_j(t)\exp[i\theta_j(t)], \tag{15.9}$$

where n_j are the position-independent pairon densities. Substituting Eq. (15.9) in Eqs. (15.6), we obtain (Problem 15.2.1.)

$$i\hbar\dot{n}_1 + \hbar n_1\dot{\theta}_1 = E_1 n_1 + K n_2 \exp[i(\theta_2 - \theta_1)],$$

$$i\hbar\dot{n}_2 + \hbar n_2\dot{\theta}_2 = E_2 n_2 + K n_1 \exp[-i(\theta_2 - \theta_1)], \tag{15.10}$$

Equating real and imaginary parts of Eqs. (15.10), we find that (Problem 15.2.2.)

$$\hbar\dot{n}_1 = K n_2 \sin(\theta_2 - \theta_1) \equiv K n_2 \sin\delta, \qquad \hbar\dot{n}_2 = -K n_1 \sin\delta, \tag{15.11}$$

$$\hbar\dot{\theta}_1 = E_1 + K\frac{n_2}{n_1}\cos\delta, \qquad \hbar\dot{\theta}_2 = E_2 + K\frac{n_1}{n_2}\cos\delta, \tag{15.12}$$

where

$$\delta \equiv \theta_2 - \theta_1 \tag{15.13}$$

is the *phase difference* across the junction.

Equations (15.11)-(15.13) can be solved by a Taylor expansion method. Assume that

$$n_j(t) = n_j^{(0)}(t) + n_j^{(1)}(t) + ..., \tag{15.14}$$

where the upper indices denote the orders in K. After simple calculations, we obtain (Problem 13.6.3)

$$n_1^{(0)} = n_2^{(0)} \equiv n_0, \tag{15.15}$$

$$\hbar \dot{n}_1^{(1)} = n_0 \sin \delta, \qquad \hbar \dot{n}_2^{(2)} = -n_0 \sin \delta, \tag{15.16}$$

$$\boxed{\hbar\, d\delta/dt = 2eV,} \tag{15.17}$$

where we assumed that the initial pairon densities are the same and they are equal to n_0. The electric current I is proportional the charge ($2e$) and the rate $\dot{n}_2$. Thus we obtain

$$\boxed{I = I_0 \sin \delta}, \qquad I_0 \equiv 2eK\hbar^{-1} n_0. \tag{15.18}$$

The last two equations are the basic equations, called the *Josephson-Feynman equations*, governing the tunneling supercurrent. Their physical meaning will become clear after solving them explicitly, which we do in the next section.

As an application of Eq. (15.18), we complete the discussion of the SQUID with quantitative calculations. The basic set-up of the SQUID is shown in Figs. 1.11 and 15.4. Consider first the current path ACB. The phase difference $(\Delta\phi)_{ACB}$ along this path is

$$(\Delta\phi)_{ACB} = \delta_C + 2e\hbar^{-1} \int_{ACB} \mathbf{A} \cdot d\mathbf{r}. \tag{15.19}$$

Similarly for the second path ADB, we have

$$(\Delta\phi)_{ADB} = \delta_D + 2e\hbar^{-1} \int_{ADB} \mathbf{A} \cdot d\mathbf{r}. \tag{15.20}$$

Now the phase difference between A and B must be independent of path:

$$(\Delta\Phi)_{ACB} = (\Delta\Phi)_{ADB}, \tag{15.21}$$

from which we obtain

$$\delta_C - \delta_D = 2e\hbar^{-1} \oint_{ADBCA} \mathbf{A} \cdot d\mathbf{r} = 2e\hbar^{-1}\Phi, \tag{15.22}$$

where Φ is the magnetic flux through the loop. The total current I is the sum of I_C and I_D. For convenience, we write

$$\delta_C \equiv \delta_0 + e\hbar^{-1}\Phi, \qquad \delta_D \equiv \delta_0 - e\hbar^{-1}\Phi, \tag{15.23}$$

where δ_0 is a constant. Note: This set satisfies Eq. (15.22). Then, using Eq. (15.18) we obtain

$$I \equiv I_C + I_D = I_0\{\sin(\delta_0 + e\hbar^{-1}\Phi) + \sin(\delta_0 - e\hbar^{-1}\Phi)\} = I_0 \sin\delta_0 \cos(e\hbar^{-1}\Phi). \tag{15.24}$$

The constant δ_0 introduced in Eq. (15.23) is an unknown parameter and it may depend on the applied voltage and other condition. However, $|\sin\delta_0|$ is bounded by unity. Thus the maximum current has amplitude $I_{\max} = I_0\,|\sin\delta_0|$, and the total current I is expressed as

$$I = I_{\max}\cos(e\hbar^{-1}\Phi) = I_{\max}\cos(\pi\Phi/\Phi_0), \qquad \Phi_0 \equiv \pi\hbar/e, \tag{15.25}$$

proving Eq. (15.1).

Problem 15.2.1. Derive Eqs. (15.10).

Problem 15.2.2. Verify Eqs. (15.11) and (15.12).

Problem 15.2.3. Verify Eqs. (15.16)-(15.18).

15.3 ac Joshepson Effect and Shapiro Steps.

Josephson predicted [1] that upon the application of a constant voltage V_0 the supercurrent passing through a junction should have a component oscillating with the *Josephson (*angular*) frequency*

$$\omega_J = 2e\hbar^{-1}V_0. \tag{15.26}$$

We discuss this ac *Josephson effect* in the present section.

Assume that a small voltage V_0 is applied across junction. Solving Eq. (15.17), we obtain

$$\delta(t) = \delta_0 + 2e\hbar^{-1}V_0 t = \delta_0 + \omega_J t, \tag{15.27}$$

where δ_0 is the initial phase. The phase difference δ changes linearly with the time t. Using Eq. (15.27) we calculate the supercurrent I_s from Eq. (15.18) and obtain

$$I_s(t) = I_0 \sin(\delta_0 + \omega_J t). \tag{15.28}$$

For the laboratory voltage and time, the sine oscillates very rapidly, and the time-averaged current vanishes:

$$\langle I_s \rangle_{time} \equiv \lim_{T\to\infty} \frac{1}{T} \int_0^T dt\, I_s(t) = 0. \tag{15.29}$$

This is quite remarkable. The supercurrent does *not obey* Ohm's law familiar in the normal conduction: I_n(normal current) $\propto V_0$ (voltage). According to Eq. (15.28) the supercurrent flows *back and forth* across the junction with the (high) frequency ω_J under the action of the dc bias. Most importantly the supercurrent passing through the junction does not gain energy from the bias voltage. Everything is consistent with our physical picture that the supercondensate is composed of independently moving $\pm$ pairons and is electrically neutral. In practice, since temperature is not zero, there is a small current I_n due to the moving charged quasi-particles (quasi-electrons, excited pairons).

Let us now apply a small ac voltage in addition to the dc voltage V_0. The AC voltage may be supplied by a microwave. We assume that

$$V = V_0 + v\cos\omega t, \qquad (v \ll V_0) \tag{15.30}$$

where ω is the microwave frequency of the order 10^{10} cycles/sec $=$ 10 GHz. Solving Eq. (15.17) with respect to t, we obtain

$$\delta(t) = \delta_0 + \int_0^t dt\, 2e\hbar^{-1}[V_0 + v\cos\omega t] = \delta_0 + 2e\hbar^{-1}V_0 t + \frac{2ev}{\hbar\omega}\sin\omega t. \tag{15.31}$$

Because $v \ll V_0$, the last term is small compared with the rest. We use Eqs. (15.28) and (15.31) to calculate the supercurrent I (the subscript s dropped)

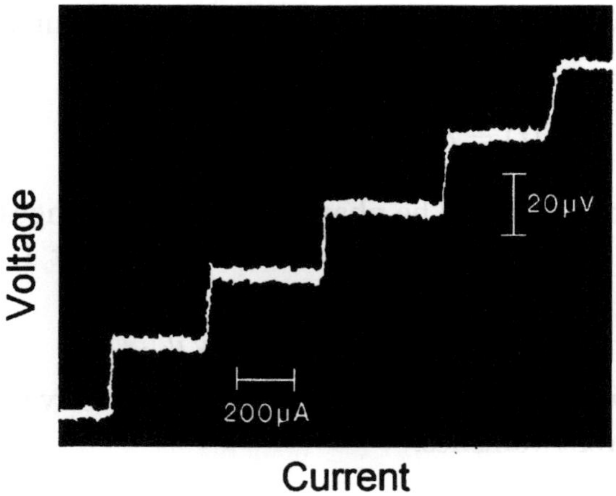

Figure 15.5: Shapiro steps in the V-I diagram, after Hartland [9].

and obtain to the first order in $2eV/\hbar\omega$ (Problem 15.3.1):

$$I = I_0 \sin \delta(t) = I_0 [\sin(\delta_0 + \omega_J t) + \frac{2ev}{\hbar\omega} \sin \omega t \cos(\delta_0 + \omega_J t)]. \qquad (15.32)$$

The first term is zero on the average as before, but the second term is nonzero if

$$\omega = 2e\hbar^{-1} V_0 \, n = \omega_J \, n, \qquad n = 1, 2, \ldots \qquad (15.33)$$

(Problem 15.3.2.). Thus there should be a dc current if the microwave frequency ω matches the Josephson frequency ω_J or its multiples $n\omega_J$. In 1963 Shapiro [8] first demonstrated such a resonance effect experimentally. If data are plotted in the V-I diagram, horizontal current strips form steps of equal height (voltage) $\hbar\omega/2e$ called *Shapiro steps.* Typical data for a Pb-PbO$_x$-Pb tunnel junction at ω = 10 GHz obtained by Hartland [9] are shown in Fig. 15.5. (Shapiro's original paper [8] contains steps but the later experimental data such as those shown here allow the more delicate interpretation of the ac Josephson effect.) The clearly visible steps are most remarkable. Fig. 15.5 indicates that horizontal current strips decrease in magnitude with increasing n, which will be explained below.

We first note that the dc bias voltage V_0 applied to the junction generates no change in the energy of the moving supercondensate. Thus we may include the effect of this voltage in the unperturbed Hamiltonian H_0 and write the total Hamiltonian H as

$$H = H_0 + 2ev\cos\omega t, \tag{15.34}$$

where the second term represents the perturbation energy due to the microwave. We may write the wavefunction corresponding to the unperturbed system in the form:

$$\Psi_0(\mathbf{r},t) = A(\mathbf{r})\exp(-iEt/\hbar), \tag{15.35}$$

where E is the energy. In the presence of the microwave we may assume a steady-state wavefunction of the form:

$$\Psi(\mathbf{r},t) = A(\mathbf{r})\exp(-iEt/\hbar)[\sum_{n=-\infty}^{\infty} B_n \exp(-in\omega t)]. \tag{15.36}$$

Substituting Eq. (15.36) into the quantum equation of motion $i\hbar\partial\Psi/\partial t = H\Psi$, we obtain (Problem 15.3.3).

$$nB_n = \frac{ev}{\hbar\omega}(B_{n-1} + B_{n+1}). \tag{15.37}$$

The solution of this difference equation can be expressed in terms of the n-th order Bessel function of the first kind [10];

$$B_n = J_n(\alpha), \qquad \alpha \equiv 2ev/\hbar\omega. \tag{15.38}$$

We thus obtain

$$\Psi(\mathbf{r},t) = A\exp(-iEt/\hbar)[\sum_{n=-\infty}^{\infty} J_n(\alpha)\exp(-in\omega t)]. \tag{15.39}$$

This steady-state solution in the presence of a microwave indicates that the condensed pairons (supercondensate) can have energies, E, $E \pm \hbar\omega$, $E \pm 2\hbar\omega$, ... Now from Eq. (15.18) we know that the supercurrent I is proportional to the condensed pairon density n_0. The amplitude of the quasi-wavefunction Ψ is also linear in this density n_0. Hence we deduce from Eq. (15.39) that the (horizontal) lengths of the Shapiro steps at n are proportional to $J_n(\alpha)$ and decrease with increasing n in agreement with experiments.

In the preceding calculation we used the quasi-wavefunction Ψ. If we adopted the G-L wavefunction Ψ' instead, we would have obtained the same solution (15.39), but the physical interpretation is different. We would have concluded that the horizontal lengths of the Shapiro steps would decrease with n in proportion to $J_n^2(\alpha)$. Experiments support the linear J_n-dependence. In other words the quasi-wavefunction Ψ, which is proportional to the supercondensate density n_0 , gives a physically correct description of pairon dynamics.

Problem 15.3.1. Verify Eq. (15.32). Use Taylor's expansion.

Problem 15.3.2. Show that the averaged current I in Eq. (15.32) is finite if Eq. (15.33) is satisfied.

Problem 15.3.3. Verify Eq. (15.37).

15.4 Discussion

15.4.1 Josephson Tunneling

If two superconductors are connected by a Josephson junction, a supercurrent can pass through the junction with no energy loss. This is the Josephson tunneling. The Josephson current is typically very small (mA), and it is very sensitive to an applied magnetic field (mG).

15.4.2 Interference and Analogy with Laser

In a SQUID two supercurrents separated up to 1 mm can exhibit an interference pattern. There is a close analogy between supercurrent and laser. Both are described by the wavefunction $A \exp i(\mathbf{k} \cdot \mathbf{r} - \omega t)$ representing a state of condensed bosons moving with a linear dispersion relation. Such a boson flux has a self-focussing power. A laser beam becomes self-focused after passing a glass plate (disperser); likewise the condensed pairon flux becomes monochromatic after passing a Josephson junction. Thus both laser and supercurrent can interfere at a macroscopic distance. Supercurrents can however carry electric currents. No self-focussing power is known for fermion flux.

15.4.3 G-L Wavefunction, Quasi-Wavefunction, and Pairon Density Operator

The G-L-wavefunction $\Psi'_\sigma(\mathbf{r})$ and the pairon density operator n are related by $\Psi'_\sigma(\mathbf{r}) = \langle \mathbf{r} | n^{1/2} | \sigma \rangle$, where σ represents the condensed pairon state. In the example of a ring supercurrent, we may simply choose $\sigma = p_m = 2\pi\hbar L^{-1} m$, $(m = 0, \pm 1, \pm 2, ...)$. For $m \neq 0$, $\Psi_m(x) = A \exp(-i\hbar^{-1} p_m x)$, $p_m \equiv 2\pi\hbar m/L$ represents a current-carrying state $p = p_m$. This state is *material-independent.* The state is qualitatively the *same for all temperatures* below T_c since there is only one quantum state. Only the density of condensed pairons changes with temperature. The quasi-wavefunction $\Psi_\sigma(\mathbf{r})$ for condensed pairons can be related to the pairon density operator n through $\Psi_\sigma(\mathbf{r}) = \langle \mathbf{r} | n | \sigma \rangle$. This $\Psi_\sigma(\mathbf{r})$ and the G-L wavefunction $\Psi'_\sigma(\mathbf{r})$ are different in the normalization. Both functions (Ψ', Ψ) can represent the state of the supercondensate. The density operator $n(t)$ changes in time, following a quantum Liouville equation, from which it follows that both (Ψ', Ψ) obey the Schrödinger equation of motion (if the repulsive interpairon interaction is neglected).

15.4.4 Josephson-Feynman Equations

The wavelength $\lambda = h/p_n = L/n$ characterizing a supercurrent is much greater than the Josephson junction size (10 Å). Thus the phase of the quasi-wavefunction $\phi(x) = kx = 2\pi\lambda^{-1}x$ measured in units of the junction size is a very slowly varying function of x. We may assume that superconductors right and left of the junction have position-independent phase θ_j. Josephson proposed two basic equations governing the supercurrent running through the junction [1]:

$$I = I_0 \sin\delta, \qquad \delta \equiv \theta_2 - \theta_1$$

$$\hbar \frac{d\delta}{dt} = 2eV.$$

The response of the supercurrent to the bias voltage V differs from that of the normal current (Ohm's law). Supercurrent does not gain energy from a dc bias. This behavior is compatible with our picture of a neutral moving supercondensate.

15.4.5 ac Josephson Effect and Shapiro Steps

On applying a dc voltage V_0, the supercurrent passing through a junction has a component oscillating with the Josephson frequency: $\omega_J \equiv 2e\hbar^{-1}V_0$. This ac Josephson effect was dramatically demonstrated by Shapiro [8]. By applying a microwave of a matching frequency $\omega = n\omega_J$, for $n = \pm 1, \pm 2$, ... step-like currents were observed in the V-I diagrams as in Fig. 15.5. The voltage step is

$$V_0 = (\hbar/2e)\omega_J. \tag{15.40}$$

Since the microwave frequency ω can accurately be measured, and $(\hbar, e)$ are constants, Eq. (15.40) can be used to define a *voltage standard.* The horizontal (current) strips in the V-I diagram decrease in magnitude with increasing n. We found that the quasi-wavefunction Ψ whose amplitude is proportional to the pairon density gives the correct pairon dynamics.

15.4.6 Independent Pairon Picture

In the present treatment of the supercurrent we assumed that $\pm$ bosonic pairons having linear dispersion relations move independently of each other. Thus the analogy between supercurrent and laser is nearly complete except that pairons have charges $\pm 2e$ and hence interact with electromagnetic fields. (Furthermore the pairons can stop with zero momentum while photons run with the speed of light and cannot stop.) The supercurrent interference at macroscopic distances is the most remarkable; it supports a B-E condensation picture of free pairons having linear dispersion relations. The excellent agreement between theory and experiment also supports our starting point, the generalized BCS Hamiltonian.

Chapter 16

Compound Superconductors

Compound superconductors exhibit type II magnetic behaviors, and they tend to have higher critical fields than type I superconductors. Otherwise they show the same superconducting behavior. The Abrikosov vortex lines, each consisting of a quantum flux and circulating supercurrents, are explained in terms of a supercondensate made up of condensed pairons. Exchange of optical and acoustic phonons are responsible for type II behavior.

16.1 Introduction

A great number of compounds superconductors have been discovered by Matthias and his group [1]. A common feature of these superconductors [2] is that they exhibit type-II magnetic behavior, which we explain fully in Section 16.2. Briefly these superconductors show the five main properties: zero resistance below the upper critical field H_{c2}, Meissner state below the lower critical field H_{c1}, flux quantization, Josephson effects, and gaps in the elementary excitation energy spectra. The critical temperatures T_c tend to be higher for compounds than for type I elemental superconductors, the highest T_c ($\sim$ 23 K) being found for Nb_3Ge. The upper critical fields H_{c2} for some compounds including Nb_3Sn can be very high, some reaching $2 \times 10^5\,G \equiv 20\,T$, compared with typical 500 G for type I superconductors. This feature makes compound superconductors very useful in devices and applications. In fact large-scale application and technology are carried out by using type II compound superconductors. Physics of compound superconductors are more complicated because of their compound lattice structures

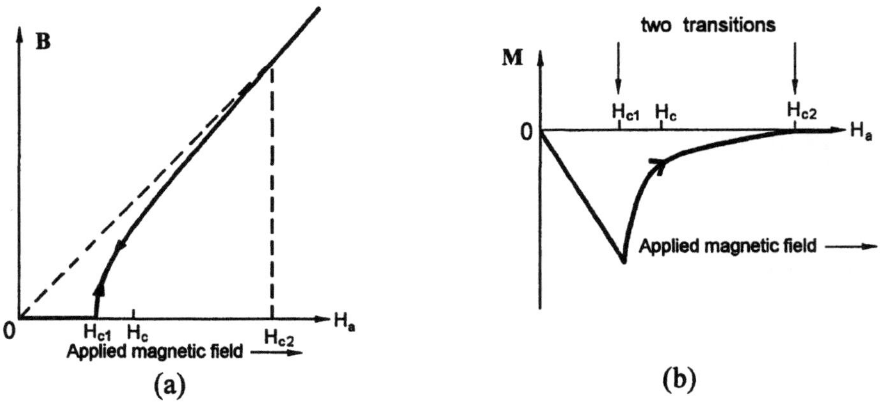

Figure 16.1: Magnetization curves of type II superconductors. Below the lower critical field H_{c1}, type II exhibit the same Meissner state $(B = 0)$ as type I.

and associated electron and phonon band structures. But the superconducting state characterized by the five main properties, is the same for both elemental and compound superconductors. By this reason the microscopic theory can be developed in a unified manner.

16.2 Type II Superconductors

The magnetic properties of type I and type II superconductors are quite different. A type *I* superconductor repels a weak magnetic field from its interior at $T < T_c$, while a type II superconductor allows a partial penetration of the field below T_c. This behavior was shown in Figs. 1.12. Because of the penetration of the magnetic field **B**, the superconducting state is more stable, making the upper critical fields in type II higher. The magnetization M and the B-field versus the H-field for type II are shown in Fig. 16.1. The (dia)magnetization M (< 0) is continuous at the lower critical field H_{c1} and also at the upper critical field H_{c2}. When the applied field H_a is between H_{c1} and H_{c2}, there is a penetration of magnetic flux B, and a complicated structure having both normal and super region are developed within the sample. In this so-called *mixed state*, a set of quantized flux lines penetrate, each flux line being surrounded by diamagnetic supercurrents. See Fig. 16.2. A quantum flux line and the surrounding supercurrents is called a *vortex line.*

The B-field is nonzero within each vortex line, and it is nearly zero excluding where the vortices are. Such a peculiar structure was predicted by Abrikosov in 1957 [3] on the basis of an extension of the G-L theory [4]. This Abrikosov vortex structure was later confirmed by experiments [5] as shown in Fig. 16.3. Each vortex line contains a flux quantum $\Phi_0 = \pi\hbar/e$. Penetration of these flux lines does not destroy the superconducting state if the flux density is not too high. If the applied magnetic field H_a is raised further, it eventually destroys the superconducting state at H_{c2}, where flux densities inside and outside become the same, and the magnetization M vanishes.

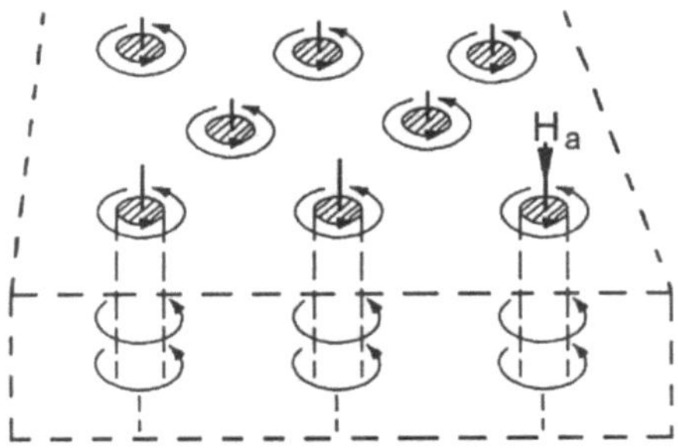

Figure 16.2: Quantized flux lines are surrounded by diamagnetic supercurrents.

Why does such a structure occurs only for a type II superconductor? To describe the actual vortex structure, we need the concept of a coherence length ξ, which was first introduced by Ginzburg and Landau [4]. In fact the distinction between the two types are made by the relative magnitudes

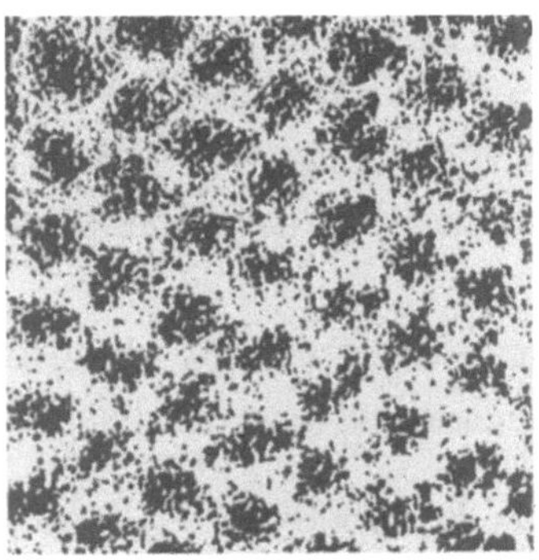

Figure 16.3: Abrikosov structure in Nb.

of the coherence length ξ and the penetration depth λ:

$$\xi > 2^{1/2}\lambda \quad \text{(type I)}, \qquad \xi < 2^{1/2}\lambda \quad \text{(type II)}. \tag{16.1}$$

Both ξ and λ depend on materials, temperature and concentration of impurities. The penetration depth λ at 0 K is about 500 Å in nearly all superconductors. The BCS coherence length $\xi \equiv \hbar v_F / \pi\Delta$ has a wider range: $25 - 10^4$ Å.

We now explain the phenomenon piece by piece.

1 We note the importance of flux quantization. Otherwise there is no vortex line.

2 Any two magnetic fluxes repel each other, as is well-known in electromagnetism; hence each vortex line contains one flux quantum $\Phi_0 = \pi\hbar/e$.

3 Each magnetic flux in a superconductor is surrounded by a supercurrent with no energy loss, forming a vortex line. As a result the B-field practically vanishes outside the vortex line.

4 Because of the supercurrent at the surface within the penetration depth λ, the B-field vanishes everywhere in the background.

5 The supercondensate is composed of condensed pairons, and hence the minimum distance over which the supercondensate density can be defined is the pairon size. Because of the Meissner pressure, each vortex may be compressed to this size. To see this, let us consider a set of four circular disks representing two + (white) and two − (light dotted) pairons, as shown in Fig. 16.4. The four disks rotate around the fixed flux line, keeping the surrounding background stationary. From Fig. 16.4 we see that the vortex line represented by the flux line and the moving disks (pairons) has a radius of the order of the pairon size.

6 Vortices are round, and they weakly repel each other. This generates a hexagonal closed-pack structure as observed, see Fig. 16.3

7 If the applied field H_a is raised to H_{c2}, the number density of vortex lines increases (the dark part in Fig. 16.3 increases), so the magnetic flux density eventually becomes equal inside and outside of the superconductor, when the super part is reduced to zero.

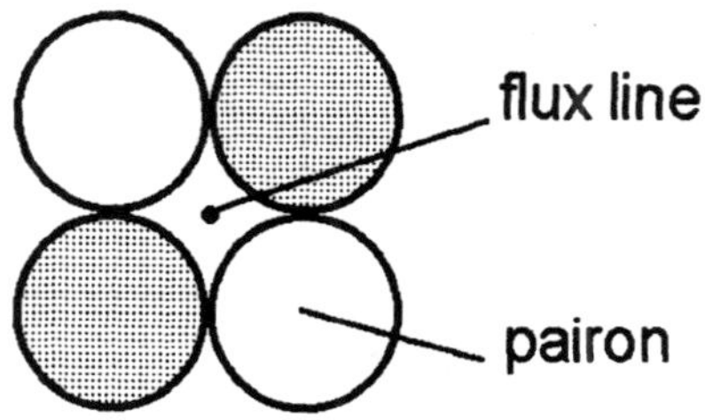

Figure 16.4: Closely-packed circular disks (pairons) rotate around the flux.

8 If the field H is lowered to H_{c1}, the number of vortex lines decreases to zero (the dark part in Fig. 16.3 decreases), and the magnetic fluxes pass through the surface layer only. These fluxes and the circulating supercurrents maintain the Meissner state in the interior of the superconductor. This perfect Meissner state is kept below H_{c1}.

9 The surface region where the supercurrents run, is the same for $H < H_{c1}$. This region remains the same for all fields $0 < H < H_{c2}$, and it is characterized by the penetration depth λ.

10 Creation of vortex lines (normal part) within a superconductor lowers the magnetic energy, but raises the Meissner energy due to the decrease in the super part. This thermodynamic competition generates a phase transition of the first order at H_{c1}, see Fig. 16.5.

11 Penetration of the vortices lightens the magnetic pressure. Hence the superconducting state is much more stable against the applied field, making H_{c2} much greater than the ideal *thermodynamic critical field* H_c defined by

$$\frac{1}{2}\mu_0 H_c^2 \equiv \frac{G_N - G_S}{V}. \tag{16.2}$$

12 Since the magnetization approaches zero near $H = H_{c2}$, [see Fig. 16.1 (b)], the phase transition is of the second order. This is in contrast with type I, where the phase transition at $H = H_c$ is of the first order. Data on heat capacity C in Nb by McConville and Serin [6], shown in Fig. 16.5, indicate the second-order phase transition. The heat capacity at the upper critical temperature T_{c2} has a jump just like the heat capacity at T_c for a type I superconductor at $B = 0$, see Fig. 16.5.

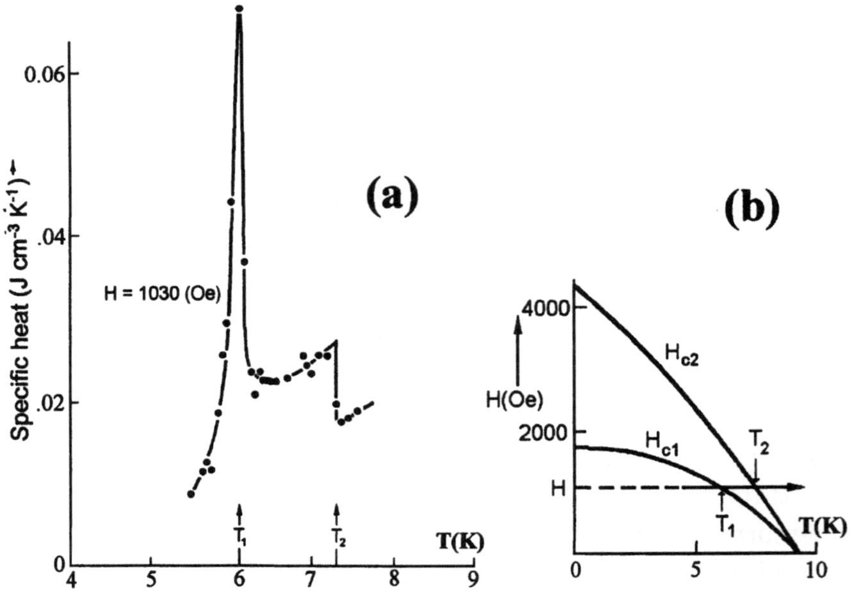

Figure 16.5: (a) Specific heat of type II superconductor (Nb) in a constant applied magnetic field H. (b) The phease diagram, after McConville and Serin [6].

13 Why can a type I superconductor develop no Abrikosov structure? The type I supercondensate is made up of large-size pairons on the order of 10^4 Å . Vortex lines having such a large size would cost too much Meissner energy to compensate the possible gain in the magnetic energy.

In summary a type II superconductor can, and does, develop a set of vortex lines of a radius of the order ξ_0 in its interior. These vortices lower the magnetic energy at the expense of the Meissner energy. They repel each other weakly and form a two-dimensional hexagonal lattice. At Abrikosov's time of work in 1957, flux quantization and Cooper pairs, which are central to the preceding arguments, were not known. The coherence length ξ_0 in the G-L theory is defined as the minimum distance below which the G-L wavefunction Ψ' cannot change appreciably. This ξ_0 is interpreted here as the Cooper pair size. The distinction between type I and type II as represented by Eq. (16.1) is equivalent to different signs of the interface energy between normal and super parts. Based on such nonmicroscopic physical ideas, Abrikosov predicted the

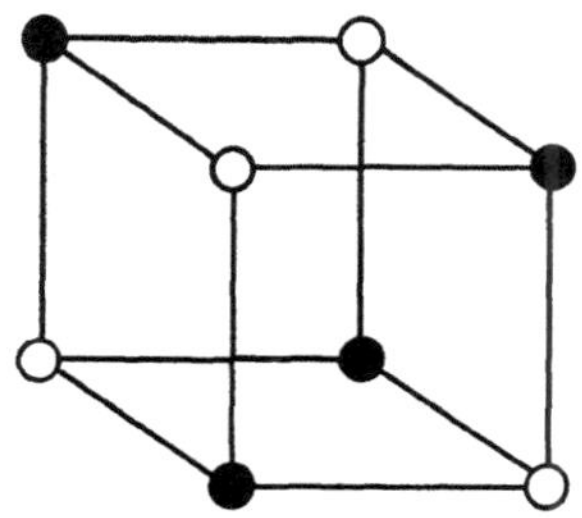

Figure 16.6: In rock salt, Na^+ and Cl^- occupy the simple cubic lattice sites alternately.

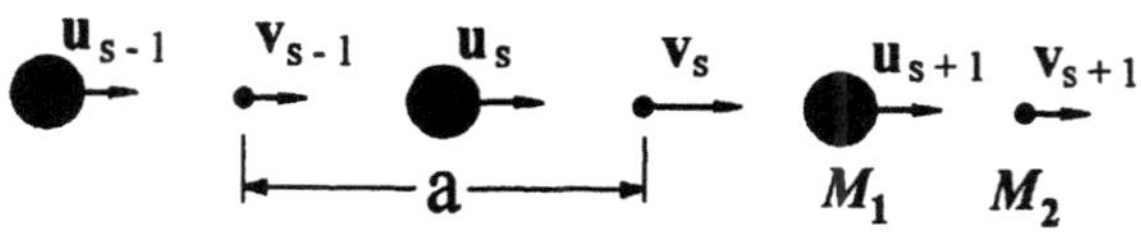

Figure 16.7: A diatomic one-dimensional lattice with masses M_1, M_2 bound by force constant C.

now-famous Abrikosov structure.

16.3 Optical Phonons

A compound crystal has two or more atoms in a unit cell; hence it has optical modes of lattice vibration. We discuss this topic in the present section.

Consider a rock salt, (NaCl) crystal whose lattice structure is shown in Fig. 16.6. In the $\langle 111 \rangle$ directions, planes containing Na ions and planes containing Cl ions alternate with a separation equal to $\sqrt{3}/2$ times the lattice constant. Thus we may imagine a density wave proceeding in this direction. This condition is similar to what we saw earlier in Section 4.2 for the lattice-vibrational modes in a sc crystal. We assume that each plane interact with its nearest neighbor planes, and that the force constants C are the same between any pairs of nearest neighbor planes. We may use a one-dimensional representation as shown in Fig. 16.7. The displacements of atoms with mass M_1 are denoted by u_{s-1}, u_s, $u_{s+1}, ...$, and those of atoms with mass M_2 by

$v_{s-1}, v_s, v_{s+1}, \ldots$ From the figure, we obtain

$$M_1 \frac{d^2 u_s}{dt^2} = C(v_s + v_{s-1} - 2u_s), \qquad M_2 \frac{d^2 v_s}{dt^2} = C(u_{s+1} + u_s - 2v_s). \quad (16.3)$$

We look for a solution in the form of a traveling wave with amplitudes (u, v):

$$u_s = u \exp i(sKa - \omega t), \qquad v_s = v \exp i(sKa - \omega t), \quad (16.4)$$

where a is the lattice distance. Introducing Eqs. (16.4) into Eqs. (16.3), we obtain

$$\begin{aligned} -\omega^2 M_1 u &= C\,v[1 + exp(-iKa)] - 2Cu, \\ -\omega^2 M_2 v &= C\,u[1 + exp(+iKa)] - 2Cv. \end{aligned} \quad (16.5)$$

From the secular equation:

$$\begin{vmatrix} 2C - M_1\omega^2 & -C[1 + \exp(-iKa)] \\ -C[1 + \exp(+iKa)] & 2C - M_2\omega^2 \end{vmatrix} = 0, \quad (16.6)$$

we obtain

$$M_1 M_2\, \omega^4 - 2C(M_1 + M_2)\,\omega^2 + 2C^2(1 - cos\,Ka) = 0, \quad (16.7)$$

which can be solved exactly. The dependence of ω on K is shown in Fig. 16.8 for $M_1 > M_2$. Let us consider two limiting cases: $K = 0$ and $K = K_{\text{max}} = \pi/a$. For small K, we have cos $Ka \simeq 1 - K^2a^2/2$, and the two roots from Eq. (16.7) are (Problem 16.3.1)

$$\omega_1 = \sqrt{\frac{2C(M_1 + M_2)}{M_1 M_2}} \left(1 - \frac{M_1 M_2\, a^2}{8(M_1 + M_2)^2} K^2\right), \quad \text{(optical branch)} \quad (16.8)$$

$$\omega_2 = \sqrt{\frac{C}{2(M_1 + M_2)}}\, Ka \quad \text{(acoustic branch)}. \quad (16.9)$$

At $K = \pi/a$, the roots are (Problem 16.3.2)

$$\omega_1 = \left[\frac{2C}{M_2}\right]^{1/2} + C_1\left(\frac{\pi}{a} - K\right)^2, \quad \text{(optical branch)} \quad (16.10)$$

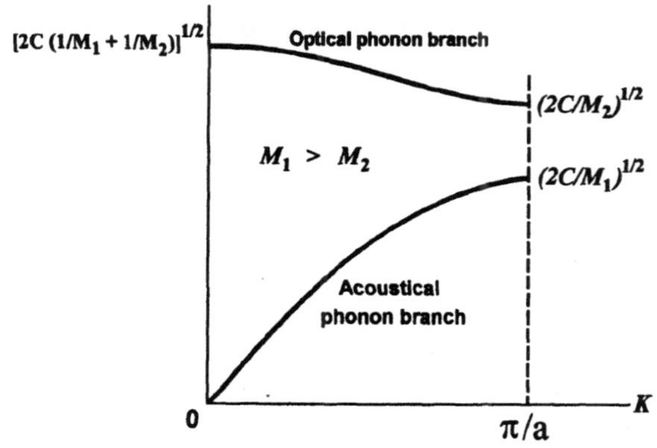

Figure 16.8: Optical and acoustic branches of the dispersion relation. The limiting frequencies at $K = 0$ and $K = \pi/a$ are shown.

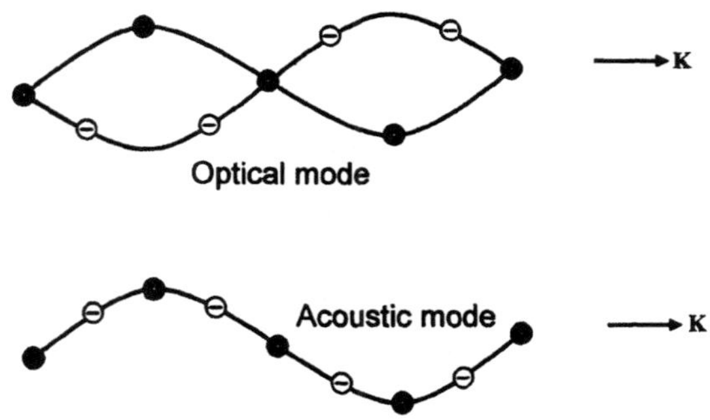

Figure 16.9: TO and TA waves in a linear diatomic lattice; $\lambda = 4a$.

$$\omega_2 = \left[\frac{2C}{M_1}\right]^{1/2} + C_2\left(\frac{\pi}{a} - K\right)^2, \qquad \text{(acoustic branch).} \tag{16.11}$$

where C_1 and C_2 are constants. These limiting cases are indicated in Fig. 16.8. Note: the dispersion relation is linear only for the low-K limit of the acoustic mode. Otherwise the dispersion relations have constants plus quadratic terms.

For a real 3D crystal there are transverse and longitudinal wave modes. The particle displacements in the *transverse acoustic* (TA) and *optical* (TO) modes are shown in Fig. 16.9. The ω-k or dispersion relations can be probed by neutron-scattering experiments, [7] whose results are in good agreement with those of the simple theory discussed here.

Problem 16.3.1. Verify Eq. (16.9).

Problem 16.3.2. Verify Eq. (16.11). Find (C_1, C_2) explicitly.

16.4 Discussion

Compound superconductors show all of the major superconducting properties found in elemental superconductors. The superconducting state is characterized by the presence of a supercondensate, and the superconducting transition is a B-E condensation of pairons. We may assume the same generalized BCS Hamiltonian and derive all the properties based on this Hamiltonian. From its lattice structure, a compound conductor provides a medium in which optical phonons as well as acoustic phonons are created and annihilated. Moreover it is most likely to have two or more sheets of the Fermi surface; one of the sheets is "electron"-like (of a negative curvature) and the other "hole"-like. If other conditions are right, a supercondensate may be formed from "electrons" and "holes" on the *different* Fermi-surface sheets mediated by optical phonons. Pair-creation and pair-annihilation of $\pm$ pairons can be done only by an optical phonon having a momentum (magnitude) greater than $\hbar$ times the minimum k-distance between "electron" and "hole" Fermi-surface sheets. Acoustic phonons of small k-vectors will not do the the intermediary.(See Section 17.1 where a 2D analogue is discussed and demonstrated). Attraction by exchange of optical phonons having a quadratic energy-momentum relation [see Eqs. (16.9) and (16.11)] is short-ranged just as the attraction by the exchange of a massive π-meson is short-ranged as shown by Yukawa [8]. Hence the pairon size should be on the order of the lattice constant (2 Å) or greater. In fact compound superconductors have correlation lengths of the order 50 Å, much shorter than the penetration depths ~500 Å. They are therefore type II superconductors.

Chapter 17

Lattice Structures of Cuprates

Cuprate superconductors have layered structures containing the copper planes (CuO_2). The electric conduction occurs in the copper plane. There are several types of cuprate superconductors.

17.1 Introduction

In 1986 Bednorz and Müller [1] reported the first discovery of the high-T_c cuprate superconductor (La-Ba-Cu-O, $T_c > 30\,\mathrm{K}$). Since then many investigations [2-5] have been carried out on the high-T_c superconductors including Y-Ba-Cu-O with $T_c \approx 94\,\mathrm{K}$ [6]. These compounds possess all of the main superconducting properties: zero resistance, Meissner effect, flux quantization, Josephson effects, and gaps in the excitation energy spectra. This means that there is the same superconducting state in high-T_c as in elemental superconductors. In addition these cuprate superconductors are characterized by [7] 2D conduction, short zero-temperature coherence length ξ_0 ($\sim 10\,\text{Å}$), high critical temperature T_c ($\sim 100\,\mathrm{K}$), type II magnetic behavior d-wave Cooper pair, two energy gaps, unusual transport and magnetization behaviors above T_c, and the dome-shaped doping dependence of T_c. In this chapter we discuss the lattice structures and the 2D conduction.

17.2 Layered Structures and 2-D Conduction

Cuprate superconductors have *layered structures* such that the copper planes (CuO_2), shown in Fig. 17.1, are periodically separated by a great distance

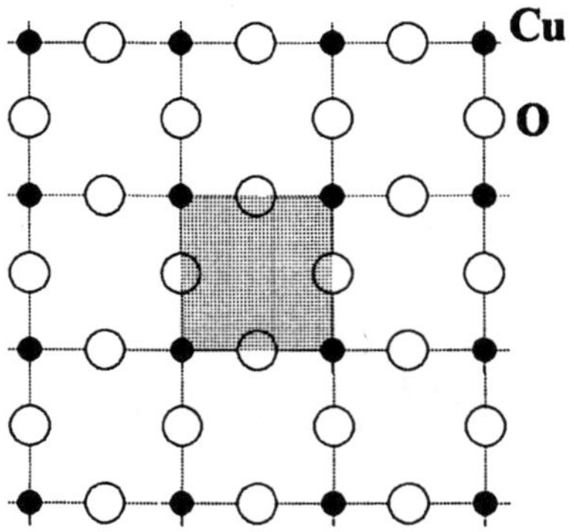

Figure 17.1: Copper plane (CuO_2). The shade area represents a unit cell.

(e.g. $a = 3.88\,\text{Å}$, $b = 3.82\,\text{Å}$, $c = 11.68\,\text{Å}$ for $YBa_2Cu_3O_{7-\delta}$). The lattice structure of YBCO is shown in Fig. 17.2. The succession of layers along the c-axis can be represented by CuO-BaO-CuO_2-Y-CuO_2-BaO-[CuO-... .

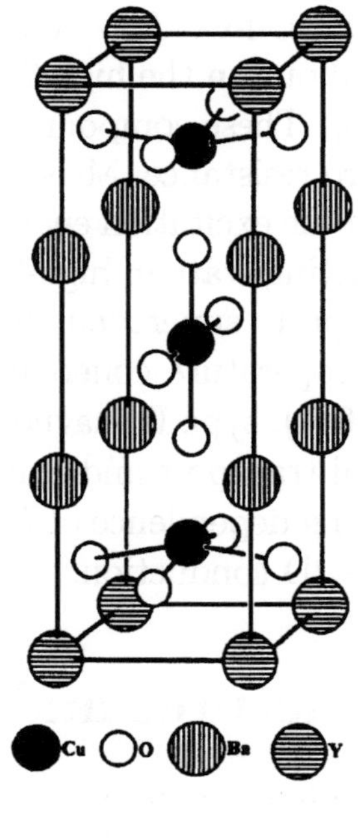

Figure 17.2: Arrangement of atoms in a crystal of $YBa_2Cu_3O_7$; superconducting $YBa_2Cu_3O_{7-\delta}$ has some missing oxygens (δ).

The buckled CuO_2 plane, where Cu-subplane and O-subplane are separated by a short distance is shown. The two copper planes separated by yttrium (Y) are about 3 Å apart, and they are responsible for the conduction.

The conductivity measured is a few orders of magnitude smaller along the c-axis than perpendicular to it [8]. This appears to contradict the prediction based on the naive application of the Bloch theorem. This puzzle may be solved as follows [9]. Suppose an electron jumps from one conducting layer to its neighbor. This generates a change in the charge states of the layers involved. If each layer is macroscopic in dimension, the charge state Q_n of the n-th layer can change without limits: $Q_n = ..., -2, -1, 0, 1, 2, ...$ in units of the electron charge e. Because of unavoidable short circuits between layers due to the lattice imperfections, Q_n may not be large. At any rate if Q_n are distributed *at random* over all layers, the periodicity of the potential for the electron along the c-axis is lost. Then the Bloch theorem based on the electron potential periodicity does not apply even though the lattice is crystallographically periodic. As a result there are no k-vectors along the c-axis. This means that the effective mass in the c-axis direction is infinity, so that the Fermi surface for the layered conductor is a right cylinder with its axis along the c-axis.

The torque-magnetometry experiment by Farrell *et al.* [8] in Tl_2Ba_2-$CaCu_2O_x$ indicates an effective-mass anisotropy of at least 10. Other experiments [10] in thin films and single crystals also indicate a high anisotropy. The most direct way of verifying the 2D structure however is to observe the orientation dependence of the cyclotron resonance (CR) peaks. The peak position (ω) in general follows Shockleys's formula

$$\frac{\omega}{eB} = \left(\frac{m_2 m_3 \cos^2(\mu, x_1) + m_3 m_1 \cos^2(\mu, x_3) + m_1 m_2 \cos^2(\mu, x_3)}{m_1 m_2 m_3} \right)^{1/2}, \tag{17.1}$$

where (m_1, m_2, m_3) are effective masses in the Cartesian axes (x_1, x_2, x_3) taken along the (a, b, c) crystal axes, and $\cos(\mu, x_j)$ is the direction cosine relative to the field $\mathbf{B}$ and the axis x_j. If the electron motion is plane-restricted, so that $m_3 \rightarrow \infty$, Eq. (17.1) is reduced to the *cosine law* formula:

$$\omega = eB(m_1 m_2)^{-1/2} \cos\theta, \tag{17.2}$$

where θ is the angle between the field and the c-axis. A second and much easier way of verifying a 2D conduction is to measure the de Haas-van Alphen

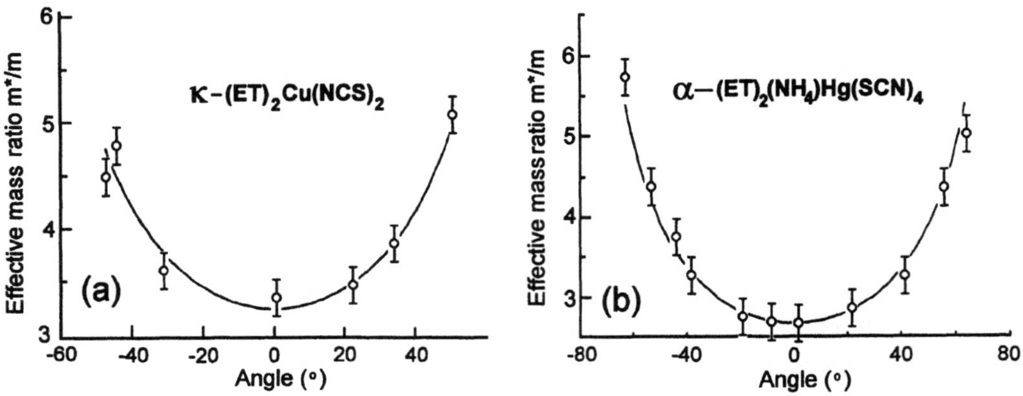

Figure 17.3: Angular dependence of the reduced effective mass in (a) κ-$(ET)_2Cu(NCS)_2$ and (b) α-$(ET_2)(NH_4)Hg(SCN)_4$. An angle of $0°$ means H is perpendicular to the conducting plane. The solid fits are obtained using Eqs. (17.2) and (17.3). After Wosnitza et al. [12].

(dHvA) oscillations and analyze the orientation dependence of the dHvA frequency with the help of Onsager's formula: [11],

$$\Delta \left[\frac{1}{B}\right] = \frac{2\pi e}{\hbar}\frac{1}{A}, \tag{17.3}$$

where A is the extremum intersectional area of the Fermi surface and the planes normal to the applied magnetic field **B**. Recently Wosnitza *et al.* [12] reported the first direct observation of the orientation dependence ($\cos\theta$ law) of the dHvA oscillations in κ-$(ET)_2Cu(NCS)_2$ and α-$(ET)_2(NH_4)Hg(SCN)_4$, both layered organic superconductors, confirming a right cylindrical Fermi surface. Their data and theoretical curves are shown in Fig. 17.3. Notice the excellent agreement between theory and experiment. Measurements of orientation-dependent magnetic or magneto-optical effects in high-T_c superconductors are highly desirable, since no transport measurements alone can give a conclusive test for a 2D conduction because of unavoidable short circuits between layers.

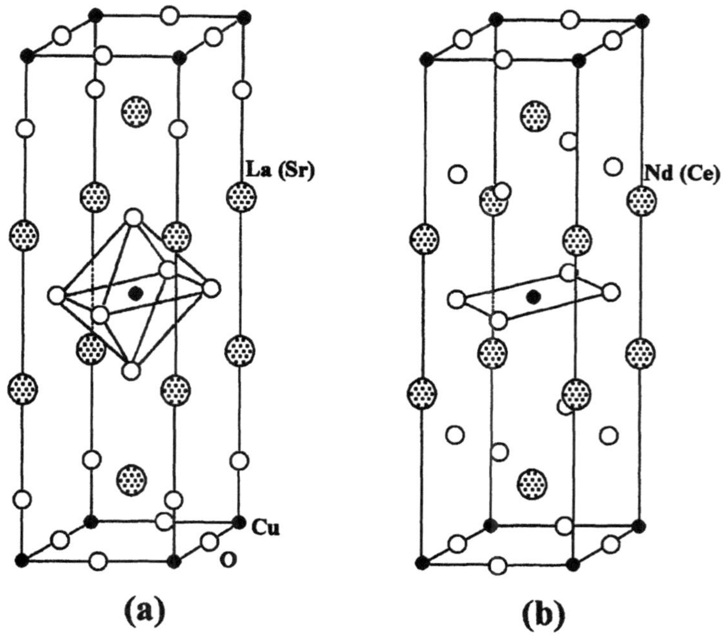

Figure 17.4: (a) $La_{2-x}Sr_xCuO_4$. (b) $Nd_{2-x}Ce_xCuO_{4-\delta}$. Small black • (large white ◦) balls represent Cu (O).

17.3 Selected Cuprate Superconductors

A number of cuprate superconductors have been discovered. Some important ones are discussed here.

(a) La_2CuO_4 and $La_{2-x}Sr_xCuO_4$.

La_2CuO_4 (parent) and $La_{2-x}Sr_xCuO_4$ (daughters) are a simplest cuprate (superconductor) system, whose lattice structure is shown in Fig. 17.4 (a). Each Cu (•) is surrounded by six O's (◦) as shown. This structure is called the perovskite (K_2NiF_4) structure. The parent La_2CuO_4 is an antiferromagnetic insulator with Néel temperature $T_a = 270$ K. The crystal as a whole must be electrically neutral. We assume that Cu (O) in the copper plane are ionized in $+2e$ $(-2e)$. Hence this plane, having twice as many O's as Cu's, has ionicity per unit cell $-2e$. The plane is sandwiched by the positively ionized blocks La_2O_2 with the ionicity of $+2e$, assuming that the trivalent La is ionized in $+3e$. The doped divalent strontium (Sr) replaces trivalent lanthanum (La). The doping takes away electrons from the copper plane through the perovskite structure and generates "holes", making a p-type conductor. The daughters $La_{2-x}Sr_xCuO_4$ are superconductors in the concentration (x) range ($0.06 < x < 0.26$). The highest T_c is 43 K at $x = 0.15$. Although T_c is not very high, this system is the simplest cuprate and has been studied extensively.

(b) $Nd_2CuO_{4-\delta}$ and $Nd_{2-x}Ce_xCuO_{4-\delta}$.

The lattice structure of Nd_2CuO_4 is shown in Fig. 17.4 (b). Each Cu is surrounded by four O's. The doped quadrivalent cerium (Ce) replaces trivalent neodymium (Nd). The parent Nd_2CuO_4 is an antiferromagnetic insulator with Néel temperature $T_a = 220$ K. The doping introduce electrons in the copper plane, making an n-type conductor. The daughters $Nd_{2-x}Ce_xCuO_{4-\delta}$ with small oxygen defect (δ) are superconductors in the range ($0.13 < x < 0.18$). The highest T_c is 42 K. The dominant carrier in the supercurrent is $-$ pairon while that in most other cuprates is $+$ pairon.

(c) $YBa_2Cu_3O_{7-\delta}$ (YBCO or Y system).

The Y in $YBa_2Cu_3O_{7-\delta}$ can be replaced by La, Nd, Sm, Eu, Gd, Dy, Ho, Er, Tm, Yb, Lu. This system with optimum O-concentration δ can attain a T_c ($\sim$ 90 K) higher than the liquid nitrogen temperature (77 K). The materials in the Y system show good superconducting properties and have been studied extensively.

The lattice structure of $YBa_2Cu_3O_7$ ($\delta = 0$) is shown in Fig. 17.5, where O's are represented by networks and others (L, Ba, Cu) by balls. There are two copper planes per unit cell separated by a Y-plane, which has no O's. Each Cu in the copper plane therefore is surrounded by five O's at the pyramidal sites. Along the c-axis Y and Ba form a series: Ba, Y, Ba, Ba, Y, Ba, ... The lattice contains Cu's besides those in the copper planes. Those Cu's are in the CuO plane sandwiched between the Ba-planes. They form a linear chain Cu, O, Cu, ... along the b-axis, see Fig. 17.2. The O's in the CuO plane are liable to dissociation by means of the oxygen partial pressure and also by changing temperatures. The excess oxygen content δ can be varied from 0 to 1. This content δ affects effectively the "hole" concentration in the copper plane.

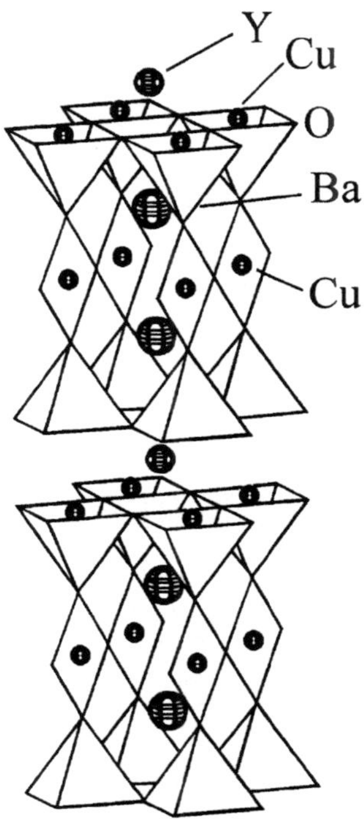

Figure 17.5: Lattice structure of $YBa_2Cu_3O_7$. The lines represent O-networks.

(d) $Bi_2Sr_2Ca_{n-1}Cu_nO_{4+2n+\delta}$, $Tl_2Ba_2Ca_{n-1}Cu_nO_{4+2n+\delta}$

The n in these systems can be changed: $n = 1, 2, 3, \ldots$. The lattice structures of $Bi_2Sr_2Ca_{n-1}Cu_nO_{4+2n}$ ($n = 1, 2, 3$), which are often denoted by Bi-2201, Bi-2212, Bi-2223, are shown in Fig. 17.6.

Special features are that there are Bi_2O_2 blocks. Note: the number of the copper planes changes with n. The three phases (a), (b) and (c) have $T_c =$ 20, 80, and 110 K, respectively. This feature of the variable n is important in basic theoretical studies and applications. Without excess oxygens ($\delta =$ 0) parent materials are antiferromagnetic insulators. Excess oxygens are thought to enter the Bi_2O_2 blocks and supply "holes" in the copper plane.

$Tl_2Ba_2Ca_{n-1}Cu_nO_{4+2n+\delta}$ have similar lattice structures and properties as $Bi_2Sr_2Ca_{n-1}Cu_nO_{4+2n+\delta}$.

(e) $HgBa_2Ca_{n-1}Cu_nO_{2+2n+\delta}$ (Hg system)

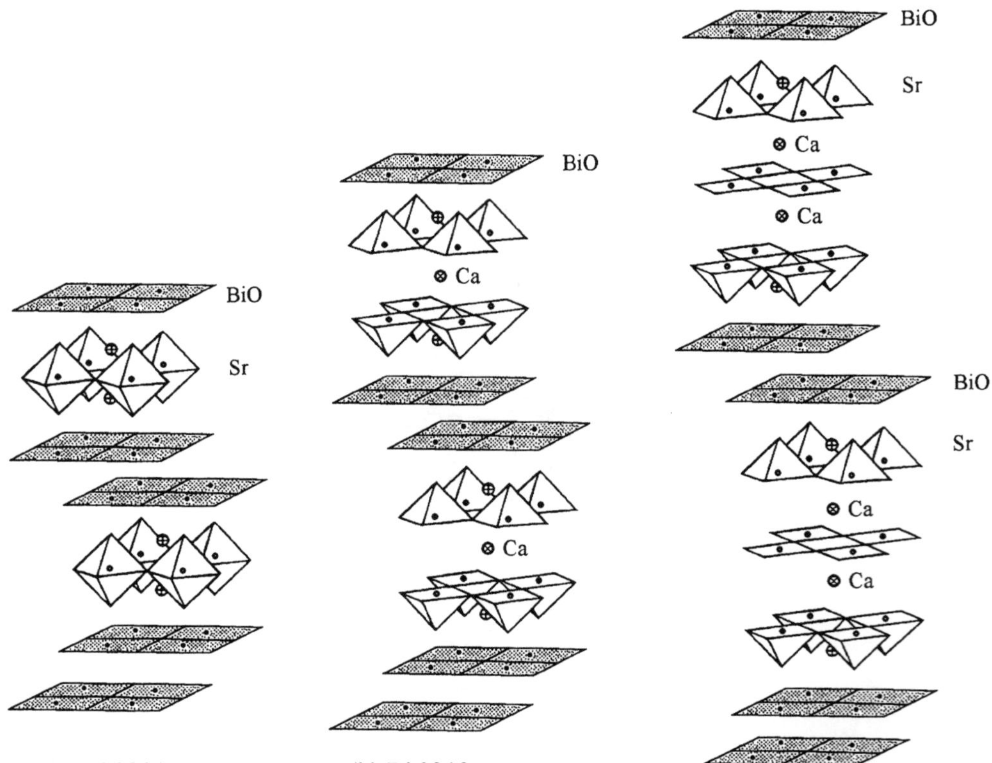

(a) Bi 2201 (b) Bi 2212 (c) Bi 2223

Figure 17.6: Lattice structures of $Bi_2Sr_2Ca_{n-1}Cu_nO_{4+2n}$ ($n = 1$, 2, 3).

The Mercury (Hg) system is similar to the Bi system discussed in (d) except that Bi_2O_2 blocks are replaced by HgO. The lattice structure of $HgBa_2CuO_4$ ($n = 1$, $\delta = 0$) is shown in Fig. 17.7. $HgBa_2Ca_2Cu_3O_{8+\delta}$ ($n = 3$) under high pressure has $T_c = 164$ K, the highest recorded [13], and has been studied widely. Unfortunately the Hg system is chemically active, and single crystals are very difficult to make. Each Cu in the copper plane is surrounded by four O's at the planar sites, which is thought to be the most favorable for high T_c.

$TlBa_2Ca_{n-1}Cu_nO_{3+2n+\delta}$ has similar lattice structures and properties.

(f) Other cuprates and Sr_2RuO_4.

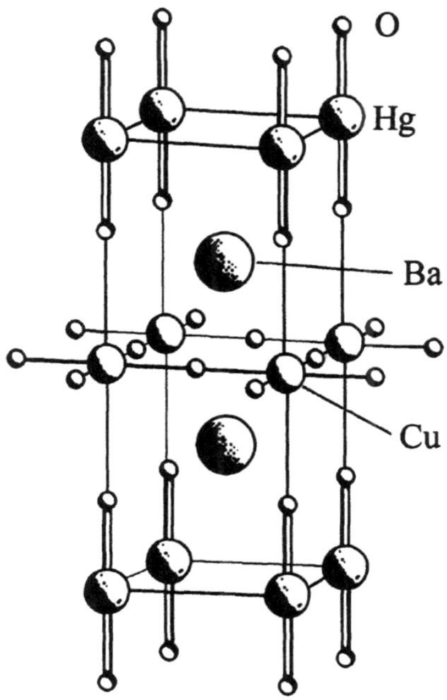

Figure 17.7: The lattice structure of $HgBa_2CuO_4$.

Other types of cuprate superconductors have been found. Common to all cuprates are the existence of the copper planes as the name indicates. In all of them the electrical conduction occurs in the copper planes, and hence they are 2D superconductors.

Are copper ions important in 2D superconductivity? The answer to this question was found negative. Maeno *et al.* [13] discovered that Sr_2RuO_4, which has the same lattice structure as La_2CuO_4, see Fig. 17.4 (a), but without Cu, is superconducting at $T_c = 0.93$ K. Although T_c is low, this case indicates that the conducting character of the CuO_2 plane plays the essential role for superconductivity rather than the actual composition.

Chapter 18

High-T$_C$ Superconductors Below T$_C$

High-T$_c$ superconductors exhibit five main superconducting properties below T_c: zero resistance, Meissner effect, flux quantization, Josephson effects, and gaps in the excitation energy spectra. In addition they are characterized by a 2D conduction, a short coherence length ξ_0 ($\sim 10\,\text{Å}$), a high T_c ($\sim 100\,\text{K}$), type II magnetic behavior two energy gaps, and *d*-wave Cooper pair wave function. These properties are discussed based on the generalized BCS model with 2D band structures, optical phonon exchange attraction, and different interaction strengths V_{ij}.

18.1 The Hamiltonian

Since the electric current flows smoothly in the copper planes, there are continuous k-vectors and the Fermi energy ε_F. Many experiments [1-3] indicate that singlet pairs with antiparallel spins (pairons) form a supercondensate whose motion generates supercurrent.

Let us examine the cause of the electron pairing. We first consider the attraction via the longitudinal acoustic phonon exchange. Acoustic phonons of lowest energies have the linear dispersion relation:

$$\varepsilon = c_s \hbar K, \qquad K \equiv 2\pi/\lambda, \tag{18.1}$$

where c_s is the sound speed. The attraction generated by the exchange of longitudinal acoustic phonons is long-ranged. This mechanism is good for a type I superconductor. This attraction is in action also for a high-T_c superconductor, but it alone is unlikely to account for the much smaller pairon size.

Second, we consider the optical phonon exchange. Each copper plane has Cu and O, and 2D lattice vibrations of optical modes are important. Optical phonons of lowest energies have short wavelengths, and they have a quadratic dispersion relation:

$$\varepsilon = \varepsilon_0 + A_1(K_1 - \frac{\pi}{a_1})^2 + A_2(K_2 - \frac{\pi}{a_2})^2, \tag{18.2}$$

where ε_0, A_1, and A_2 are constants. The attraction generated by the exchange of a massive boson is short-ranged just as the short-ranged nuclear force between two nucleons generated by the exchange of massive pions. Lattice constants for YBCO (a_1, a_2) are (3.88, 3.82) Å, and the limit wavelengths $(\lambda_{\rm min})$ at the Brillouin boundary are twice these values. The observed coherence length ξ_0 has the same order of magnitude as $\lambda_{\rm min}$:

$$\xi_0 \sim \lambda_{\rm min} \cong 8 \text{ Å}. \tag{18.3}$$

Thus the electron-optical-phonon interaction is a viable candidate for the cause of the electron pairing. [4]

To see this in more detail, let us consider the copper plane. With the neglect of small difference in the lattice constants along the a- and b-axes, Cu atoms form a square lattice of a lattice constant $a_0 = 3.85$ Å, as shown in Fig. 18.1 (a). Oxygen atoms (O) occupy mid-points of the nearest neighbors (Cu, Cu) in the plane. The unit cell (dotted area) is located at the center. Observe that Cu's line up in the [110] and $[1\bar{1}0]$ directions with a period $\sqrt{2}a_0$ while O's line up in [100] and [010] with the lattice constant a_0. The first Brillouin zone is shown in Fig. 18.1 (b). The compound copper plane is likely to contain "electrons" and "holes". In equilibrium the "electrons", each having charge $-e$, are uniformly distributed over the sublattice of *positive* ions Cu^{+2} due to Bloch's theorem. If the number density of "electrons" is small, the Fermi surface should then be a small circle as shown in the central part in (b). The "holes", having charge $+e$, are uniformly distributed over the sublattice of the *negative* ions O^{-2}. If the number of "holes" is small, the Fermi surface should consists of the four small pockets shown in (b). The Fermi surface constructed here will play a very important role in our microscopic theory. We shall examine it from a different angle.

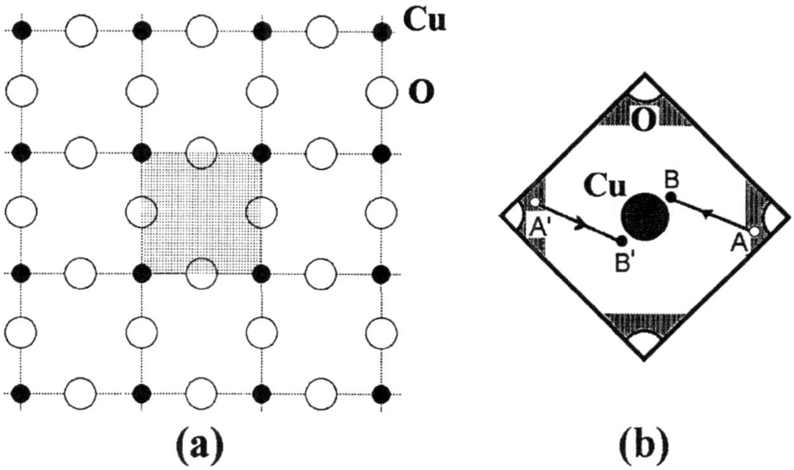

Figure 18.1: (a) Copper plane. (b) The Fermi surface of a cuprate model has a small circle ("electrons") at the center and a set of four small pockets ("holes") at the Brillouin boundary. Exchange of a phonon can create the $-$ pairon at (B, B') and the $+$pairon at (A, A'). The phonon must have a momentum equal to $\hbar$ times the k-distance AB, which is greater than the minimum k-distance between the"electron" circle and the "hole" pockets.

First let us look at the motion of an "electron" wave packet that extends over a unit cell. This wave packet ("electron") may move easily in [110] or $[1\bar{1}0]$ because the O-sublattice charged uniformly and weakly favors the motion over the possible motion in [100] and [010]. In other words the easy axes of motion for the "electron" are the second-nearest (Cu-Cu) neighbor directions [110] and $[1\bar{1}0]$ rather than the first neighbor directions [100] and [110]. The Bloch wave packets are superposable; hence the "electron" can move in any direction characterized by the 2D k-vectors with bases taken along [110] and $[1\bar{1}0]$. Second we consider a "hole" wave packet which extends over a unit cell. It may move easily in [100] or [010] because the Cu-sublattice of a uniform charge distribution favors such a motion.

Under the assumption of the Fermi surfaceshown, pair-creation of $\pm$ pairons by the exchange of an optical phonon may occur as indicated in (b). Here a single-phonon exchange generates an electron transition from A in the O-Fermi sheet to B in the Cu-Fermi sheet and another electron transition from A' to B', creating the $-$pairon at (B, B') and the +pairon at (A, A'). From momentum conservation the momentum (magnitude) of a phonon must be equal to $\hbar$ times the k-distance AB, which is approximately equal to the momentum of an optical phonon of the smallest energy. Thus the Fermi surface comprising a small "electron" circle and small "hole" pockets are quite favorable for forming a supercondensate by exchanging an optical phonon. Note: the pairon formation via optical phonon exchange is anisotropic, yielding the d-wave Cooper pairs, which will be discussed in Chapter 25.

Generally speaking any and every possible cause for the electron pairing, including spin-dependent one must be enumerated, and its importance be evaluated. However if the net interaction between two electrons due to all causes is attractive, pairons will be formed. Then a BCS-like Hamiltonian can be postulated generically irrespective of specific causes. Because of this nature of theory, we may set up a generalized BCS Hamiltonian as follows. [5] We assume that:

1 The conduction electrons move in 2D.

2 There exists a well-defined Fermi energy ε_F for the normal state.

3 There are "electrons" and "holes" with different effective masses:

$$m_1 \neq m_2. \tag{18.4}$$

4 The electron-phonon attraction generates pairons near the Fermi surface within a distance (energy) $\varepsilon_c = \hbar\omega_D$.

5 The interaction strengths v_{ij} satisfy

$$v_{11} = v_{22} < v_{12} = v_{21}, \tag{18.5}$$

since the Coulomb repulsion between two electrons separated by 10 Å is not negligible due to the incomplete screening.

Under these conditions we write down a generalized BCS Hamiltonian:

$$\begin{aligned} H &= \sum_{\mathbf{k},s} \varepsilon_k^{(1)} n_{\mathbf{k},s}^{(1)} + \sum_{\mathbf{k},s} \varepsilon_k^{(2)} n_{\mathbf{k},s}^{(2)} \\ &\quad - \sum_{\mathbf{k}}{}' \sum_{\mathbf{k}'}{}' [v_{11} b_{\mathbf{k}'}^{(1)\dagger} b_{\mathbf{k}}^{(1)} + v_{12} b_{\mathbf{k}'}^{(2)\dagger} b_{\mathbf{k}}^{(1)\dagger} + v_{21} b_{\mathbf{k}}^{(2)} b_{\mathbf{k}'}^{(1)} + v_{22} b_{\mathbf{k}}^{(2)} b_{\mathbf{k}'}^{(2)\dagger}] \\ &\quad - \sum_{\mathbf{k}}{}' \sum_{\mathbf{q}}{}' \sum_{\mathbf{k}'}{}' [v_{11} B_{\mathbf{kq}}^{(1)\dagger} B_{\mathbf{k}'\mathbf{q}}^{(1)} + v_{12} B_{\mathbf{kq}}^{(1)\dagger} B_{\mathbf{k}'\mathbf{q}}^{(2)\dagger} + v_{21} B_{\mathbf{kq}}^{(2)} B_{\mathbf{k}'\mathbf{q}}^{(1)} + v_{22} B_{\mathbf{kq}}^{(2)} B_{\mathbf{k}'\mathbf{q}}^{(2)\dagger}]. \end{aligned} \tag{18.6}$$

Assumption 3 will be explained next section. All other assumptions are essentially the same as those for elemental superconductors, and detailed explanations were given in Section 7.2. In summary we assume the same generalized BCS Hamiltonian for cuprates. Only 2D electron motion, optical-phonon exchange attraction and inequalities (18.5) are newly introduced.

18.2 The Ground State

In section 7.2 we studied the ground state of the generalized BCS system. We can extend our theory to the cuprate model straightforwardly. We simply summarize methods and results. At 0 K there are only *stationary* pairons described in terms of $(b, b^\dagger)$. The ground state Ψ for the system can be described by the *reduced* Hamiltonian per unit volume given by

$$\begin{aligned} H_0 &= \sum_{\mathbf{k}}\sum_{s} \varepsilon_k^{(1)} n_{\mathbf{k}s}^{(1)} + \sum_{\mathbf{k}}\sum_{s} \varepsilon_k^{(2)} n_{\mathbf{k}s}^{(2)} \\ &\quad -{\sum_{\mathbf{k}}}'{\sum_{\mathbf{k}'}}' [v_{11} b_{\mathbf{k}}^{(1)\dagger} b_{\mathbf{k}'}^{(1)} + v_{12} b_{\mathbf{k}}^{(1)\dagger} b_{\mathbf{k}'}^{(2)\dagger} + v_{21} b_{\mathbf{k}}^{(2)} b_{\mathbf{k}'}^{(1)} + v_{22} b_{\mathbf{k}}^{(2)} b_{\mathbf{k}'}^{(2)\dagger}]. \end{aligned} \tag{18.7}$$

Following BCS [7], we assume the normalized ground-state ket

$$|\Psi\rangle \equiv {\prod_{\mathbf{k}}}' (u_{\mathbf{k}}^{(1)} + v_{\mathbf{k}}^{(1)} b_{\mathbf{k}}^{(1)\dagger}) {\prod_{\mathbf{k}'}}' (u_{\mathbf{k}'}^{(2)} + v_{\mathbf{k}'}^{(2)} b_{\mathbf{k}'}^{(2)\dagger}) \,|0\rangle, \tag{18.8}$$

where u's and v's are probability amplitudes satisfying

$$u_{\mathbf{k}}^{(j)2} + v_{\mathbf{k}}^{(j)2} = 1. \tag{18.9}$$

We now determine u's and v's such that ground state energy

$$\begin{aligned} W &\equiv \langle \Psi \,| H_0 |\, \Psi \rangle = {\sum_{\mathbf{k}}}' 2\varepsilon_k^{(1)} v_{\mathbf{k}}^{(1)2} + {\sum_{\mathbf{k}}}' 2\varepsilon_{k'}^{(2)} v_{\mathbf{k}'}^{(2)2} \\ &\quad -{\sum_{\mathbf{k}}}'{\sum_{\mathbf{k}'}}' \sum_{i}\sum_{j} v_{ij}\, u_{\mathbf{k}}^{(i)} v_{\mathbf{k}}^{(i)}\, u_{\mathbf{k}'}^{(j)} v_{\mathbf{k}'}^{(j)} \end{aligned} \tag{18.10}$$

have a minimum value. After variational calculations, we obtain

$$2\varepsilon_k^{(j)} u_{\mathbf{k}}^{(j)} v_{\mathbf{k}}^{(j)} - (u_{\mathbf{k}}^{(j)2} - v_{\mathbf{k}}^{(j)2}) {\sum_{\mathbf{k}'}}' [v_{j1} u_{\mathbf{k}'}^{(1)} v_{\mathbf{k}'}^{(1)} + v_{j2} u_{\mathbf{k}'}^{(2)} v_{\mathbf{k}'}^{(2)}] = 0. \tag{18.11}$$

To simply treat these equations subject to Eqs. (18.9), we introduce a set of energy-parameters:

$$\Delta_{\mathbf{k}}^{(j)}, \qquad E_k^{(j)} \equiv \left[\varepsilon_k^{(j)2} + \Delta_{\mathbf{k}}^{(j)2}\right]^{1/2} \tag{18.12}$$

such that

$$u_{\mathbf{k}}^{(j)2} - v_{\mathbf{k}}^{(j)2} = \frac{\varepsilon_k^{(j)}}{E_k^{(j)}}, \qquad u_{\mathbf{k}}^{(j)} v_{\mathbf{k}}^{(j)} = \frac{\Delta_{\mathbf{k}}^{(j)}}{2E_k^{(j)}}. \tag{18.13}$$

Then, Eq. (18.11) can be re-expressed as

$$\Delta_j \equiv \Delta_{\mathbf{k}}^{(j)} = \sum_{\mathbf{k}'}{}' \sum_i v_{ij} \frac{\Delta_i}{2E_{\mathbf{k}'}^{(i)}}, \tag{18.14}$$

which are the generalized energy gap equations. These equations can alternatively be obtained by the equation-of-motion method.

Using Eqs. (18.13) and (18.14), we calculate the ground-state energy W and obtain [Eq. (7.28)]

$$\begin{aligned} W &\equiv \sum_{\mathbf{k}}{}' \sum_j 2\varepsilon_k^{(j)} v_{\mathbf{k}}^{(j)2} - \sum_{\mathbf{k}}{}' \sum_{\mathbf{k}'}{}' \sum_i \sum_j v_{ij} u_{\mathbf{k}}^{(i)} v_{\mathbf{k}}^{(i)} u_{\mathbf{k}'}^{(j)} v_{\mathbf{k}'}^{(j)} \\ &= \sum_{\mathbf{k}}{}' \sum_{j=1}^{2} \left\{ \varepsilon_k^{(j)} \left[1 - \frac{\varepsilon_k^{(j)}}{E_k^{(j)}}\right] - \frac{\Delta_j^2}{2E_k^{(j)}} \right\}. \end{aligned} \tag{18.15}$$

The ground-state ket $|\Psi\rangle$ in Eq. (18.8) is a superposition of many-pairon states. Each component state can be reached from the physical vacuum state $|0\rangle$ by pair creation and/or pair annihilation of $\pm$ pairons and by pair stabilization via a succession of phonon exchanges. Since phonon exchange processes are charge-conserving, the supercondensate is composed of equal numbers of $\pm$ pairons. In the bulk limit we obtain

$$\begin{aligned} W &= \sum_{j=1}^{2} \mathcal{N}(0) \int_0^{\hbar\omega_D} d\varepsilon \left[\varepsilon - \frac{\varepsilon^2}{(\varepsilon^2 + \Delta_j^2)^{1/2}} - \frac{\Delta_j^2}{2(\varepsilon^2 + \Delta_j^2)^{1/2}}\right] \\ &\equiv \frac{1}{2} N_0 (w_1 + w_2), \end{aligned} \tag{18.16}$$

$$w_i \equiv \hbar\omega_D \{1 - \left[1 + (\Delta_i/\hbar\omega_D)^2\right]^{1/2}\} \; (< 0), \tag{18.17}$$

$$N_0 \equiv \hbar\omega_D \mathcal{N}(0). \tag{18.18}$$

The binding energies $|w_i|$ for $\pm$ pairons are different. To proceed further we must find Δ_j from the gap equations (18.14). In the bulk limit, these equations are simplified to

$$\begin{aligned}\Delta_j &= \frac{1}{2}v_{j1}\mathcal{N}(0)\int_0^{\hbar\omega_D} d\varepsilon \frac{\Delta_1}{(\varepsilon^2+\Delta_1^2)^{1/2}} + \frac{1}{2}v_{j2}\mathcal{N}(0)\int_0^{\hbar\omega_D} d\varepsilon \frac{\Delta_2}{(\varepsilon^2+\Delta_2^2)^{1/2}} \\ &= \frac{1}{2}v_{j1}\mathcal{N}(0)\Delta_1 \sinh^{-1}(\hbar\omega_D/\Delta_1) + \frac{1}{2}v_{j2}\mathcal{N}(0)\Delta_2 \sinh^{-1}(\hbar\omega_D/\Delta_2),\end{aligned} \tag{18.19}$$

whose solutions will be discussed in Section 18.5.

18.3 High Critical Temperature

In a cuprate superconductor, pairons move in the copper plane with the linear dispersion relation:

$$\varepsilon = (2/\pi)v_F p = cp. \tag{18.20}$$

Earlier in Section 9.1, we saw that free bosons moving in 2D with the dispersion relation $\varepsilon = cp$ undergoes a B-E condensation at [Eq. (9.11)] [6]

$$T_c = 1.954\,\hbar c n^{1/2} k_B^{-1}. \tag{18.21}$$

After setting $c = (2/\pi)v_F$, we obtain

$$k_B T_c = 1.24\,\hbar v_F n_0^{1/2} = 1.24\,\hbar v_F r_0^{-1}, \tag{18.22}$$

where n_0 represents the number density of pairons in the superconductor and $r_0 \equiv n_0^{-1/2}$ is the average interpairon distance.

Let us compare our results with the case of elemental (type I) superconductors. The critical temperature T_c for 3D superconductors is

$$k_B T_c = 1.01\,\hbar v_F n_0^{1/3} = 1.01\,\hbar v_F r_0^{-1}, \tag{18.23}$$

[Eq. (9.38)]. The similarity between Eqs. (18.22) and (18.23) is remarkable; in particular T_c depends on $(v_F,\ r_0)$ nearly in the same way. Now the interpairon distance r_0 is different by the factor $10^2 \sim 10^3$ between type I and cuprate. The Fermi velocity v_F is different by the factor $10 \sim 10^2$. Hence the *high* critical temperature is explained by the very short interpairon distance, partially compensated by a smaller Fermi velocity.

The critical temperature T_c is much lower than the Fermi temperature T_F. The ratio T_c/T_F computed from Eq. (18.22) is

$$k_B T_F = \frac{1}{2} m^* v_F^2 = \frac{\hbar^2 (2\pi n_{el})}{2m^*} = \frac{\pi \hbar^2}{m^* R_0^2}, \tag{18.24}$$

yielding

$$\frac{T_c}{T_F} = 0.99 \frac{R_0}{r_0}, \tag{18.25}$$

which represents the law of corresponding states for T_c in 2D superconductors. Note: Eq. (18.25) is remarkably close to the 3D formula (9.48).

If we assume a 2D Cooper system with a circular Fermi surface, we can calculate the ratio R_0/r_0 and obtain

$$\frac{R_0}{r_0} = \left(\frac{\Theta_D}{2T_F} \right)^{1/2}. \tag{18.26}$$

Introducing the pairon formation factor α, (see Section 9.4), we rewrite Eq. (18.25) as

$$T_c = 0.70\, \alpha (\Theta_D T_F)^{1/2}, \tag{18.27}$$

which indicates that T_c is high if T_F and Θ_D are both high. But the power laws are different compared with the corresponding 3D formula (9.54). The observed pairon formation factor α for cuprates is in the range $\sim 10^{-2}$, much greater than that for elemental superconductors. This is reasonable since the 2D Fermi surface has an intrinsically more favorable symmetry for the pairon formation.

We saw earlier that the interpairon distance r_0 in 3D is several times greater than the coherence length $\xi_0 = \hbar v_F / \pi \Delta$. [7] For 2D, we obtain from Eq. (18.22)

$$r_0 = 6.89\, \xi_0. \tag{18.28}$$

Thus the 2D pairons do not overlap in space. Hence the superconducting temperature T_c can be calculated based on the free-moving pairons. Experiments indicate that $\xi_0 = 14\,\text{Å}$ and $T_c = 94\,\text{K}$ for YBCO. Using these values, we estimate the value of the Fermi velocity v_F from Eq. (18.22):

$$v_F = 10^5\,\text{ms}^{-1}, \tag{18.29}$$

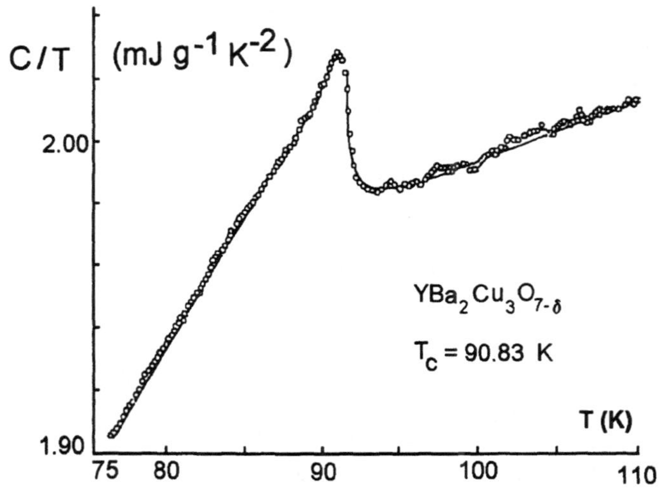

Figure 18.2: The electronic heat capacity near the critical temperature in polycrystal YBCO, after Fisher et al. [8].

which is reasonable.

The smallness of v_F partly explains the high thermodynamic critical field H_c of these materials, since $H_c \propto v_F$.

18.4 The Heat Capacity

We first examine the heat capacity C near the superconducting transition. Since T_c is high, the electron contribution is very small compared with the phonon contribution. The systematic studies by Fisher *et al.* [8] of the heat capacities of high-T_c materials (polycrystals) with and without applied magnetics fields indicate that there is a distinct maximum near T_c. A summary of the data is shown in Fig. 18.2. Since materials are polycrystals with a size distribution, the maximum observed is broader compared to those observed for pure metals. But the data are in agreement with what is expected of a B-E condensation of free massless bosons in 2D, a peak with no jump at T_c with the T^2-law decline on the low temperature side. Compare Fig. 18.2 with Fig. 8.2.

Loram *et al.* [9] extensively studied the electronic heat capacity of YBa_2CuO_{6+} with varying oxigen concentration. A summary of their data are shown in

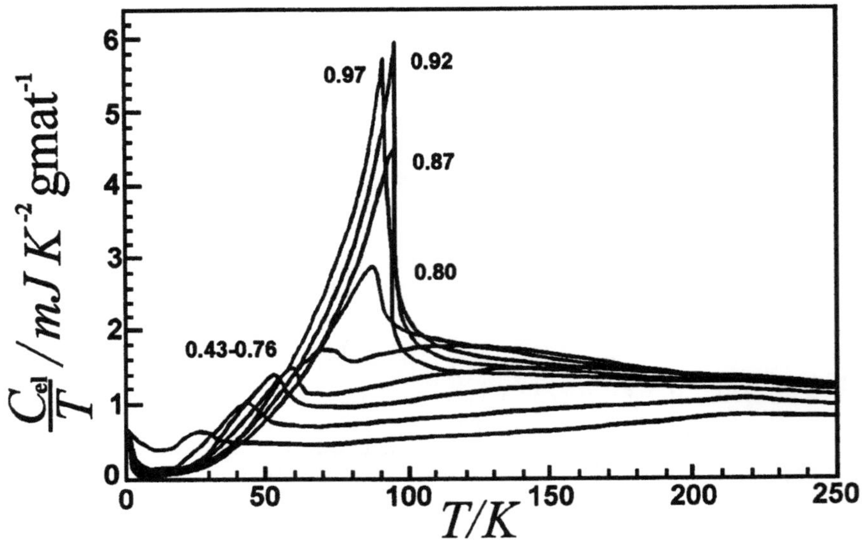

Figure 18.3: Electronic heat capacity plotted as C_{el}/T after Loram et al. [9] for $YBa_2Cu_3O_{6+x}$ with the x values shown.

Fig. 18.3. The maximum heat capacity at T_c with a shoulder on the high temperature side can only be explained naturally from the view that the superconducting transition is a macroscopic change of state generated by the participation of a great number of pairons with no dissociation. The standard BCS model predicts no features above T_c.

18.5 Two Energy Gaps; Quantum Tunneling

The pairon size represented by the coherence length ξ_0 for YBCO is 14 Å. The density of conduction electrons that controls the screening effect is not high. Then the Coulomb repulsion between the constituting electrons is not negligible, so the interaction strengths satisfy the inequalities:

$$v_{11} = v_{22} < v_{12} = v_{21}. \tag{18.30}$$

As a result there are *two* quasi-electron energy gaps (Δ_1, Δ_2) satisfying

$$\Delta_j = \frac{1}{2}\sum_i v_{ji}\mathcal{N}(0)\sinh^{-1}(\hbar\omega_D/\Delta_i). \tag{18.31}$$

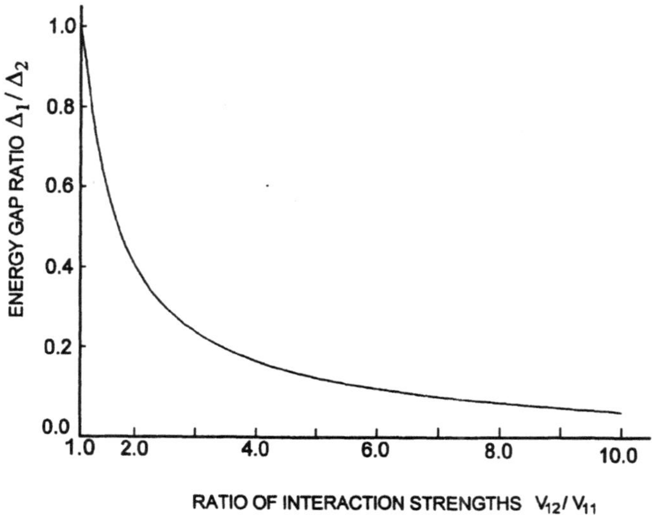

Figure 18.4: Variation of the ratio Δ_1/Δ_2 as a function of the ratio v_{11}/v_{12}.

The ratio Δ_1/Δ_2 and its inverse Δ_2/Δ_1 satisfy the same equations since $v_{11} = v_{22}$ and $v_{12} = v_{21}$. This means that we cannot determine from Eqs. (18.31) alone which types of pairons have the higher energy gaps $\varepsilon_j \equiv \varepsilon_{gj}$. To answer this question, we must examine the behavior of excited pairons. At any rate the ratio Δ_1/Δ_2 (or Δ_2/Δ_1) varies with the ratio of the interaction strengths, v_{12}/v_{11}. This behavior is shown in Fig. 18.4. The values of the ratio Δ_1/Δ_2 change significantly near $v_{12}/v_{11} = 1$. Examining Eqs. (18.17) reveals that the greater the gaps Δ_j, the greater is the binding-energy $|w_j|$. Hence, pairons with greater Δ_j are major contributors to ground-state energy. Pairons with smaller gaps ε_j are easier to excite, and these come from smaller gaps with smaller Δ_j. The situation is thus rather complicated. In any event there are two energy gaps (ε_1, ε_2). Using this fact we now discuss the behavior of pairon energy gaps in quantum tunneling experiments. The appearance of the energy gaps is one of the most important signatures of a superconducting state. A great number of quantum-tunneling studies have been made. [10] Since cuprate superconductors are ceramics and contain many imperfections, a wide range of scattered data were reported. The following general features however stand out: asymmetric I-V curve for

S-I-N systems, a wide scatter of energy gaps data, and complicated conductance (dI/dV)-voltage (V)curves. We comment on these features separately.

18.5.1 Asymmetric I-V Curves for S-I-N

In Section 11.2 we showed that the I-V curve for a S_1-I-S_2 (N) must in general be asymmetric. Briefly if the bias voltage is reversed, different charge carriers are involved in the quantum tunneling. Since moving pairons have two energy gaps (ε_1, ε_2), threshold energies (V_{t1}, V_{t2}) are different, generating easily recognizable asymmetries. (For a type I superconductor, there is a single energy gap ($\varepsilon_1 = \varepsilon_2$) and a single threshold voltage V_t for both polarities.) In actual data for cuprate superconductors, the difference between V_{t1} and V_{t2} is less than 3%, indicating

$$\frac{|\varepsilon_1 - \varepsilon_2|}{\varepsilon_1} < 0.03. \tag{18.32}$$

18.5.2 Scattered Data for Energy Gaps

In Section 10.3, we showed that threshold voltage V_t for an S-I-N depends on the nature of the metal N. There are two cases: a normal state superconductor and a true normal metal like Na. The threshold voltages are then different by the factor of 3:

$$3V_a = V_c, \tag{18.33}$$

which is consistent with most of the data on high-T_c superconductors. For S=YBCO,

$$2V_a = 8 \sim 10\,\text{mV} \qquad \text{and} \qquad 2V_c(\text{Pt}) = 20 \sim 25\,\text{mV}. \tag{18.34}$$

All experiments were done without knowledge of Eq. (18.33). Therefore no efforts were made to use different metals intentionally for N. Further systematic experiments are required to see how well Eq. (18.33) is observed.

18.5.3 Complicated I-V Curves

Energy gaps (ε_1, ε_2) have a small difference. Hence pairons of both charge types are excited, generating complicated I-V curves. In summary the data for quantum tunnelings in high-T_c superconductors show more complicated features compared to the case of type-I superconductors. All of the features coming from two pairon energy gaps (ε_1, ε_2) are in qualitative agreement with experiment.

Chapter 19

Doping Dependence of T_c

The critical temperature T_c in $La_{2-x}Sr_xCuO_4$ has a dome-shaped concentration (x) dependence. This is microscopically explained, starting with a generalized BCS Hamiltonian and using the band structures of electrons and phonons. The disappearance of T_c at the end of the overdoping region, which coincides with the sign change of the Hall coefficient R_H is shown to arise from the curvature inversion of the O-(Cu-)Fermi surface.

19.1 Introduction

The doping dependence of the critical temperature T_c has been recognized ever since the first discovery of a high-T_c superconductor by Bednorz and Müller [1]. The systematic studies of T_c in $La_{2-x}Sr_xCuO_4$ [2-3] indicate that T_c has a maximum about 40 K at concentration $x = 0.15$ and decreases on both sides in the range $(0.06 < x < 0.25)$, see Fig. 19.1(a). Similar doping behaviors were observed for other systems ($YBa_2Cu_3O_{7-\delta}$, $Tl_2Ba_2CuO_{6+\delta}$). In these materials each Cu in the copper plane (CuO_2) is surrounded by six (6) O's, see Fig. 17.3 (a), and the conduction is p-type. In contrast each Cu in the copper plane of $Nd_{2-x}Ce_xCuO_{4-\delta}$ is surrounded by four (4) O's, see Fig. 17.3 (b) and the conduction is n-type [3]. This material exhibits a similar doping dependence of T_c on x in the range $(0.13 < x < 0.18)$ [3]. In this chapter we show that the doping dependence of T_c can be explained based on the generalized BCS model which incorporates the electron energy bands and optical-phonon exchange interaction.

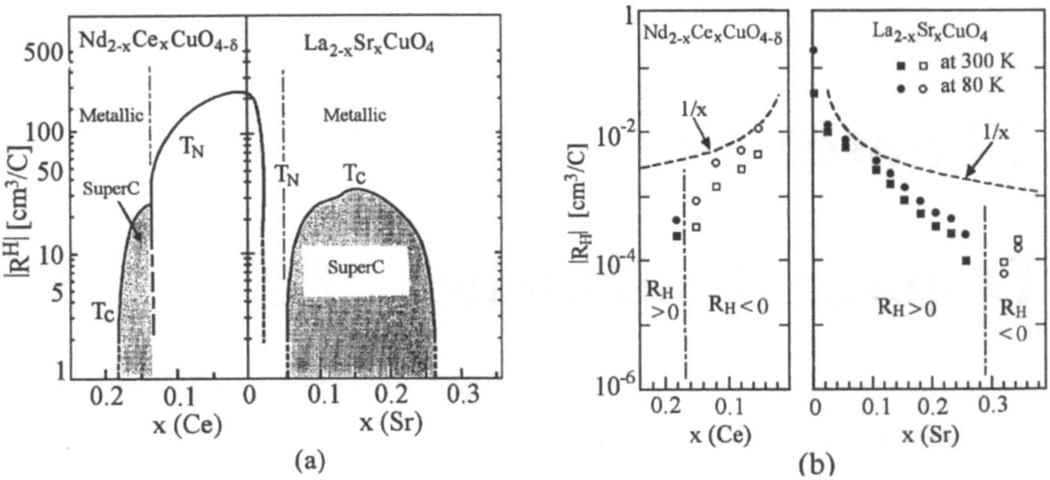

Figure 19.1: (a) Phase diagrams for $Nd_{2-x}Ce_xCuO_{4-\delta}$ and $La_{2-x}Sr_xCuO_4$. (b) The doping dependence of Hall coefficient R_H for the two systems, after Takagi [3].

19.2 Critical Temperature T_c

BCS introduced electron-pair operators [4]:

$$b_k \equiv c^\dagger_k c^\dagger_{-k}, \qquad b_k \equiv c_{-k} c_k, \tag{19.1}$$

where $(c, c^\dagger)$ are electron operators (spin indices omitted) satisfying the Fermi anticommutation rules. They investigated the commutators among b and $b^\dagger$, which do not satisfy the usual Bose commutation rules. Based on this and $b_k^2 = 0$, BCS did not consider the bosonic nature of pairons. In fact, BCS used $b_k^2 = 0$ to construct the ground state to show that the ground-state energy is lower than that of the Bloch system without the pairing interaction. Based on the wavefunction symmetry argument Feynman [5] noted that the pairons, whose CM motion generate a supercurrent, move as bosons, and proceeded to derive the Josephson-Feynman equations [6]:

$$\hbar \frac{d}{dt}\delta = 2eV, \qquad i = i_0 \sin\delta, \qquad V = \text{voltage difference}, \qquad i_0 = \text{constant}, \tag{19.2}$$

governing the current i across a Josephson junction generating the phase difference δ. As we saw in section 8.2, the eigenvalues for $n_{12} \equiv c^\dagger_{k_1} c^\dagger_{k_2} c_{k_2} c_{k_1} \equiv$

$c_1^\dagger c_2^\dagger c_2 c_1$ in the pair-states (k_1, k_2) are limited to 0 or 1 (fermionic property) while the eigenvalues of the pair number operator

$$n_0 \equiv \sum_k b_k^\dagger b_k \tag{19.3}$$

are non-negative integers (bosonic property):

$$n_0' = 0, 1, 2, \ldots \tag{19.4}$$

Both fermionic and bosonic natures of the pairons must be used in the total description of superconductivity. The most significant signature of bosonic pairons is the Bose-Einstein Condensation (BEC).

The flux (Φ) quantization experiments [7] show that

$$\Phi = n\Phi, \qquad n = 0, 1, 2, \ldots, \tag{19.5}$$

$$\Phi_0 \equiv \frac{\hbar}{2e} = \frac{\pi\hbar}{e}, \tag{19.6}$$

indicating that the carriers in the supercurrent are pairons having charge magnitude $2e$. The supercurrent runs persistently. Hence the momentum state along the ring,

$$p_n = \frac{2\pi\hbar}{L} \tag{19.7}$$

with L being the ring circumference, is also an energy-eigenstate so that

$$\varepsilon_n = \varepsilon(p_n), \tag{19.8}$$

meaning that the condensed pairons move independently. The Josephson interference in a SQUID [8] indicates that two supercurrents macroscopically separated up to 1 mm can interfere just as two laser beams coming from the same source. This means that the condensed pairons move with a non-dispersive linear dispersion relation: $\varepsilon = cp$, (c = constant). Cooper solved the Cooper equation (19.1), and obtained a *linear* dispersion relation for moving pairons (unpublished).

$$w_p = w_o + (1/2)v_F p, \qquad \text{(3D)}, \tag{19.9}$$

where w_o is a negative pairon ground-state energy and v_F the Fermi velocity. Formula (19.9) is recorded in Schrieffer's book [9], Eq. (2-15). If the pairon moves in 2D, the relation changes slightly:

$$w_p = w_o + (2/\pi)v_F p, \qquad \text{(2D)}. \tag{19.10}$$

The BEC temperature T_c for free massless bosons moving in D dimension can be found from

$$n_0 = \frac{1}{(2\pi\hbar)^D} \int d^D p \frac{1}{\exp(\varepsilon/k_B T_c) - 1}. \qquad (\varepsilon = cp) \tag{19.11}$$

Using (19.10)-(19.11), we obtain

$$T_c = \begin{cases} 1.01\, \hbar v_F\, n_0^{1/3}/k_B & (3D) \\ 1.24\, \hbar v_F\, n_0^{1/2}/k_B & (2D). \end{cases} \tag{19.12}$$

The 2D BEC is noteworthy since the BEC of massive bosons ($\varepsilon = p^2/2m$) is known to occur in 3D only. This is not a violation of Hohenberg's theorem [10] that there can be no long range order in 2D, the theorem derived with the assumption of an f-sum rule representing the mass conservation law. The theorem does not hold for massless bosons. There is no BEC in 1D. Formulas (19.12) for the BEC temperature are obtained with the assumption of free-moving pairons, which can be justified as follows. The interpairon distance r_0 computed from (19.12),

$$r_0 \equiv \begin{cases} n_0^{-1/3} = 1.01\, \hbar v_F\, /(k_B T_c) & (3D) \\ n_0^{-1/2} = 1.24\, \hbar v_F\, /(k_B T_c) & (2D), \end{cases} \tag{19.13}$$

are compared with the zero-temperature BCS pairon size [4]

$$\xi_0 = \frac{\hbar v_F}{\pi \Delta} = 0.18\, \frac{\hbar v_F}{k_B T_c}. \tag{19.14}$$

From the last two equations we obtain

$$r_o/\xi_0 = \begin{cases} 5.6 & (3D) \\ 6.9 & (2D), \end{cases} \tag{19.15}$$

indicating that the condensed pairons do not overlap in space, whence the free pairon model can be used to evaluate T_c.

In the present work we regard the BEC temperature T_c as the superconducting temperature. Such a view was proposed previously by Ogg and others [11]. Since superconductivity and superfluidity are remarkably similar, a unified BEC approach is natural and desirable [12]. In particular Schafroth, Blatt and Butler [13] (SBB) proposed that the superconducting transition can be regarded as a BEC of *massive* (M) bound pair of electrons having a quadratic relation:

$$\varepsilon = \frac{p^2}{2M}. \tag{19.16}$$

A significant difficulty of the SBB model is that the interpair distance r_0 calculated from the critical temperature

$$T_c = 3.31\,\hbar^2 (Mk_B)^{-1} n_0^{2/3} \tag{19.17}$$

for a system of free pairs with the mass M ($\sim 2m_e$) moving in 3D, is smaller than the pairon size ξ_0, and hence the pairs must overlap in space and cannot move independently. Attempts for overcoming this difficulty have been made by Leggett [14], Nozierers and Schmitt-Rink [15] and others with limited successes. We stress that our quantum statistical theory using the linear dispersion relation (19.9) has no such difficulty. Besides, our theory yields an exact expression for T_c for a 2D system (cuprate superconductor).

We stress that formulas (19.12) were independent of the famous BCS formula (in the weak coupling limit):

$$T_c = 1.13\,\hbar\omega_D \exp\left[\frac{1}{v_0 \mathcal{N}(0)}\right]. \tag{19.18}$$

The pairon density n_0 and the Fermi velocity v_F appearing in (19.12) can be experimentally determined from the data of the resistivity, the Hall coefficient, the Hall angle, the specific heat, and the superconducting temperature, while the BCS formula (19.18) contains the pairing strength v_0, which appears together with the density of sates at ε_F, $\mathcal{N}(0)$, and is hard to determine unambiguously.

19.3 Doping Dependence of T_c

Let us consider $La_{2-x}Sr_xCuO_4$. The parent material La_2CuO_4 is an antiferromagnetic insulator with transition temperature $T_a = 270$ K. Substitution of trivalent La by divalent Sr changes the ionicity in the La_2O_2 blocks neighboring the copper plane, see Fig. 18.2 (a). Since O's are connected in the perovskite, this doping changes the electron density at O's in the copper plane, and hence the O-Fermi surface. This density change adversely affects the antiferromagnetic state and hence the Néel temperature T_N declines. The

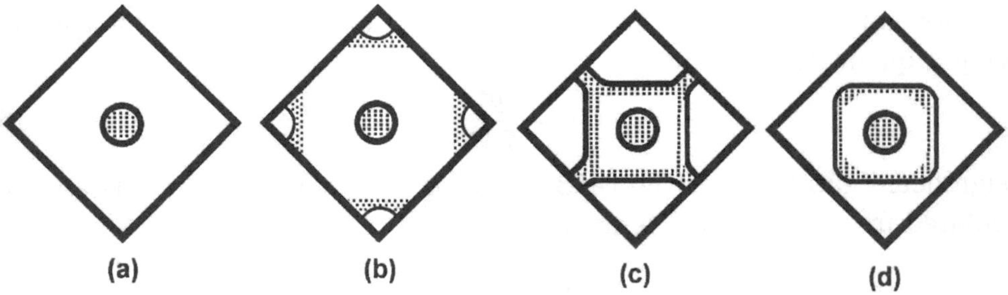

Figure 19.2: The doping in $La_{2-x}Sr_xCuO_4$ reduces the electron density at O's and hence changes the O-Fermi surface. The Fermi surface at optimum doping is shown in (b). Further doping generates a curvature-inversion of the O-Fermi surface: (c)→(d). The "holes"-like Fermi surface disappears in (a) as x is reduced to 0.06.

doping destroys the antiferromagnetic phase at $x = 0.02$. A further doping changes the "hole" density and the O-Fermi surface so that $\pm$ pairons are created by optical-phonon-exchange attraction, generating a superconducting state ($0.06 < x < 0.25$). For comparison consider $Nd_{2-x}Ce_xCuO_4$. The parent material Nd_2CuO_4, see Fig. 17.4 (b), has no apical oxygen and each Cu in the copper plane is surrounded by four O´s. Substitution of trivalent Nd by quadrivalent Ce changes the ionicity in Nd_2O_2 blocks. This doping changes the electron density at Cu's in the copper plane and the Cu-Fermi surface.

A summary of the data [2] for the dependence of T_c for $La_{2-x}Sr_xCuO_4$ on the concentration x is shown in Fig 19.1 (a). The dome-shaped curve can be explained as follows. We assume that the Fermi surface at the optimum doping (highest T_c) is the same as that in Fig. 17.2, which is reproduced in Fig. 19.2 (b). The doping decreases the electron density at O's and the curvature of the O-Fermi surface eventually changes its nature from "hole"-like in (c) to "electron"-like in (d), where no "holes" are present and hence no pairons can be generated. This corresponds to the fall of T_c at the end of the overdoping ($x = 0.25$). The curvature change can be checked by observing the Hall coefficient R_H. Experiments [16] show that

$$\begin{aligned} R_H &> 0 \quad \text{for } x < 0.25 \\ R_H &< 0 \quad \text{for } x > 0.25. \end{aligned} \tag{19.19}$$

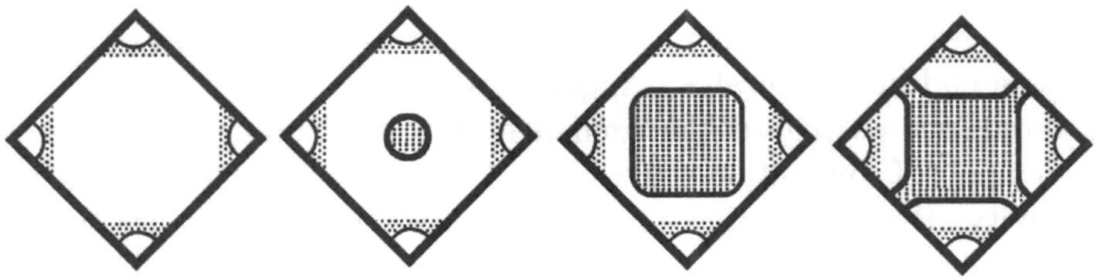

Figure 19.3: The doping in $Nd_{2-x}Ce_xCuO_4$ increases the electron density at Cu's and hence changes the Cu-Fermi surface. The Fermi surface at optimum doping is shown in (b). Further doping induces a curvature-inversion of the Cu-Fermi surface (c)→(d) as x changes from 0.13 to 0.18.

The absolute values $|R_H|$ ($\sim 10^{-4}$ cm^3/C) near the inversion point $x = 0.25$ are relatively small, and they are nearly equal to each other on both sides of this point. These experimental data are in accord with the physical picture that the Fermi surface is large and nearly equal in size near the inversion point, see Fig. 19.2 (c) and (d). Reduction of the doping increases the electron density at O's and makes the "hole" pocket size smaller and eventually the "hole"-like Fermi surface disappears at $x = 0.06$ as indicated in (a), where no "holes" are present and hence no pairons can be generated. The change in the curvature of the Fermi surface originates in the fact that the surface must approach the Brillouin boundary at right angles, which arises from the mirror symmetry possessed by the CuO_2 lattice.

Consider now $Nd_{2-x}Ce_xCuO_4$. The doping increases the electron density at Cu's in the copper plane. The Cu-Fermi surface in the center grows and it is "electron"-like as the electron density increases as indicated in Fig. 19.3 (a)-(b). Further doping (increased electron density) eventually causes an curvature inversion of the Cu-Fermi surface from "electron"-like in (c) to "hole"-like in (d). In (d) ($x = 0.18$) the system contains "holes" only, and cannot pair-create $\pm$ pairons, and hence T_c must vanish. Fig. 19.1 shows a summary of data for both systems, indicating the connection between the fall of T_c and the inversion of the Fermi surface (the sign change of R_H).

Other cuprates including $YBa_2Cu_3O_{7-\delta}$, $BaPb_{1-x}Bi_xO_3$, show also dome-shaped doping dependence. In our theory the fall of T_c at the overdoping end is interpreted in terms of the curvature inversion of the Fermi surface.

To test this feature experimental observation of the sign change of R_H at the overdoping end in all cuprates is highly desirable.

In summary we regard the exchange of a longitudinal optical phonon between a pair of conduction electrons in a cuprate as the cause of the superconductivity. By the acoustic phonon exchange, the pairons are stabilized. The pairon moves in the copper plane with a linear dispersion relation. Since the pairon is composed of two electrons, its CM motion is bosonic. We regard the BEC temperature of independently moving pairons as the superconducting temperature T_c, which depends on the pairon density n_0. This density n_0 depends critically on the Fermi surface of electrons. The doping can change the electron density at O (Cu) and the Fermi surface. The highest T_c temperature corresponds to the greatest pairon density. The overdoping eventually causes the curvature inversion of the Fermi surface and leads to a vanishing superconductivity. We note that the same model can account for the d-wave pairon formation with strong binding along the a-and b-axis of the copper plane, see Chapter 25.

Chapter 20

Transport Properties Above T_c

Magnetotransport properties in $Nd_{2-x}Ce_xCuO_4$, and $La_{2-x}Sr_xCuO_4$ above T_c show unusual behaviors with respect to the temperature (T)- and doping concentration (x)-dependence. The resistivity ρ in optimum samples shows a T-linear behavior while that in highly overdoped samples exhibits a T-quadratic behavior. The $\cot\theta_H$, θ_H = the Hall angle, shows a T-quadratic behavior. The Hall coefficient R_H changes the sign as the concentration x passes the super-to-normal phase. We shall explain these properties based on the independent pairon model in which $\pm$ pairons and conduction electrons are carriers in the superconductor.

20.1 Introduction

A summary of data for the resistivity ρ in various cuprates at optimum doping is shown as solid lines in Fig. 20.1 [1]. The T-linear behavior:

$$\rho \propto T \tag{20.1}$$

has been observed ever since the discovery of the first high-T_c superconductor by Bednorz and Müller [2]. Data for ρ in highly overdoped samples are shown in dotted lines; they show a T-quadratic behavior. The $\cot\theta_H$, θ_H = the Hall angle, follows a T-quadratic behavior

$$\cot\theta_H \propto T^2. \tag{20.2}$$

Data for the Hall coefficient R_H in $Nd_{2-x}Ce_xCuO_4$, and $La_{2-x}Sr_xCuO_4$ versus the "electron" ("hole")-concentration are shown in Fig. 20.2 (b) [3, 4]. Note

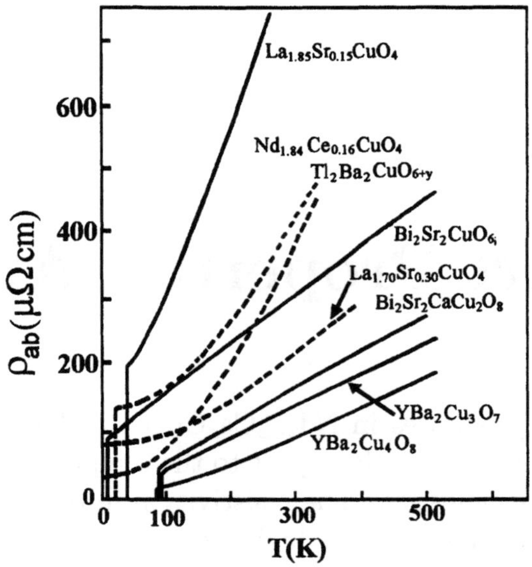

Figure 20.1: Resistivity in the *ab* plane, ρ_{ab} versus temperature T. Solid lines represent data for cuprates at optimum doping and dashed lines data for highly overdoped samples, after Iye [1]

that the sign change in R_H at the end of the overdoping region (dash-dot lines). Also note that $|R_H|$ is smaller for higher T and for greater x in superconductor samples. These data clearly indicate that at least two (charge) carriers contribute to the electrical conduction. We shall show that the unusual magnetotransport properties can quantitatively be explained based on a two-carrier model, in which fermionic electrons and bosonic pairons are scattered by acoustic phonons.

20.2 Simple Kinetic Theory

20.2.1 Resistivity

We use simple kinetic theory to describe the transport properties. [5] Kinetic theory originally was developed for a dilute gas. Since a conductor is far from being the gas, we shall discuss the applicability of kinetic theory. The Bloch

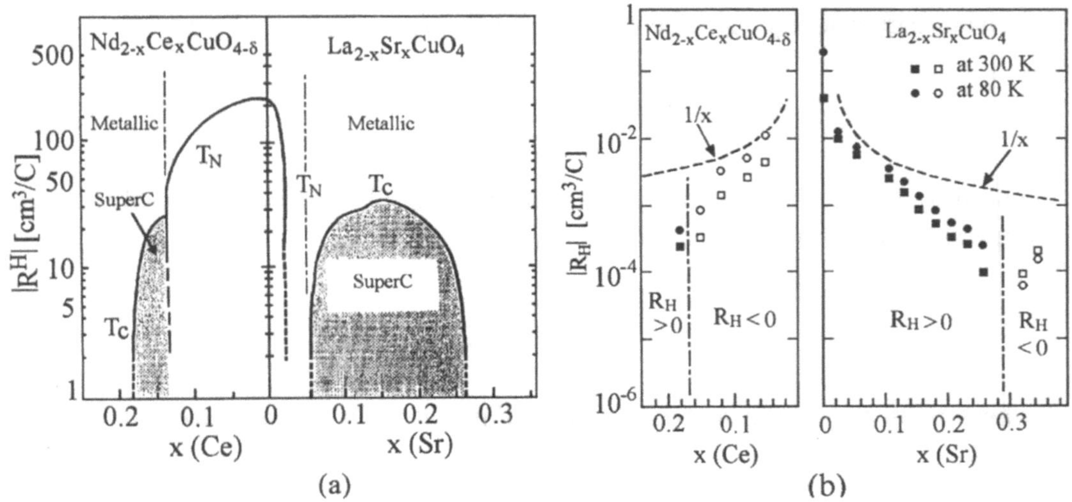

Figure 20.2: (a) Phase diagrams for $Nd_{2-x}Ce_xCuO_4$, and $La_{2-x}Sr_xCuO_4$. (b) Doping dependence of Hall coefficient R_H, after Takagi [3].

wave packet in a crystal lattice extends over one unit cell or more, and the lattice-ion force averaged over unit cell vanishes, see Eq. (3.36). Hence the electron ("electron", "hole") runs straight and changes direction if it meets an impurity or phonon (wave packet). The electron-electron collision conserves the net momentum and hence its contribution to the current correlation function [6] is zero. Upon the application of a magnetic field, the system develops a Hall electric field so as to balance out the Lorentz magnetic force, see Fig. 20.3 (a). Thus the electron still moves straight, which makes the kinetic theory applicable.

Consider a system of "holes", each having effective mass m_1 and charge e, scattered by phonons. Assume a weak electric field $\mathbf{E}$ applied along the x-axis. Newton's equation of motion for the "hole" with the neglect of the scattering is $m_1 dv_x/dt = eE$. Solving it for v_x and assuming that the acceleration persists in the mean free time τ_1, we obtain

$$v_d = \frac{eE}{m_1}\tau_1 \tag{20.3}$$

for the drift velocity v_d. The current density (x-component) j is

$$j = en_1 v_d = \frac{e^2}{m_1} n_1 \tau_1 E, \tag{20.4}$$

where n_1 is the "hole" density. Assuming Ohm's law: $j = \sigma E$, we obtain an expression for the electrical conductivity:

$$\sigma_1 = \frac{e^2}{m_1} n_1 \frac{1}{\gamma_1}, \tag{20.5}$$

where $\gamma_1 \equiv \tau_1^{-1}$ is the scattering rate. This rate can be computed, using

$$\gamma_1 = n_{ph} v_F S_1, \tag{20.6}$$

where S_1 is the scattering diameter. If acoustic phonons having average energies: $\langle \hbar\omega_q \rangle \equiv \alpha_0 \hbar\omega_D \ll \mathrm{k}_B T$, $\alpha_0 \sim 0.20$ are assumed, the phonon number density is [6]

$$n_{ph} = n_a [exp(\alpha_0 \hbar\omega_D / \mathrm{k}_B T) - 1]^{-1} \simeq n_a \frac{\mathrm{k}_B T}{\alpha_0 \hbar\omega_D}, \tag{20.7}$$

where $n_a \equiv (2\pi)^{-2} \int d^2k$ is the small k-space area where the acoustic phonons are located.

Using (20.5)-(20.7), we obtain

$$\sigma_1 = \frac{ae^2 n_1}{T}, \qquad a \equiv \frac{\alpha_0 \hbar\omega_D}{n_a m_1 \mathrm{k}_B v_F S_1}. \tag{20.8}$$

Thus resistivity ρ (conductivity σ) is (inversely) proportional to T.

Let us now consider a system of +pairons, each having charge $+2e$ and moving with the linear dispersion relation: $\varepsilon = cp$. Since $v_x = (d\varepsilon/dp)(\partial p/\partial p_x)$ $= c(p_x/p)$, Newton's equation of motion is [7]

$$\frac{p}{c}\frac{dv_x}{dt} = \frac{\varepsilon}{c^2}\frac{dv_x}{dt} = 2eE, \tag{20.9}$$

yielding $v_x = 2e(c^2/\varepsilon)Et$ + initial velocity. Hence we obtain

$$v_d^{(2)} = 2ec^2 \tau_2 E \, \langle \varepsilon^{-1} \rangle, \tag{20.10}$$

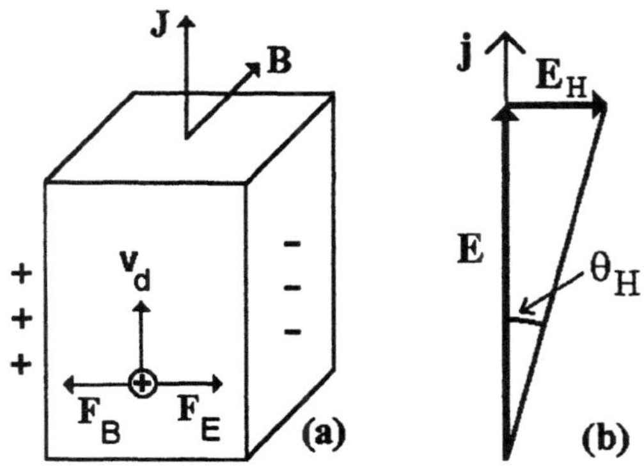

Figure 20.3: (a) The magnetic and electric forces (F_B, F_E) balance out to zero in the Hall effect measurement. (b) The Hall angle θ_H.

where τ_2 is the pairon mean free time and the angular brackets denote a thermal average. Using this and Ohm's law, we obtain

$$\sigma_2 = (2e)^2 c \langle \varepsilon^{-1} \rangle n_2 \gamma_2^{-1}, \qquad \gamma_2 \equiv \tau^{-1}, \tag{20.11}$$

where n_2 is the pairon density. If we assume a Boltzmann distribution for bosonic pairons above T_c, $(T > T_c)$, we obtain

$$\langle \varepsilon^{-1} \rangle \equiv \frac{2\pi}{(2\pi\hbar)^2} \int_0^\infty dp\, p \frac{1}{\varepsilon} e^{-\beta cp} \Big/ \frac{2\pi}{(2\pi\hbar)^2} \int_0^\infty dp\, p\, e^{-\beta cp} = (\mathrm{k}_B T)^{-1}. \tag{20.12}$$

The rate γ_2 is calculated with the assumption of a phonon scattering. We then obtain

$$\sigma_2 = \frac{4e^2 c^2 n_2}{\mathrm{k}_B T \gamma_2} = \frac{2e^2 b n_2}{T^2}, \qquad b \equiv \frac{8}{\pi^2} \frac{\alpha_0 \hbar \omega_D v_F}{n_a \mathrm{k}_B^2 S_2}. \tag{20.13}$$

The total conductivity σ is $\sigma_1 + \sigma_2$. Taking the inverse, we obtain

$$\rho \equiv \frac{1}{\sigma} = \frac{T^2}{e^2(a n_1 T + 2b n_2)}. \tag{20.14}$$

20.2.2 Hall Coefficient

We take a rectangular sample having only "holes", see Fig. 20.3 (a). The current $\mathbf{j}$ runs in the z-direction. Experiments show that if a magnetic field

$\mathbf{H}$ is applied in the y-direction, the sample develops a cross (Hall) electric field $\mathbf{E}_H$ so that the magnetic force $\mathbf{F}_B \equiv e(\mathbf{v}_d \times \mathbf{B})$ be balanced out to zero:

$$e(\mathbf{E}_H + \mathbf{v}_d \times \mathbf{B}) = 0 \qquad \text{or} \qquad E_H = -v_d B. \tag{20.15}$$

If the sample contains $+$pairons only, Eq. (20.15) also holds. Since the "hole" and the $+$pairon have like charges $(e, 2e)$ the two components ("holes", $+$pairons) separately maintain drift velocities $(v_d^{(1)}, v_d^{(2)})$ and Hall fields $(E_H^{(1)}, E_H^{(2)})$ so that

$$E_H^{(j)} = -v_d^{(j)} B, \qquad j = 1, 2. \tag{20.16}$$

The total Hall field E_H is

$$E_H = E_H^{(1)} + E_H^{(2)}. \tag{20.17}$$

Let us check the validity. Eqs. (20.16) and (20.17) are reduced to the familiar one-component equations if either component is absent and also if the two components are assumed identical.

The negative signs in Eqs. (20.16) mean that the Hall electric force $\mathbf{F}_E \equiv e\mathbf{E}_H$ opposes the Lorentz-magnetic force $\mathbf{F}_B$. In the present geometry the induced field $\mathbf{E}_H$ for the "holes" points in the positive x-direction. We take the convention that $\mathbf{E}_H$ is measured relative to this direction.

The Hall coefficient R_H is defined and calculated as

$$R_H \equiv \frac{E_H}{jB} = \frac{v_d^{(1)} + v_d^{(2)}}{en_1 v_d^{(1)} + 2en_2 v_d^{(2)}} = \frac{aT + b}{e(an_1 T + 2bn_2)}, \tag{20.18}$$

where Eqs. (20.3) and (20.10) were used . Formula (20.18) is obtained with the assumption that the carriers have like charges. If carriers have unlike charges, e.g. $\pm e$, R_H has a far more complicated expression. [8]

20.2.3 Hall Angle

The Hall angle θ_H is the angle between the current $\mathbf{j}$ and the combined field $\mathbf{E} + \mathbf{E}_H$, see Fig. 20.3 (b). This angle θ_H is very small under normal experimental conditions. We now consider

$$\cot\theta_H = \frac{E}{E_H} = \frac{E}{v_d B} = \frac{\rho}{BR_H}. \tag{20.19}$$

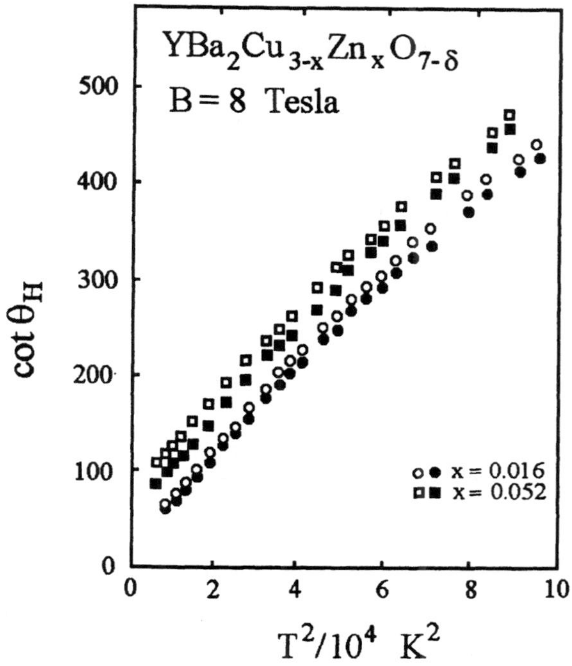

Figure 20.4: $\cot\theta_H$ versus $T^2/10^4$ K^2 for $YBa_2Cu_{3-x}Zn_xO_{7-\delta}$, $B = 8$ tesla, after Chien et. al. [9].

Using (20.14) and (20.18), we obtain

$$\cot\theta_H = \frac{T^2}{eB(aT + b)}. \tag{20.20}$$

Formulas (20.14), (20.18) and (20.20) for ρ, R_H and $\cot\theta_H$, respectively, will be used for data analysis in the following section.

20.3 Data Analysis

A summary of the data for $\cot\theta_H$ in $YBa_2Cu_{3-x}Zn_xO_{7-\delta}$ [9] is shown in Fig. 20.4. The doped divalent Zn substitute for divalent Cu and the doping therefore introduces additional scatterers in the copper plane. The data may be represented by

$$\cot\theta_H = \alpha T^2 + C(x), \qquad \alpha = \text{constant}, \tag{20.21}$$

$$C(x) \to 0 \quad \text{as } x \to 0. \tag{20.22}$$

The term $C(x)$ (> 0) in Eq. (20.21) can be regarded as the impurity (Zn) scattering contribution. Thus Matthiessen's rule holds.

From now on we consider the undoped sample ($x = 0$, $C = 0$):

$$\cot\theta_H = \alpha T^2. \tag{20.23}$$

This T-quadratic behavior can be obtained from formula (20.20) if

$$\frac{aT}{b} = \frac{v_d^{(1)}}{v_d^{(2)}} = \frac{\pi^2}{16}\frac{S_2}{S_1}\frac{\mathrm{k}_B T}{\varepsilon_F} \ll 1, \tag{20.24}$$

which is reasonable since $T \ll T_F$. Physically from Eq. (20.9) low-energy pairons are accelerated strongly in proportion to ε^{-1}, and hence $v_d^{(2)}$ dominates $v_d^{(1)}$, causing the strong inequality (20.24).

Consider now the T-linear law for as expressed in Eq. (20.1). This behavior can be obtained from formula (20.14) if

$$an_1 T \gg 2bn_2, \tag{20.25}$$

which is reasonable since the "hole" density n_1 is much greater than the pairon density n_2. Assuming inequality (20.24), we obtain from (20.18)

$$R_H \approx \frac{b}{e(an_1T + 2bn_2)}. \tag{20.26}$$

The "hole" density n_1 monotonically increases with increasing x ($x < 0.25$) while the pairon density, calculated from Eq. (18.22),

$$n_2 = 0.65\,\frac{\mathrm{k}_B^2}{\hbar^2 v_F^2}\,T_c^2 \tag{20.27}$$

has a dome-shaped maximum in the superconducor phase: ($0.05 < x < 0.25$), see Fig. 20.2. (a). The critical temperature T_c is a simple function of the pairon density n_2 and the Fermi speed v_F. We may use this relation to make a numerical estimate of n_2. Experiments indicate that $T_c = 40\,\mathrm{K}$, ξ_0(coherence length) = 30 Å, $x = 0.15$. Using these data, we obtain $n_{2,\max} = 2.3 \times 10^{11}$ cm^{-2}, $v_F = 8.8 \times 10^6$ cm sec^{-1}, which are reasonable. Experiments show that R_H decrease significantly in the range ($0.06 < x < 0.25$), see Fig. 20.2 (b). This behavior arises from the pairon term $2bn_2$ in the denominator in Eq. (20.26). To see the temperature behavior more explicitly, we assume the two inequalities (20.24) and (20.25). We then obtain from Eq. (20.26)

$$R_H \approx \frac{b}{en_1 aT} = \frac{16 S_1 \varepsilon_F}{\pi^2 S_2 k_B T} \frac{1}{en_1}, \tag{20.28}$$

indicating that R_H is smaller if T is high and if n_1 is high, in good agreement with the experimental data, see Fig. 20.2 (b).

Let us now consider the overdoped sample. As noted earlier, the Hall coefficient R_H in $La_{2-x}Sr_xCuO_4$ changes sign at the end of the overdoping ($x = 0.25$), which coincides with vanishing T_c. We interpret this change in terms of the curvature inversion of the O-Fermi surface: (c)→(d) in Fig. 19.2. Near the inflexion point the Fermi surface becomes large and hence the "hole" (or "electron") density is high. Hence $|R_H|$ should have a minimum (10^{-4} cm^3/C). If we use $|R_H| = (n_1 e)^{-1}$, we obtain $n_{1,\max} = 10^{23}$ cm^{-3}, a remarkably high electron density comparable to that in silver (Ag). Near the inflexion point, the Fermi surface is large and its shape changes significantly with the energy, meaning that the density of states is very large and the "hole" effective mass m_1 is extremely large. Then the "hole" contribution to σ ($\sigma_1 \propto m_1^{-1} \ll \sigma_2$) becomes negligible. Hence the pairon contribution yields

$$\rho \approx \frac{1}{\sigma_2} = \frac{T^2}{2e^2 b n_2}. \tag{20.29}$$

Thus the T^2-law behavior in highly overdoped sample, $La_{1.70}Sr_{0.30}CuO_4$, see Fig. 20.1, is explained.

20.4 Discussion

The unusual magnetotransport in $La_{2-x}Sr_xCuO_4$ are explained based on the model in which "holes" and +pairons are carriers scattered by phonons. We note that no adjustable parameters were introduced in the theory.

Because of the non-perovskite structure, the substitution of trivalent Nd by quadrivalent Ce increases the electron density at Cu in the copper plane. Hence the doping in $Nd_{2-x}Ce_xCuO_{4-\delta}$ changes the electron density and the

Cu-Fermi surface. The phase diagram in Fig. 20.2(a) shows that T_c falls to zero as x approaches 0.17 from below, where R_H changes the sign. This behavior can be interpreted in terms of the curvature inversion of the Cu-Fermi surface occurring in the reversed sense. See Fig. 19.3. This is corroborated by the T^2-law resistivity in highly overdoped sample $Nd_{1.84}Ce_{0.16}CuO_4$, see Fig. 20.1, indicating that the "electrons" move as *heavy-fermions* and do not contribute much to the conduction, and hence the pairon contribution generates the T-quadratic behavior, see Eq. (20.29).

Parent materials ($x = 0$) of $La_{2-x}Sr_xCuO_4$, and $Nd_{2-x}Ce_xCuO_4$, are antiferromagnetic insulators at 0 K, see the phase diagram in Fig. 20.2 (a). First consider $La_{2-x}Sr_xCuO_4$. If the electrons at O-sites are taken away by doping, "holes" are created and the "hole" density initially increases. This density increase adversely affects the antiferromagnetic state and hence the Néel temperature T_N is reduced. The doping destroys the antiferromagnetic phase at $x = 0.02$. A further doping changes the "hole" density and the O-Fermi surface so that $\pm$ pairons are created by optical-phonon-exchange attraction, generating a superconducting state ($0.06 < x < 0.25$). A further doping increase eventually causes the curvature inversion of the O-Fermi surface, which terminates the superconducting phase ($x = 0.25$). Second, consider $Nd_{2-x}Ce_xCuO_4$. The doping increases the electron density at Cu, which adversely affects the antiferromagnetic state and T_N therefore decreases. A further doping makes the Cu-Fermi surface to grow so that $\pm$ pairons are generated by phonon-exchange, generating a superconducting state ($0.13 < x < 0.17$). From the diagram $T_N > T_c$, meaning that the exchange energy is greater than the pairon binding energy $|w_0|$. This explains the suppression of the underdoping part of the otherwise dome-shaped T_c curve as observed in $La_{2-x}Sr_xCuO_4$. Near the phase change point ($x = 0.13$) the antiferromagnetic and superconducting tendencies compete with each other. We predict that there will be a finite change in the "electron" density, making the phase change to be of the first order. An experimental check is called for here. We discussed (T_N, T_c) separately, which means that the coupling between conduction electrons and spins is weak.

Chapter 21

Out-of-Plane Transport

The out-of-plane resistivity ρ_c in single-crystal $YBa_2Cu_3O_y$, follows the experimental law: $\rho_c = C_1\rho_{ab} + C_2/T$, C_1, C_2 = constants, ρ_{ab} = in-plane resistivity, in the concentration range ($6.6 < y < 6.92$) with the highest T_c (90 K) occurring at $y = 6.88$ and in the temperature range ($T_c < T <$ room temperature). This behavior is microscopically explained under the assumption that the charge carriers are a mixture of "electrons", "holes" and $\pm$ pairons. The second term C_2/T arises from the quantum tunneling between copper planes of $-$ pairons, moving with a linear dispersion relation, and the first term $C_1\rho_{ab}$ from the in-plane currents due to the "holes" and $+$ pairons.

21.1 Introduction

Terasaki *et al.*[1] systematically investigated the temperature (T)- and the oxygen concentration (y)-dependence of the out-of-plane resistivity ρ_c in $YBa_2Cu_3O_y$. They found that the data, reproduced in Fig. 21.1, can be fitted by

$$\rho_c = C_1\rho_{ab} + C_2/T, \qquad C_1 = 35, \qquad C_2 = 4.3 \times 10^{-8}\ \Omega^2\ \mathrm{cm}^2. \tag{21.1}$$

Here single crystal $YBa_2Cu_3O_y$ is a superconductor in the measured range $6.60 < y < 6.92$, with the highest T_c (90 K) occurring at $y = 6.88$; ρ_{ab} is the in-plane resistivity. We notice quite different behaviors between ρ_c and ρ_{ab}. First of all, ρ_c is greater than ρ_{ab} by a few orders of magnitude. The decreasing T-dependence of ρ_c in $YBa_2Cu_3O_{6.6}$ is against conventional wisdom but such behavior is also observed in $La_{2-x}Sr_xCuO_4$ ($x = 0.06, 0.11$, underdoped

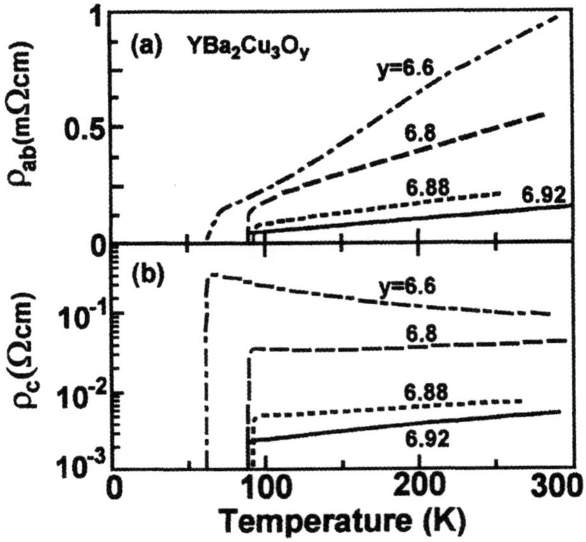

Figure 21.1: Anisotropic resistivities of $YBa_2Cu_3O_y$ single crystals for various oxygen content y: (a) in-plane resistivity ρ_{ab} and (b) out-of-plane resistivity ρ_c after Terasaki et al. [1].

samples) [2]. In the present chapter we give a microscopic foundation to formula (21.1) under the assumption that the charge carriers are a mixture of "electrons", "holes" and $\pm$ pairons. Analyses of the data for ρ_c, the in-plane resistivity ρ_{ab}, the Hall coefficients R_c^H (induced voltage along the c-axis) and R_{ab}^H (current and induced voltage in the ab plane) indicate that: **(a)** the second term C_2/T in (21.1) arises from the quantum tunneling between planes of $-$ pairons, and that **(b)** the first term $C_1\rho_{ab}$ from the in-plane currents with "holes" and $+$ pairons being the main carriers. [3] Using the same independent pairon model, we have discussed earlier the transport properties in the ab plane (chapter 20) and the doping dependence of T_c (chapter 19).

21.2 Theory

Cuprate superconductors have layered structures, and the electrical conduction occurs in the copper plane. There are no k-vectors along the c-axis, and the effective mass along the c-axis is infinity so that the Fermi surface is a

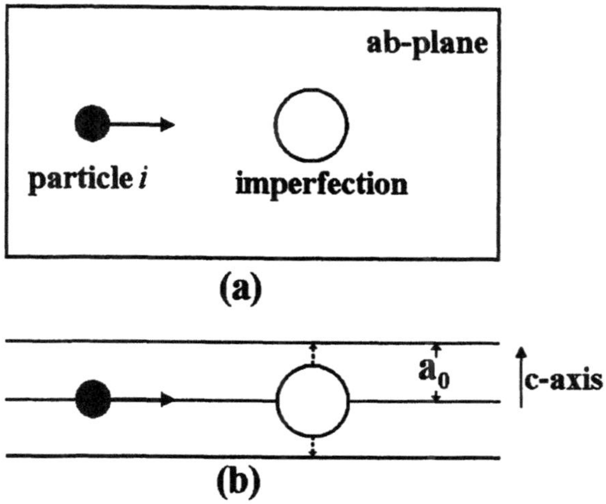

Figure 21.2: Particle i in the standard plane (layer) arrives at an imperfection and quantum-jumps to the neighboring layer. (a) the ab-plane, (b) the side view.

right cylinder with its axis along the c-axis. [4]

We consider a cuprate subject to a voltage applied along the c-axis. Charged particles ("electrons", "holes", pairons) may run in the copper (ab) plane in lengths much greater than the mean free path and occasionally quantum-jump (or tunnel) from plane to plane. According to Ohm's second law, the resistance R for a homogeneous metallic bar of length L is proportional to the length L and the resistivity ρ:

$$R = \rho L. \tag{21.2}$$

Generalizing this to the case of the inhomogeneous anisotropic medium, we may assume [5]

$$R = R_{ab} + R_{jump}, \tag{21.3}$$

where R_{ab} is the conductive resistance when the charges travel in the ab plane, and R_{jump} is the jump resistance.

The jump resistance R_{jump} may be calculated as follows. A charged particle arrives at a certain lattice-imperfection (impurity, lattice defect, etc.) and quantum-jumps to a neighboring layer, see Fig. 21.2, with the jump rate given by Fermi's golden rule:

$$\omega = \frac{2\pi}{\hbar} \left| \langle \mathbf{p}_i | \, U \, | \mathbf{p}_f \rangle \right|^2 \delta(\varepsilon_i - \varepsilon_f), \tag{21.4}$$

where $\mathbf{p}_i$ ($\mathbf{p}_f$) and ε_i (ε_f) are respectively initial (final) momentum and energy, and U is the imperfection-perturbation. The jump rate is finite only if the final-states are in the continuous energy bands, which is implied by $\delta(\varepsilon_i - \varepsilon_f)$.

The current density $j_c^{(i)}$ along the c-axis due to a set of particles i, having charge $q^{(i)}$ and momentum-energy $(\mathbf{p}_i, \varepsilon_i)$, will be calculated by

$$j_c^{(i)} = j_{c,jump}^{(i)} - j_{c,down}^{(i)} = q^{(i)} n^{(i)} a_0 (\omega_{up}^{(i)} - \omega_{down}^{(i)}), \tag{21.5}$$

where $n^{(i)}$ is the number density and a_0 the interlayer distance. When a small voltage is applied, the jump rates upwards and downwards are different, causing a steady current.

We first consider the "hole" with the effective mass m^* moving in the plane. We assume that the voltage is applied upward. Using Eqs. (21.4) and (21.5), we obtain

$$\begin{aligned} j &= e(\frac{2\pi}{\hbar})K_1 \frac{M_1^2 a_0}{A}\{\int_0^{\varepsilon_0 - eV} d\varepsilon\, \mathcal{N}(\varepsilon) f_1(\varepsilon)[1 - f_2(\varepsilon)] \\ &\quad - \int_{eV}^{\varepsilon_0} d\varepsilon\, \mathcal{N}(\varepsilon - eV) f_2(\varepsilon - eV)[1 - f_1(\varepsilon - eV)]\}, \end{aligned} \tag{21.6}$$

were $\mathcal{N}$ represents the density of states, and A the normalization area; f_1 and f_2 are the Fermi distribution functions at the standard (1) and upper (2) planes. Because of the voltage difference the chemical potential at the two planes are different:

$$\begin{aligned} f_1(\varepsilon; \beta, \mu) &\equiv [e^{\beta(\varepsilon - \mu)} + 1]^{-1} \equiv f(\varepsilon), \\ f_2(\varepsilon; \beta, \mu) &\equiv [e^{\beta(\varepsilon - \mu + eV)} + 1]^{-1} = f(\varepsilon + eV, \mu). \end{aligned} \tag{21.7}$$

In Eq. (21.6) we included the quantum statistical factors $[1 - f(\varepsilon_f)]$ for the final states. We assumed a continuous energy bands in the range $(0, \varepsilon_0)$; V is the positive voltage difference. The role of the lattice-imperfection perturbation U is to trigger a tunneling. Because of this incidental nature we assumed a constant absolute squared matrix element:

$$M_1^2 \equiv |\langle p_i| U |p_f\rangle|^2 . \tag{21.8}$$

Not all "holes" arriving at an imperfection may be triggered into tunneling, whence we added a correction factor K_1, which also takes care of the physical dimension of the current density j. The first (second) term corresponds to the upward (downward) particle flow. If the voltage sign is reversed, the upward (downward) contribution is given by the second (first) term with the reversed sign. Thus the current density j is antisymmetric with respect to the voltage.

For a very small V, we obtain from Eqs. (21.6) and (21.7)

$$j = e(\frac{2\pi}{\hbar})K_1 M_1^2 \frac{a_0}{A}(-eV) \int_0^{\varepsilon_0} d\varepsilon\, \mathcal{N}(\varepsilon)\, \frac{df}{d\varepsilon}. \tag{21.9}$$

The effects of the quantum statistical factor $1 - f$ are canceled out in this expression. The Fermi distribution function f is normalized such that

$$\int_0^{\varepsilon_0} d\varepsilon\, \mathcal{N}(\varepsilon)\, f(\varepsilon) = An_1, \qquad \mathcal{N}(\varepsilon) = \frac{m^* A}{\pi\hbar^2}, \tag{21.10}$$

where n_1 is the "hole" density in 2D. We assume a high degeneracy for the "holes". After straightforward calculations, we obtain

$$j = e^2 V(\frac{2\pi}{\hbar})K_1 M_1^2 a_0 \frac{n_1}{\varepsilon_F}. \tag{21.11}$$

Using Ohm's first law:

$$j = \sigma E, \qquad E = V/a_0, \tag{21.12}$$

we then obtain an expression for the conductivity σ_1 as

$$\sigma_1 = \alpha_1 n_1, \qquad \alpha_1 \equiv e^2 a_0^2 (\frac{2\pi}{\hbar}) \frac{K_1 M_1^2}{\varepsilon_F}. \tag{21.13}$$

This σ_1 is temperature-independent, which is reasonable since this contribution arises from the lattice-imperfection, and the jump rate is uninfluenced by the T-dependent lattice vibration.

The contribution of the "electron" can be calculated similarly, and the conductivity is given by formula (21.13) with n_1 = "electron" density.

Let us now consider + pairons, each having charge $2e$. We assume that **(a)** the pairon moves in the plane with a linear dispersion relation

$$\varepsilon = \begin{cases} (2/\pi)v_F p \equiv cp, & p < p_0 \equiv |w_0|/c \\ 0 & p > p_0, \end{cases} \tag{21.14}$$

where w_0 is the pairon ground-state energy and $v_F \equiv (2\varepsilon_F/m^*)^{1/2}$ the Fermi velocity; **(b)** the pairons move as bosons, and their population is given by the Bose distribution function $F(\varepsilon)$. Using Eqs. (21.4), (21.5) and (21.14), we obtain

$$\begin{aligned} j &= (2e)K_2(\frac{2\pi}{\hbar})\frac{2\pi M_2^2 a_0}{(2\pi\hbar)^2 c^2 A}\{\int_0^{cp_0-2eV} d\varepsilon\, \varepsilon\, F_1(\varepsilon)[1+F_2(\varepsilon)] \\ &\quad - \int_{2eV}^{cp_0} d\varepsilon\, F_2(\varepsilon-2eV)[1+F_1(\varepsilon-2eV)]\}, \end{aligned} \tag{21.15}$$

where we included the quantum statistical factors $[1+F(\varepsilon_f)]$ for the final states and the correction factor K_2.

Assuming a very small voltage, we obtain

$$j = -(2e)^2 V(\frac{2\pi}{\hbar})K_2 M_2^2 \frac{2\pi}{(2\pi\hbar)^2}\frac{a_0}{c^2 A}\int_0^{cp_0} d\varepsilon\, \varepsilon\, \frac{dF(\varepsilon)}{d\varepsilon}. \tag{21.16}$$

The effect of the quantum statistical factor $1-F$ is cancelled out in this expression. The function F is normalized such that

$$\frac{2\pi}{(2\pi\hbar)^2 c^2 A}\int_0^{cp_0} d\varepsilon\, \varepsilon\, F(\varepsilon) = n_2 = \text{ pairon density.} \tag{21.17}$$

The Bose distribution function F above T_c will be approximated by the Boltzmann distribution function:

$$F \sim F_c \equiv e^{-\beta(\varepsilon-\mu)}, \tag{21.18}$$

which is normalized by the same form (21.17). Using

$$-\int_0^{\infty} d\varepsilon\, \varepsilon\, \frac{dF_c}{d\varepsilon} = \int_0^{\infty} d\varepsilon\, F_c = e^{\beta\mu}\beta^{-1}, \qquad \int_0^{\infty} d\varepsilon\, \varepsilon\, F_c = e^{\beta\mu}\beta^{-2}, \tag{21.19}$$

and Ohm's law, we then obtain from Eq. (21.16)

$$\sigma_2 = \beta_2 n_2 T, \qquad \beta_2 \equiv (2e)^2 a_0^2(\frac{2\pi}{\hbar})K_2 M_2^2 k_B. \tag{21.20}$$

The case of $-$ pairons, each having charge $-2e$, can be treated similarly. Formulas (21.13) and (21.20) will be used to analyze the data in the following section.

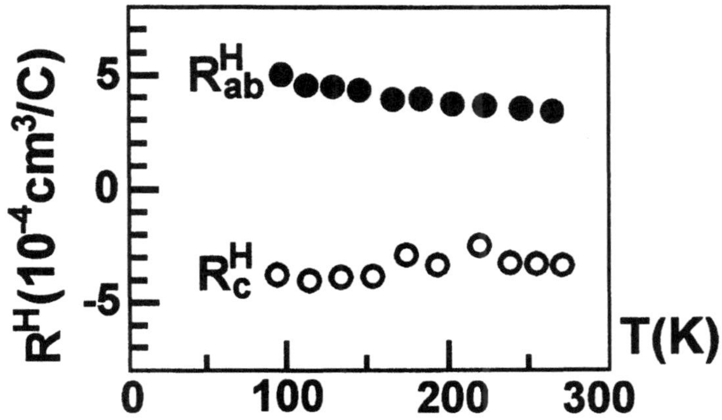

Figure 21.3: Temperature dependence of the Hall coefficients R^H_{ab} an R^H_c. for single-crystal $YBa_2Cu_3O_{6.92}$ after Terasaki et al. [1].

21.3 Data Analysis

The data by Terasaki *et al.* [1] for the *c*-axis-induced-voltage Hall coefficient R^H_c and the *ab* plane-current Hall coefficient R^H_{ab} are reproduced in Fig. 21.3. Note the different signs, indicating that the dominant carriers are different for the *c*-axis and *ab*-plane currents. The measured Hall coefficient R^H_c is negative. Hence we have either "electrons" or − pairons for the carrier. Slower particles are more likely to be trapped by the perturbation and going into tunneling. Since − pairons move with the smallest speed: $(2/\pi)v_F$, we assume that the − pairon is the dominant carrier in the *c*-axis current. We then obtain from Eq. (21.20)

$$\rho_2 = \frac{1}{\sigma_2} = C_2\frac{1}{T}, \qquad C_2 \equiv \frac{1}{\beta_2 n_2}, \tag{21.21}$$

which exhibits the experimentally observed T^{-1}-law. If "electrons" were chosen as the carriers, their contribution would have been T-independent, see Eq. (21.13). Hence we conclude that the dominant carrier in the *c*-axis current is − pairons. Experiments show, see Fig. 21.3, that R^H_c is T-independent, which also supports this conclusion. In fact, the Hall coefficient for −pairons alone,

$$R_H = -\,(2en_{-pairon})^{-1}, \tag{21.22}$$

is T-independent since the pairon density n_{pairon} is roughly constant below the Debye temperature (450 K). From the measured values: $\left|R_c^H\right| = 3.5 \times 10^{-4}$ cm^3/C we estimate $n_{pairon} = 10^{22}$ cm^{-3} (3D), which is reasonable.

We used formula: $R = R_{ab} + R_{jump}$ with the assumption that the current paths in the ab plane are macroscopic so that the carriers suffer a number of scatterings (mainly by phonons, see below) between the tunneling. To obtain the resistivity corresponding to these current paths, we imagine a unit size (1 cm^3) sample and obtain

$$\rho_{c,path} = C_1 \rho_{ab}, \qquad C_1 = 35. \tag{21.23}$$

The in-plane resistivity ρ_{ab} obeys roughly the T-linear law, see Fig. 21.1:

$$\rho_{ab} \propto T, \tag{21.24}$$

the characteristic behavior which has repeatedly been observed in many cuprates at optimum doping (highest T_c) [6]. The fact that the data for ρ_c can be fitted with two terms having different T-dependencies supports the applicability of generalized Ohm's second law (21.3), the basic assumption in the present theory.

21.4 Discussion

We have shown that the out-of-plane resistivity ρ_c has two components, one arising from the resistive conduction in the ab plane and the other from the quantum tunneling between the planes. The tunneling particles in YBCO are −pairons, the slowest moving among "electrons", "holes", and +pairons and −pairons. These tunneling pairons generate a T^{-1}-law: C_2/T. This effect is more important in the underdoped region, where there are relatively less "holes" than in the overdoped region. The experimental check is called for here in other cuprates including $Bi_2Sr_2Ca_{n-1}Cu_nO_{4+2n+\delta}$, $Tl_2Ba_2Ca_{n-1}Cu_nO_{4+2n+\delta}$ and $HgBa_2Ca_{n-1}Cu_nO_{2+2n+\delta}$, $n = 1, 2, 3, \ldots$.

The T^{-1}-law (or decreasing-in-T) behavior for ρ_c has been reported not only in cuprates but also in non-copper perovskite superconductors [7]. The phonon exchange generates ± pairons in equal numbers in the superconductor: $n_{+pairon} = n_{-pairon}$. In the low temperature region, we then have

$$\left|R_c^H\right| \approx \left|R_{ab}^H\right|. \tag{21.25}$$

which is consistent with the experimental data, see Fig. 21.3. The experimental R_{ab}^H decreases with the temperature T. This is due to the "holes" (in addition to the + pairons), whose density increases with T.

In summary the transport properties (ρ_c, R_c^H) and $(\rho_{ab}, R_{ab}^H, \cot\theta_H)$ of $YBa_2Cu_3O_y$ above T_c, are explained consistently based on independent pairon model in which there are free-moving $\pm$ pairons besides "electrons" and "holes" in the superconductor.

Chapter 22

Seebeck Coefficient (Thermopower)

Based on the idea that different temperatures generate different carrier densities and the resulting diffusion causes the thermal emf, a new formula for the Seebeck coefficient (thermopower) S is obtained: $S = (2\ln 2/3)(qn)^{-1}\varepsilon_F k_B$ $(\mathcal{N}_0/\mathcal{V})$, where q, n, ε_F, $\mathcal{N}_0$ and $\mathcal{V}$ are respectively charge, carrier density, Fermi energy, density of states at ε_F and volume. Ohmic and Seebeck currents are fundamentally different in nature. This difference can cause significantly different transport behaviors. For a multi-carrier metal the Einstein relation between the conductivity and the diffusion coefficient does not hold in general. Seebeck (S) and Hall (R_H) coefficients in noble metals have opposite signs. This is shown to arise from the Fermi surface having "necks" at the Brillouin boundary. The measured in-plane Seebeck coefficient S_{ab} in single-crystal $YBa_2Cu_3O_{7-\delta}$ is negative and T-independent. This behavior is shown to arise from the thermal diffusion of "electrons", which are the minority carriers in the ab-plane conduction. The measured out-of-plane Seebeck coefficient S_c is positive and shows a T-linear dependence. This behavior arises from the quantum tunneling between the copper planes of the − pairons.

22.1 Introduction

When a metallic bar is subjected to a voltage (V) or a temperature (T) difference, the electric current is generated. For small voltage and tempera-

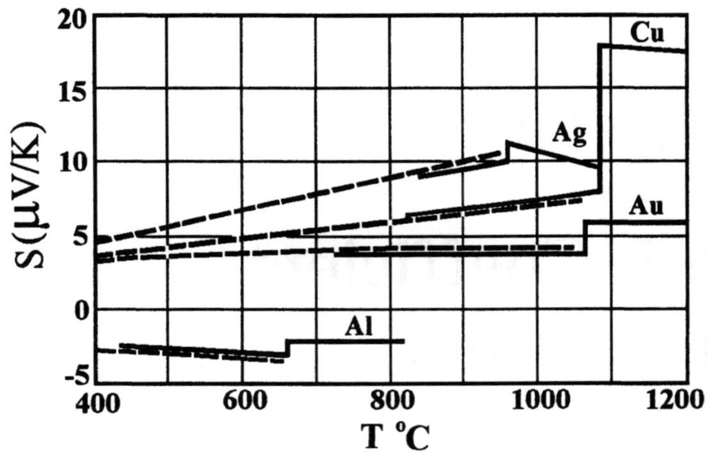

Figure 22.1: High temperature Seebeck coefficients above 400°C for Ag, Al, Au, and Cu.

ture gradients we may assume a linear relation between the electric current density $\mathbf{j}$ and the gradients:

$$\mathbf{j} = \sigma(-\nabla V) + A(-\nabla T) = \sigma \mathbf{E} - A\nabla T, \tag{22.1}$$

where σ is the conductivity. In the condition where the ends of the conducting bar are maintained at different temperatures, no electric current flows. Thus from Eq. (22.1),

$$\sigma \mathbf{E}_s - A\nabla T = 0, \tag{22.2}$$

where $\mathbf{E}_s$ is the field generated by the thermal emf. The *Seebeck coefficient*, also called the *thermopower*, S is defined through

$$\mathbf{E}_s = S\nabla T, \qquad S \equiv A/\sigma, \tag{22.3}$$

The conductivity σ is positive but the Seebeck coefficient S can be positive or negative. As shown in Fig. 22.1, the measured S in Al at high temperatures (400-700 °C) is negative while the S in noble metals (Cu, Ag, Au) are positive [1].

Based on the idea that different temperatures generate different average electron velocities, (which is reasonable under the assumption of a classical statistics) and different drift velocities, we obtain (Problem 22.1.1)

$$S = -\frac{c_v}{3ne}, \tag{22.4}$$

where c_v is the specific heat. Setting c_v equal to $3nk_B/2$, we obtain the classical formula:

$$S_{\text{classical}} = -\frac{k_B}{2e} = -0.43 \times 10^{-4} \text{ volt K}^{-1}. \tag{22.5}$$

Observed metallic Seebeck coefficients at room temperature are of the order of microvolts per degree, see Fig. 22.1., a factor of 10 smaller than $S_{classical}$. If we introduce the Fermi-statistically computed specific heat $c_v = (\pi^2/3)k_B^2 T \mathcal{N}_0$ in Eq. (22.4), we obtain

$$S_{\text{semi-quantum}} = -\frac{\pi}{6}\frac{k_B}{e}\left(\frac{k_B T}{\varepsilon_F}\right), \tag{22.6}$$

which is often quoted in materials-handbooks [1]. Formula (22.6) remedies the difficulty with respect to the magnitude. But the correct theory must explain the two possible signs of S.

Terasaki *et al.* [2] measured in-plane (ab) and out-of-plane (c) Seebeck coefficients (S_{ab}, S_c) and resistivities (ρ_{ab}, ρ_c) and Hall coefficients (R_{ab}^H, R_c^H) in single-crystal $YBa_2Cu_3O_{7-\delta}$. The data are reproduced in Fig. 22.2. We note different signs:

$$S_{ab} < 0, \qquad S_c > 0, \tag{22.7}$$

and different T-dependence:

$$S_{ab} = \text{constant}, \qquad S_c \propto T. \tag{22.8}$$

Also the Hall and Seebeck coefficients have different signs:

$$R_{ab}^H > 0, \qquad S_{ab} < 0, \tag{22.9}$$

$$R_c^H < 0, \qquad S_c > 0, \tag{22.10}$$

which is against the conventional wisdom.

We shall develop a quantum theory of the Seebeck coefficient [4] and explain the sign and the T-dependence of the Seebeck coefficient under the assumption that there are "electrons", "holes" and $\pm$ pairons in the cuprates above T_c.

Problem 22.1.1. Derive formula (22.4), using kinetic theory.

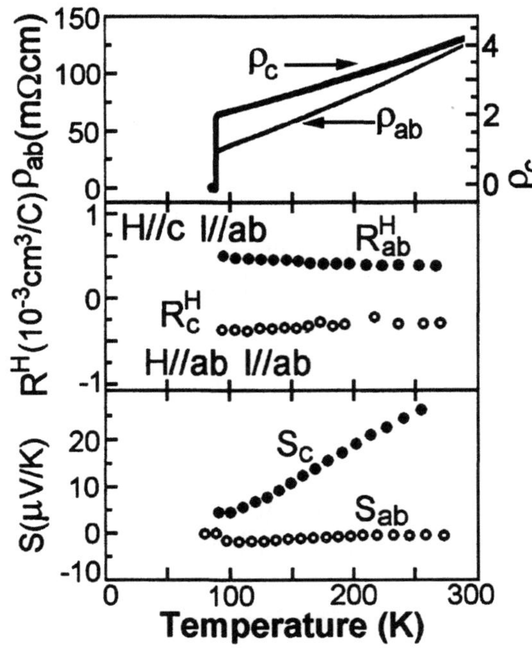

Figure 22.2: Temperature-dependent normal-state transport of highly oxygenated $YBa_2Cu_3O_y$ after Terasaki et al. [2]. In-plane (ρ_{ab}) and out-of-plane (ρ_c) resistivities are shown in the top panel; Hall coefficients with the Hall voltage in the ab plane (R_{ab}^H) and that along the c-axis (R_c^H) are shown in the middle panel; in-plane (S_{ab}) and out-of-plane (S_c) Seebeck coefficients are shown in the bottom panel.

22.2 Theory

We assume that the carriers are conduction electrons, each having charge q [$= -(+)e$ for "electron" ("hole")] and mass m^*. Assuming a one-component system, the conductivity σ after standard kinetic theory is

$$\sigma = \frac{q^2 n \tau}{m^*}, \tag{22.11}$$

where n is the carrier density and τ the mean free time. Note that σ is always positive irrespective of whether $q = -e$ or $+e$. The Fermi distribution

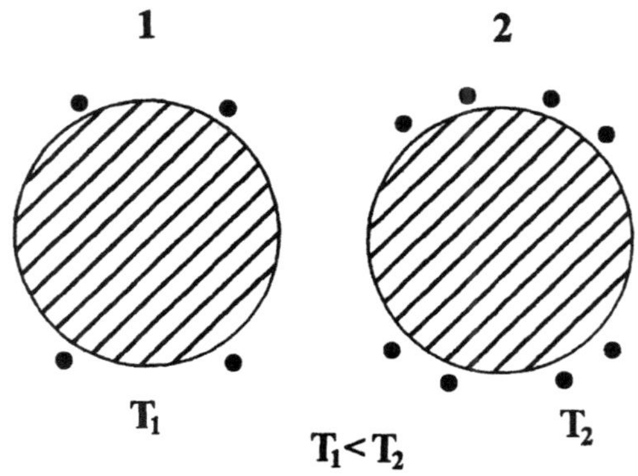

Figure 22.3: More "electrons" are excited at the high temperature end: $T_2 > T_1$. "Electrons" diffuse from 2 to 1.

function f is

$$f(\varepsilon;\beta,\mu)=\frac{1}{e^{\beta(\varepsilon-\mu)}+1},\qquad \beta\equiv(k_BT)^{-1},\tag{22.12}$$

where μ is the chemical potential whose value at 0 K is the Fermi energy ε_F. The voltage difference $\Delta V = LE$, with L being the sample length, generates the chemical potential difference $\Delta\mu$, the change in f and consequently the electric current. Similarly the temperature difference ΔT generates the change in f and the current. We assume the high Fermi degeneracy: $T_F \gg T$.

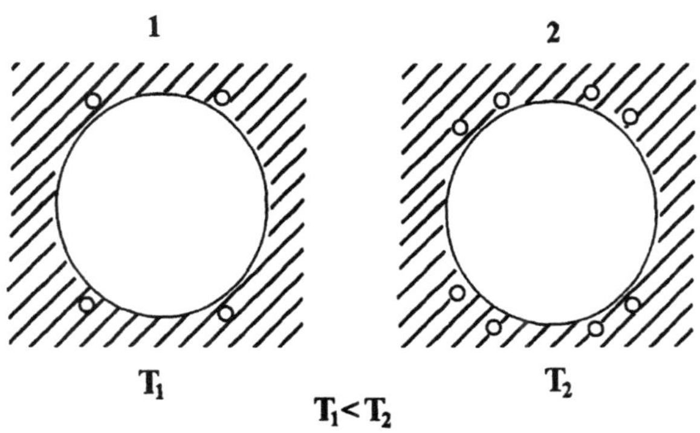

Figure 22.4: More "holes" are excited at the high temperature end: $T_2 > T_1$. "Holes" diffuse from 2 to 1.

At 0 K the Fermi surface is sharp and there are no conduction electrons. At a finite T "electrons" ("holes") are thermally excited near the Fermi surface if the curvature of the surface is negative (positive), see Figs. 22.3 and 22.4. Consider first the case of "electrons". The number of thermally excited "electrons", N_x, having energies greater than ε_F is defined and calculated as [4] (Problem 22.2.1)

$$N_x \equiv \int_{\varepsilon_F}^{\infty} d\varepsilon \mathcal{N}(\varepsilon) \frac{1}{e^{\beta(\varepsilon-\mu)}+1} = \ln 2\, k_B T \mathcal{N}_0, \qquad \mathcal{N}_0 = \mathcal{N}(\varepsilon_F). \tag{22.13}$$

where $\mathcal{N}$ is the density of states. Then the excited particle density $n \equiv N_x/V$ is higher at the high-temperature end and the particle current runs from the high- to the low-temperature end. This means that the electric current runs towards (away from) the high-temperature end in an "electron" ("hole)-rich material. Hence using Eqs. (22.1) and (22.3), we obtain:

$$\begin{aligned} S &< 0 \qquad \text{for "electrons"}, \\ S &> 0 \qquad \text{for "holes"}. \end{aligned} \tag{22.14}$$

At low temperatures the phonon drag effect due to the non-equilibrium phonon distribution is known to be important in the noble metals [1]. We neglected this contribution, which is normally important below 100 K, in the above calculation.

The Seebeck current arises from the thermal diffusion. We assume Fick's law:

$$\mathbf{j} = q\,\mathbf{j}_{\text{particle}} = -qD\boldsymbol{\nabla} n, \tag{22.15}$$

where D is the diffusion coefficient, which is computed from the standard formula:

$$D = \frac{1}{d} v\, l = \frac{1}{d} v_F^2\, \tau, \qquad v = v_F, \tag{22.16}$$

where d is the dimension. (In this chapter the dimension is denoted by d instead of D, which denotes the diffusion coefficient.) In the present case $d = 2$. The density gradient $\boldsymbol{\nabla} n$ is generated by the temperature gradient $\boldsymbol{\nabla} T$ and is given by

$$\boldsymbol{\nabla} n = \frac{\ln 2}{dV} k_B \mathcal{N}_0\, \boldsymbol{\nabla} T, \tag{22.17}$$

where Eq. (22.13) is used. Using the last three equations and Eq. (22.1), we obtain

$$A = \frac{\ln 2}{V} q v_F^2\, k_B \mathcal{N}_0 \tau. \tag{22.18}$$

Using Eqs. (22.3), (22.11) and (22.18), we get

$$S = \frac{A}{\sigma} = \frac{\ln 2}{d}(\frac{1}{qn})\varepsilon_F \, k_B \frac{\mathcal{N}_0}{V}. \tag{22.19}$$

Note that τ is compensated from numerator and denominator.

Problem 22.2.1. Verify Eq. (22.13).

22.3 Discussion

Our formula (22.19) for the Seebeck coefficient S, was derived based on the idea that the Seebeck emf arises for the thermal diffusion. We used the high Fermi degeneracy condition $T_F \gg T$. The relative errors due to this approximation and the neglect of the T-dependence of μ are both of the order $(k_BT/\varepsilon_F)^2$. Formula (22.19) can be negative or positive while the handbook formula (22.6) has the negative sign, a deficiency. The average speed v for highly degenerate electrons is equal to the Fermi velocity v_F (independent of T). Hence semi-classical formulas (22.4)-(22.6) break down. Ashcroft and Mermin [4] (AM) discussed the origin of a positive S in terms of a mass tensor $M = \{m_{ij}\}$. This tensor M is real and symmetric, and hence it can be characterized by the principal masses $\{m_j\}$. Formula for S obtained by AM, [4], Eq. (13.62), can be positive or negative, but it has a complicated structure and is hard to apply in practice. In contrast our formula (22.19) can be interpreted straightforwardly. Besides our formula for a one-carrier system is T- independent while the AM formula is linear in T. This difference arises from the fact that the thermal diffusion is the cause of the Seebeck current.

Formula (22.19) is remarkably similar to the standard formula for the Hall coefficient:

$$R_H = (qn)^{-1}. \tag{22.20}$$

Both Seebeck and Hall coefficients are inversely proportional to charge q, and hence they give important information about the carrier charge sign. In fact the measurement of the thermopower of a semiconductor can be used to see if the conductor is n-type or p-type (with no magnetic measurements). If only one kind of carrier exists in a conductor, the Seebeck and Hall coefficients must have the same sign.

Let us consider the electric current caused by a voltage difference. The current is generated by the electric force which acts on all electrons. The

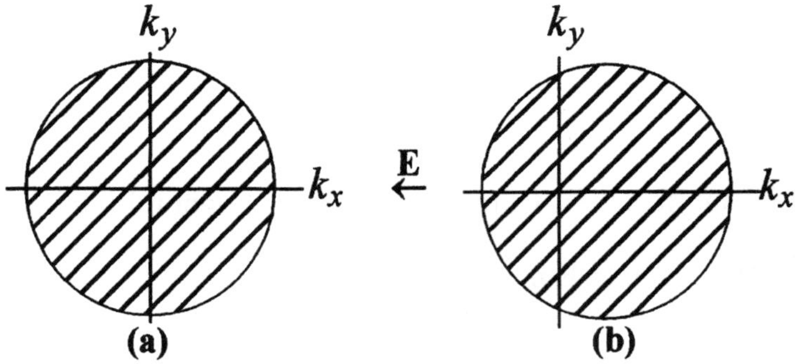

Figure 22.5: Due to the electric field E pointed in the negative x-direction, the steady-state electron distribution in (b) is generated, which is a translation of the equilibrium distribution in (a) by the amount $\hbar^{-1}eE\tau$.

electron's response depends on its mass m^*. The conductivity σ representing this response depends directly on m^*. The n-dependence of σ can be understood by examining the current-carrying steady-state (b) in Fig. 22.5. The electric field $\mathbf{E}$ displaces the electron distribution by a small amount $\hbar^{-1}qE\tau$ from the equilibrium distribution in (a). Since all the conduction electrons are displaced, the conductivity σ depends on the particle density n. The Seebeck current is caused by the density difference in the thermally excited electrons near de Fermi surface, and hence the thermal diffusion coefficient A depends on the density of states, $\mathcal{N}_0$, see Eq. (22.18). We further note that the diffusion coefficient D does not depend on m^* directly, see Eq. (22.16). Thus the Ohmic and Seebeck currents are fundamentally different in nature.

For a single-carrier metal such as an alkali metal Na, where only "electrons" exist both R_H and S are negative. The Einstein relation between the conductivity σ and the diffusion coefficient D holds: $\sigma \propto D$. In fact using Eqs. (22.11) and (22.16), we obtain

$$\frac{D}{\sigma} = \left(v_F^2\tau/3\right) / \left(q^2 n\tau/m^*\right) = \frac{2}{3}\frac{\varepsilon_F}{q^2 n}, \tag{22.21}$$

which is a material constant. For a multi-carrier system, the Einstein relation is valid for each (ideal) carrier. But the relation does not hold in general for the total coefficients. In fact, the ratio D/σ for a two-carrier system containing "electrons" (1) and "holes" (2) is given by

$$\frac{D}{\sigma} = \left(\frac{1}{3}v_1^2\tau_1 + \frac{1}{3}v_2^2\tau_2\right) / \left(q_1^2\frac{n_1}{m_1}\tau_1 + q_2^2\frac{n_2}{m_2}\tau_2\right), \tag{22.22}$$

which is a complicate function of (m_1/m_2), (n_1/n_2), (v_1/v_2) and (τ_1/τ_2). In particular the mass ratio m_1/m_2 may vary significantly for a heavy fermion condition which occurs whenever the Fermi surface just touches the Brillouin boundary, see below. Experimental check on the violation of the Einstein relation may be carried out by simply examining the T-dependence of the ratio D/σ. This ratio from Eq. (22.21) is constant for a single-carrier system while from Eq. (22.22) it depends on T since the generally T-dependent mean free times (τ_1, τ_2) arising from the electron-phonon scattering do not cancel out from numerator and denominator. Conversely, if the Einstein relation holds for a metal, the spherical Fermi surface approximation with a single effective mass m^* is valid for this single-carrier metal.

The difference between the Ohmic and diffusion currents manifests more distinctly in the behavior of the Seebeck coefficient which will be discussed in the next section.

22.4 Seebeck Coefficient in Metals

Our quantum theory of the Seebeck coefficient is applicable in any dimension. In this section we apply it to the 3D normal metals.

Fig. 22.1 shows the T-dependence of the Seebeck coefficient S for selected metals. Note that the S in the familiar noble metals (Cu, Ag, Au) is positive and each curve shows a weak T-dependence above 100 K. In particular the S in Au is positive and nearly T-independent up to the melting point (1060 °C), which cannot be explained in terms of the handbook formula (22.6). The Hall coefficient R_H for the noble metals is negative. The reason why S and R_H have opposite signs must be explained.

Formula (22.19) with $d = 3$ reads:

$$S = \frac{\ln 2}{3}\frac{1}{qn}\varepsilon_F\, k_B \frac{\mathcal{N}_0}{V}. \tag{22.23}$$

The Fermi surface in Ag is far from spherical and has a set of "necks" at the Brillouin boundary, see Fig. 22.6 [4]. The curvatures along the axes of each

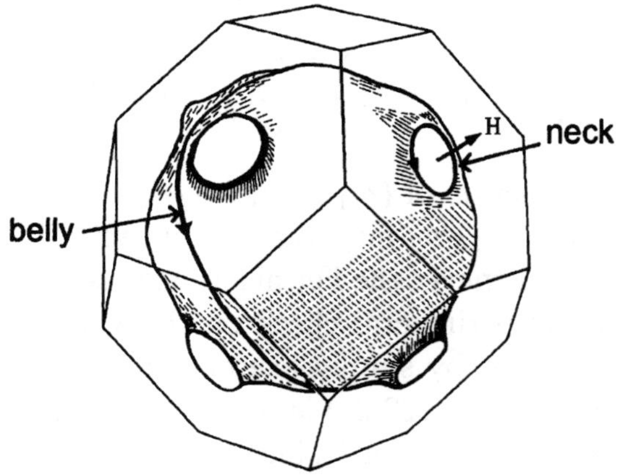

Figure 22.6: The Fermi surface of silver (fcc) has "necks", with the axes in the $\langle 111 \rangle$ direction, located near the Brillouin boundary.

"neck" are positive and hence the Fermi surface is "hole"-generating. Experiments indicate [4] that the minimum neck area A_{111}(neck) in the k-space is 1/51 of the maximum belly area A_{111}(belly), meaning that the Fermi surface just touches the Brillouin boundary (the figure exaggerates the neck area). The density of "hole"-like states, $\mathcal{N}_{\text{hole}}$, associated with the $\langle 111 \rangle$ "necks", having the heavy-fermion character due to the rapidly varying surface with energy, is much greater than that of "electron"-like states, $\mathcal{N}_{\text{electron}}$, associated with the $\langle 100 \rangle$ belly. Then the thermally excited "hole" density is higher than the "electron" density, yielding a positive S. The principal mass m_1 along the axis of a small "neck" ($m_1^{-1} = \partial^2 \varepsilon / \partial p_1^2$) is positive ("hole"-like) and it is extremely large. Hence the "hole" contribution to the conduction is small, ($\sigma \propto m^{*-1}$). So is the "hole" contribution to the Hall voltage. Then the "electrons" associated with the non-neck Fermi surface dominate and yield a negative R_H.

Our formula (22.19) indicates that thermal diffusion contribution to S is T-independent. As noted earlier, the inhomogeneous phonon distribution can account for the phonon-drag contribution, which is important below 100 K. Excluding this component, the observed S in many metals is mildly T-dependent. In particular, the coefficient S in Ag increases slightly before melting while the coefficient S in Au is nearly constant and decreases. These behaviors may come from the incomplete compensation of the scattering effects. "Electrons" and "holes" which are generated from the complicated Fermi surfaces will have different effective masses and densities, and the resulting incomplete compensation of τ's yields a T-dependence.

22.5 In-Plane Seebeck Coefficient S_{ab}

The majority carrier in the ab-plane conduction in $YBa_2Cu_3O_y$ is "hole". The "hole", having smaller m^* and higher $v_F \equiv (2\varepsilon_F/m^*)^{1/2}$, dominates the Ohmic conduction and also the Hall current, yielding positive R^H_{ab}, see Fig. 22.2. But the experiments indicate that $S_{ab} < 0$, which must come from the "electron" diffusion. This puzzle may be solved as follows. [5]

For conduction electrons in the ab plane, we assume the effective mass approximation:

$$\varepsilon = (p_x^2 + p_y^2)(2m^*)^{-1}. \tag{22.24}$$

The density of states (spin degeneracy included), $\mathcal{N}$, is then,

$$\mathcal{N} = m^* \mathcal{A}(\pi\hbar^2)^{-1}, \tag{22.25}$$

which is independent of energy. The "electron" (minority carrier), having heavier mass, contributes more to the coefficient A, and hence the Seebeck coefficient S_{ab} can be negative. In fact, considering the currents due to "electrons"(1) and "holes"(2) we obtain from Eqs. (22.18) and (22.25)

$$\begin{aligned} A &= \frac{-e}{2\pi\hbar^2} v_F^{(1)2} k_B m_1^* \tau_1 + \frac{e}{2\pi\hbar^2} v_F^{(2)2} k_B m_2^* \tau_2 \\ &= \frac{-e}{\pi\hbar^2} \varepsilon_F (\tau_1 - \tau_2). \end{aligned} \tag{22.26}$$

If the acoustic phonon scattering is assumed, and the scattering rates $\gamma_j \equiv \tau_j^{-1}$ are given by

$$\gamma_j = n_{\text{phonon}} v_F^{(j)} s = \frac{n_a k_B T s}{\alpha_0 \hbar \omega_D} v_F^{(j)}, \tag{22.27}$$

where s is the scattering diameter and α_0 a small number (~ 0.20), we obtain

$$\tau_1 - \tau_2 \equiv \frac{1}{\gamma_1} - \frac{1}{\gamma_2} > 0, \qquad v_F^{(1)} < v_F^{(2)}. \tag{22.28}$$

The total conductivity is

$$\sigma = \sigma_1 + \sigma_2 = e^2 n_1 (m_1^*)^{-1} \tau_1 + e^2 n_2 (m_2^*)^{-1} \tau_2. \tag{22.29}$$

Using this and Eq. (22.7) we obtain

$$S = \frac{A}{\sigma} = -\frac{k_B \varepsilon_F}{\pi \hbar^2 e} \left(\frac{1}{v_F^{(1)}} - \frac{1}{v_F^{(2)}} \right) / \left(\frac{n_1}{m_1^*} \frac{1}{v_F^{(1)}} + \frac{n_2}{m_2^*} \frac{1}{v_F^{(2)}} \right) < 0. \tag{22.30}$$

This S which is T-independent explains the experimentally observed behavior. See Fig. 22.2.

22.6 Out-of-Plane Seebeck Coefficient S_c

Terasaki *et al.* [7] measured the out-of-plane resistivity ρ_c, the Hall coefficient R_c^H and others in $YBa_2Cu_3O_y$. In the range $6.6 < y < 6.92$, the data for ρ_c can be fitted with

$$\rho_c = C_1 \rho_{ab} + C_2/T, \qquad C_1, C_2 = \text{constant}. \tag{22.31}$$

Fujita, Ho and Godoy [8] interpreted this behavior based on the model in which the charge carriers are a mixture of "electrons", "holes" and $\pm$ pairons. The second term C_2/T arises from the quantum tunneling between the copper planes of $-$pairons, each having charge $-2e$, moving with a linear dispersion relation

$$\varepsilon = (2/\pi) v_F p \equiv cp, \tag{22.32}$$

The first term $C_1 \rho_{ab}$ arises from the in-plane conduction due to the (predominant) "holes" and +pairons. The Hall coefficient R_c^H (induced field along the c-axis) is observed to be negative, see Fig. 22.2, indicating that the carriers have negative charges. Hence either "electrons" or $-$ pairons are carriers. Only the pairon motion can account for the resistivity component C_2/T. Hence the carriers in the c-axis current are $-$ pairons.

The tunneling current may be calculated as follows. A pairon arrives at a certain lattice-imperfection (impurity, lattice defect, etc.) and quantum-jumps to a neighboring layer with the jump rate given by Fermi's golden rule:

$$\omega = \frac{2\pi}{\hbar} \left| \langle \mathbf{p}_i | U | \mathbf{p}_f \rangle \right|^2 \delta(\varepsilon_i - \varepsilon_f) = \frac{2\pi}{\hbar} M^2 \delta(\varepsilon_i - \varepsilon_f), \tag{22.33}$$

where $\mathbf{p}_i$ ($\mathbf{p}_f$) and ε_i (ε_f) are respectively initial (final) momentum and energy, and U is the imperfection-perturbation. The jump rate is finite only if the final-states are in the continuous energy band, which is implied by $\delta(\varepsilon_i - \varepsilon_f)$. The current density $j_c^{(i)}$ along the c-axis due to a group of particles i having charge $q^{(i)}$ and momentum-energy $(\mathbf{p}, \varepsilon)$ is calculated by

$$j_c^{(i)} = j_{c,1}^{(i)} - j_{c,2}^{(i)} = q^{(i)} a_0 w n^{(i)} (K_1^{(i)} - K_2^{(i)}), \tag{22.34}$$

where $n^{(i)}$ is the number density and a_0 the interlayer distance. The pairon having the linear dispersion relation (22.32) obeys the equation of motion (Problem 22.6.1)

$$\frac{p}{c}\frac{d\mathbf{v}}{dt} = \frac{\varepsilon}{c^2}\frac{d\mathbf{v}}{dt} = q(\mathbf{v} \times \mathbf{B}). \tag{22.35}$$

Lower energy pairons are more likely to get trapped by the imperfection and go into tunneling. We represent this tendency by

$$K = B\frac{1}{\varepsilon}, \qquad B = \text{constant}. \tag{22.36}$$

When a small temperature gradient exists, the thermal average of K is different, causing a current. The temperature difference $\Delta T = T_2 - T_1$, generates the change in the Bose distribution function,

$$F(\varepsilon; T, \mu) \equiv [\exp\beta(\varepsilon - \mu_p) - 1]^{-1}, \tag{22.37}$$

where μ_p is the pairon chemical potential.

We may then compute the c-axis current density j_c as

$$j_c = -2eB\frac{2\pi}{\hbar}\frac{2\pi M^2 a_0}{(2\pi\hbar)^2 A}\int d\varepsilon\,[F(\varepsilon; T + \Delta T, \mu_p) - F(\varepsilon; T, \mu_p)], \tag{22.38}$$

Not all pairons arriving at an imperfection are triggered into tunneling. The factor B contains this correction and also takes care of the physical dimension of j_c. The lower the temperature of the initial state, the tunneling occurs more frequently and hence the particle current $j_{particle}$ runs towards the high-temperature end, the opposite direction to the case of the diffusion in the ab plane. Thus, after considering the carrier charges, we obtain

$$\begin{aligned} S_c &> 0 \quad \text{for } - \text{pairons} \\ S_c &< 0 \quad \text{for } + \text{pairons}. \end{aligned} \tag{22.39}$$

In the present case the carrier is $-$ pairon, and hence $S_c > 0$, which is in accord with the experiments, see Fig. 22.2.

We now calculate expression (22.38), assuming a small ΔT, and obtain (Problem 22.5.2),

$$j_c = \frac{2e}{\hbar^3} BM^2 \frac{a_0}{c^2} \frac{\Delta T}{k_B T^2} \int d\varepsilon \, \frac{dF(\varepsilon)}{d\beta}. \tag{22.40}$$

The function F is normalized such that

$$\frac{1}{A} \int d\varepsilon \, \mathcal{N}_p(\varepsilon) \, F(\varepsilon) = \text{pairon density}, \qquad \mathcal{N}_p(\varepsilon) = \frac{2\pi A \, \varepsilon}{(2\pi\hbar)^2 c^2}. \tag{22.41}$$

At the BE condensation temperature T_c, the pairon chemical potential μ_p vanishes: $\mu(T_c) = 0$. The number $\beta\mu_p \equiv \mu_p/(k_B T)$ is negative and small in magnitude for $T > T_c$. Hence

$$\frac{d}{d\beta} \int_0^{cp_0} d\varepsilon \, F(\varepsilon) \simeq \frac{d}{d\beta} (e^{\beta\mu_p} \frac{1}{\beta}) \simeq \frac{-1}{\beta^2}. \tag{22.42}$$

Using this, we obtain from Eq. (22.38)

$$A_c \cong -\frac{2e}{\hbar^3} BM^2 k_B a_0. \tag{22.43}$$

Note that A_c is T-independent. Experiments [2] indicate that the first term $C_1 \rho_{ab}$ in Eq. (22.31) is dominant for $y > 6.8$ in $YBa_2Cu_3O_y$:

$$\rho_c \cong C_1 \rho_{ab} \propto T \qquad \text{at} \;\; y = 7. \tag{22.44}$$

Hence we obtain

$$S_c \equiv \frac{A_c}{\sigma_c} = A_c \rho_c \propto T > 0, \tag{22.45}$$

which is in agreement with the experiments, see Fig. 22.2.

Problem 22.6.1. Verify Eq. (22.13).

Problem 22.6.2. Verify Eq. (22.40).

22.7 Discussion

Based on the idea that the Seebeck current arises from the diffusion of the thermally excited conduction electrons, formula (22.19) for the Seebeck coefficient S is derived. According to this formula S for a one carrier-system is negative (positive) if the carrier is the "electron" ("hole"), which is the same criterion applied to the Hall coefficient R_H. The Seebeck coefficient S however depends on the density of states, $\mathcal{N}(\varepsilon_F)$, while the Hall coefficient R_H does not.

The measured in-plane Seebeck coefficient S_{ab} in $YBa_2Cu_3O_{7-\delta}$ is negative and T-independent. This is explained using the new formula in terms of the thermal diffusion of "electrons", the minority carriers in the ab-plane conduction. The measured out-of-plane Seebeck coefficient S_c is positive and shows a T-linear dependence. This behavior is explained in terms of the quantum tunneling between the copper planes of the $-$pairons, which are the minority charge carriers. It is quite remarkable that all carriers ("hole", "electron", $+$pairon, $-$pairon) in $YBa_2Cu_3O_{7-\delta}$ contribute explicitly and selectively in various transport coefficients ($\rho_{ab}, \rho_c, R_{ab}^H, R_c^H, S_{ab}, S_c$).

Chapter 23

Magnetic Susceptibility

The curvature inversion of the O-Fermi surface at the end of the overdoping in $La_{2-x}Sr_xCuO_4$, $x = 0.25$, shows up distinctly in the magnetic susceptibility χ. At the inflection point, χ has a maximum value and shows a temperature behavior: $\chi = A_0 + B_0/T$. The pairon has no net spin, and hence its spin contribution to χ is zero. But its motion can contribute diamagnetically. This contribution generates a temperature behavior: $\chi = C_1 - D_1/T$ in the concentration range $(0.06 < x < 0.25)$. These two contributions generate a χ- maximum at T_m in the range $(0.15 < x < 0.25)$.

23.1 Introduction.

Takagi *et al.* [1], Torrance *et al.* [2], Terasaki *et al.* [3] and others studied transport and magnetic properties of $La_{2-x}Sr_xCuO_4$ over a wide range of concentration x including a nonsuperconducting phase beyond the overdoped region. Remarkable change in the resistivity ρ, the Hall coefficient R_H and the magnetic susceptibility χ are observed near the super to normal conductor transition at $x = 0.25$. Previously in Chapter 19 we showed that the coincidence between the disappearance of T_c and the sign change of R_H at the end of overdoping $(x = 0.25)$ can be accounted for by the curvature change of the O-Fermi surface. The susceptibility data after Torrance *et al.* [2] are reproduced in Fig. 23.1. The notable features of the data are:

(a) χ is nearly flat at $x = 0.04$ in $0 < T < 400\,\text{K}$. At this concentration the material is not a superconductor.

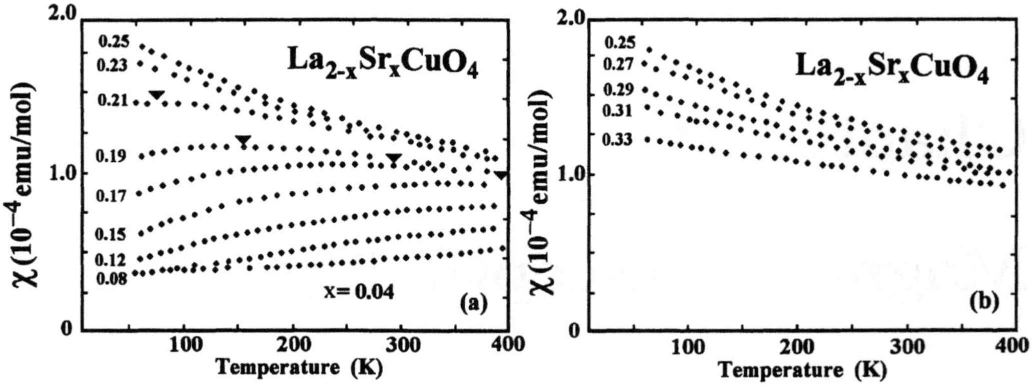

Figure 23.1: Normal-state magnetic susceptibility of $La_{2-x}Sr_xCuO_4$ which, for increasing x, (a) increases in the range $0.04 < x < 0.25$. and (b) decreases for $0.25 < x < 0.33$. The filled triangles ▼ mark the temperatures where χ is a maximum.

(b) χ at 400 K grows with increasing x in the range $(0.04 < x < 0.25)$ and decreases in the range $(0.25 < x < 0.33)$. The turning point $x = 0.25$ roughly corresponds to the concentration at which the superconducting temperature T_c vanishes. See Fig. 19.4.

(c) In the range $(0.33 > x > 0.25)$, where no superconductivity is observed at any temperatures, χ has a T-dependence:

$$\chi = A_0 + B_0/T, \qquad A_0,\ B_0 = \text{ constant.} \tag{23.1}$$

(d) In the range $(0.06 < x < 0.15)$, χ has the following T-dependence:

$$\chi = A_1 - B_1/T, \qquad A_1,\ B_1 = \text{ constant.} \tag{23.2}$$

The end point $x = 0.15$ coincides where the highest T_c is observed, see Figs. 23.1 and 19.4.

(e) In the range $(0.13 < x < 0.25)$, χ has a maximum, indicated by ▼ in Fig. 23.1, at T_m. This T_m disappears at $x = 0.25$, where T_c vanishes.

23.2 Theory.

The feature (a) suggests a Pauli paramagnetism: [4, 5]

$$\chi = \frac{2\mu_B^2}{\mathcal{A}}\mathcal{N}_0(\varepsilon_F), \qquad \mu_B = \text{Bohr magneton} \tag{23.3}$$

where $\mathcal{N}_0(\varepsilon_F)$ is the density of states per spin, and $\mathcal{A}$ the normalization area. Note that this χ is T-independent. We assume the high Fermi degeneracy

$$T_F \gg T. \tag{23.4}$$

If we assume an effective mass (m^*) approximation:

$$\varepsilon = \frac{1}{2m^*}(p_x^2 + p_y^2), \tag{23.5}$$

the density of states per spin in 2D is

$$\mathcal{N}_0(\varepsilon) = m^*\mathcal{A}(2\pi\hbar^2)^{-1}, \tag{23.6}$$

which is independent of energy ε.

At high temperatures (400 K), the electron spin contribution minus the Landau diamagnetism [5], which is ideally one third of the Pauli paramagnetism can account for the behavior (b) as follows.

The doping in $La_{2-x}Sr_xCuO_4$ reduces the number of electrons in the O-Fermi sheet with increasing x, and changes the nature of the Fermi surface from "hole"-like (a) to "electron"-like (b) as shown in Fig. 23.2. Observe the curvature sign change. The density of states $\mathcal{N}$ has a maximum at the inflection point ($x = 0.25$). In this neighborhood χ is T-dependent, which will be shown below. [6]

In the range ($0.25 < x < 0.33$) the Hall coefficient R_H is negative: $R_H < 0$, see Fig. 19.4, indicating that the "electron" is the majority carrier in the conduction. We may explain the behavior (23.1) as follows.

Assume that the density of states per spin at $B = 0$ is given by

$$\mathcal{N}(\varepsilon) = \mathcal{N}_0 + c_0\,\delta(\varepsilon - \varepsilon_0), \qquad c_0,\, \varepsilon_0 = \text{constants}. \tag{23.7}$$

The density of states in an isotropic-mass 2D system is energy-independent, see Eq. (23.6), and this part is represented by the constant $\mathcal{N}_0$. By examining the changes (a)$\rightleftarrows$(b), we see that $\mathcal{N}(\varepsilon)$ grows indefinitely at the inflection

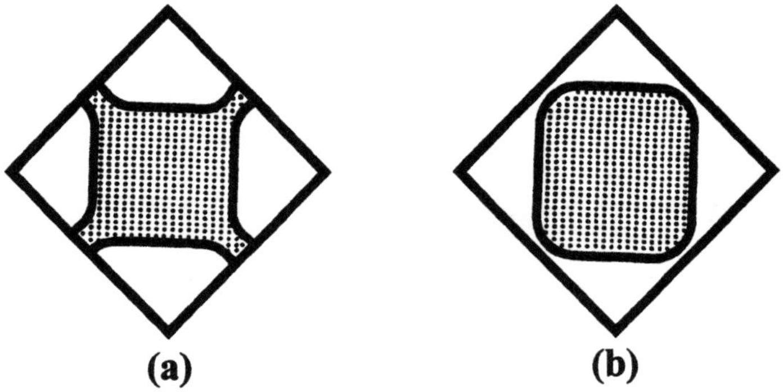

Figure 23.2: The Fermi surface changes its character from "hole"-like in (a) to "electron"-like in (b) as the number of electrons is reduced.

point $\varepsilon = \varepsilon_0$ and it is symmetric in the neighborhood of this point. We can represent this component by the delta-function term $c_0\delta(\varepsilon - \varepsilon_0)$, see Fig. 23.3.

When a magnetic field is applied, the electron energy ε is split, and the up-spin (+) and down-spin (−) electrons have different energies:

$$\varepsilon_{\pm} = \varepsilon_p \mp \mu_B B = \varepsilon \mp x, \qquad x \equiv \mu_B B. \tag{23.8}$$

The magnetization (magnetic moment per unit area) I is

$$I = \frac{\mu_B}{\mathcal{A}}(N_+ - N_-) \equiv \frac{\mu_B}{\mathcal{A}}\Delta N, \tag{23.9}$$

where N_+ (N_-) is the number of up (down) spin electrons. The electron-number difference $\Delta N \equiv N_+ - N_-$ can be read from Fig. 23.4, and is

$$\Delta N = \int_{-x}^{\varepsilon_1 - x} d\varepsilon_+ \mathcal{N}_+(\varepsilon_+) f_+(\varepsilon_+, \mu_+) - \int_{x}^{\varepsilon_1 + x} d\varepsilon_- \mathcal{N}_-(\varepsilon_-) f_-(\varepsilon_-, \mu_-), \tag{23.10}$$

where ε_1 is the upper band edge, see Fig. 23.3. The spin-dependent density of states, $\mathcal{N}_\pm(\varepsilon)$, can be related to that of the no-field system, $\mathcal{N}(\varepsilon)$, by

$$\begin{aligned} \mathcal{N}_+(\varepsilon_+) &= \mathcal{N}(\varepsilon + x) \qquad \varepsilon_+ > -x, \\ \mathcal{N}_-(\varepsilon_-) &= \mathcal{N}(\varepsilon - x), \qquad \varepsilon_- > x. \end{aligned} \tag{23.11}$$

see Fig. 23.4. The two systems of the up- and down- spin electrons are in equilibrium, and hence they have the same chemical potential (the same

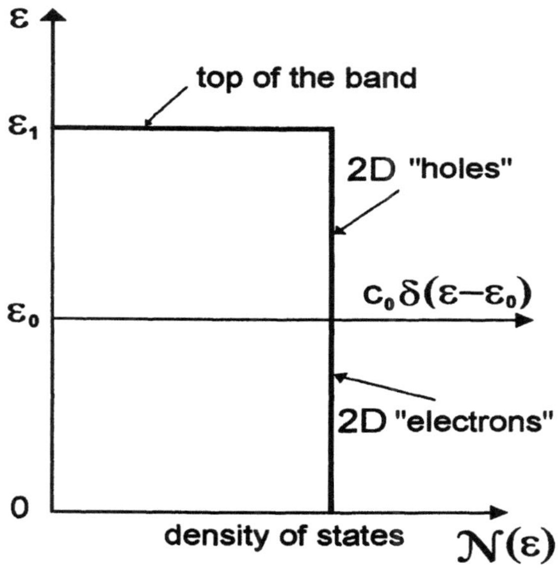

Figure 23.3: The density of states per spin, $\mathcal{N}(\varepsilon)$, at zero B-field.

dotted level in Fig. 23.4.) Using Eqs. (23.11), we obtain from (23.10)

$$\Delta N = \int_0^{\varepsilon_1} d\varepsilon\, \mathcal{N}(\varepsilon)[f(\varepsilon; \varepsilon_F + x) - f(\varepsilon; \varepsilon_F - x)]. \tag{23.12}$$

In general if $\mathcal{N}$ depends on ε, μ_0 must be determined from the normalization condition:

$$N = N_+ + N_- = \int_0^{\varepsilon_1} d\varepsilon\, \mathcal{N}(\varepsilon)[f(\varepsilon; \mu_0 + x) + f(\varepsilon; \mu_0 - x)]. \tag{23.13}$$

In the present model a constant density of states $\mathcal{N}_0$ is assumed so that $\mu_0 = \varepsilon_F$. We now use $\mathcal{N}(\varepsilon)$, given in Eq. (23.7), and evaluate Eq. (23.12). The constant term ($\mathcal{N}_0$) reproduces Pauli's formula (23.3). The delta-function term generates the contribution linear in B:

$$(\Delta N)_\delta = c_0 \mu_B B \frac{1}{2k_B T} \cosh^{-2}\left[(\varepsilon_0 - \varepsilon_F)/(2k_B T)\right]. \tag{23.14}$$

The magnetization I is obtained by multiplying this expression by $\mu_B \mathcal{A}^{-1}$. The susceptibility χ, defined through $I = \chi B$, is

$$\chi_\delta = \frac{c_0 \mu_B^2}{2\mathcal{A}} \frac{1}{k_B T} \cosh^{-2}\left[(\varepsilon_0 - \varepsilon_F)/(2k_B T)\right]. \tag{23.15}$$

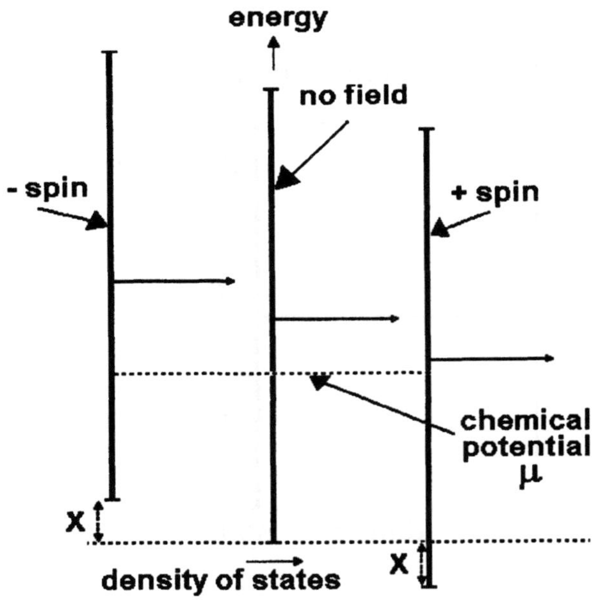

Figure 23.4: The spin dependent density of states $\mathcal{N}_{\pm}(\varepsilon)$ in the presence of a B-field.

This result indicates that: (a) if $\varepsilon_F = \varepsilon_0$, χ_δ behaves like B_0/T, (b) if the Fermi energy ε_F is away from ε_0, this contribution becomes smaller because of $\cosh^{-2}[(\varepsilon_0 - \varepsilon_F)/(2k_BT)]$. Both behaviors are in good agreement with the experimental data, see Fig. 23.1.

The pairon has zero net spin and its spin does not contribute to χ. Its motion in the ab plane can contribute diamagnetically to χ. The pairon density n_2 and the critical temperature T_c are related by

$$T_c = 1.24\,\hbar k_B^{-1} v_F\, n_2^{1/2}. \tag{23.16}$$

Hence the pairon density n_2 is proportional to T_c^2:

$$n_2 \propto T_c^2. \tag{23.17}$$

From Fig. 19.4, we see that n_2 has a maximum at $x = 0.15$. Since

$$v_x = \frac{\partial \varepsilon}{\partial p_x} = c\frac{p_x}{p}, \tag{23.18}$$

each pairon having charge q in the presence of a static B-field moves, following the equation of motion:

$$\frac{p}{c}\frac{d\mathbf{v}}{dt} = q(\mathbf{v} \times \mathbf{B}), \qquad \mathbf{v} \equiv \frac{\partial \varepsilon}{\partial \mathbf{p}}. \tag{23.19}$$

The cyclotron frequency ω_0 is

$$\omega_0 = \frac{c}{p}\,|q|\,B = \frac{c^2}{\varepsilon}\,|q|\,B. \tag{23.20}$$

The magnitude of the magnetic moment μ for a circulating pairon is

$$\mu = (\text{current}) \times (\text{area}) = (\frac{|q|\,v}{2\pi r})\,(\pi r^2) = \frac{1}{2}\,|q|\,r^2\omega_0, \tag{23.21}$$

where r is the cyclotron radius. Thus the magnetic moment μ is proportional to charge magnitude $2e$ and the cyclotron frequency ω_0.

As Landau's treatment indicates [5], the motional diamagnetism cannot be treated in terms of the classical orbits but must be treated quantum mechanically. We take Onsager's quantum flux view of the magnetization. [7] If a magnetic field is applied perpendicular to the ab plane, a set of flux quanta penetrate the plane. A charged particle circulates about the flux, occasionally scattered by impurities, phonons, ... The particle motion acts such that it always reduces the magnetic field energy $\int d^3r\, B^2/(2\mu_0)$. That is, the motion diamagnetically (with $-$ sign) contributes to the magnetization I. We postulate that each pairon contribute an amount proportional to $|q|\,\omega_0$. We then obtain

$$I_{\text{pairon}} = -\alpha \int_0^\infty d\varepsilon\, \mathcal{N}_p(\varepsilon)\,|q|\,\omega_0(\varepsilon)\,F(\varepsilon;\beta,\mu), \qquad \alpha = \text{constant}, \tag{23.22}$$

where F is the Bose distribution function:

$$F(\varepsilon) = \frac{1}{e^{\beta(\varepsilon-\mu_p)} - 1} \tag{23.23}$$

with μ_p being the pairon chemical potential, normalized such that

$$\int d\varepsilon\, \mathcal{N}_p(\varepsilon) F(\varepsilon) = N_2 = \text{pairon number}. \tag{23.24}$$

We note that

$$|q|\,\omega_0 = \frac{c^2}{\varepsilon} q^2 B > 0. \tag{23.25}$$

The magnetization I_{pairon} is negative irrespective of whether $q = 2e$ or $-2e$.

The density of states for pairon, $\mathcal{N}_p(\varepsilon)$, is

$$\mathcal{N}_p(\varepsilon) = \frac{\mathcal{A}\varepsilon}{2\pi\hbar^2 c^2}. \tag{23.26}$$

Far above T_c, the Bose distribution function F can be approximated by a Boltzmann distribution function:

$$F \simeq F_c = e^{-\beta(\varepsilon-\mu)}, \qquad T \gg T_c, \tag{23.27}$$

normalized in the form (23.24). Carrying out the ε-integration in Eq. (23.22) we obtain

$$\chi_{\text{pairon}} = -\Delta\, n_2/T, \qquad n_2 \equiv N_2/\mathcal{A}, \qquad \Delta \equiv \alpha c^2 q^2 k_B^{-1}, \tag{23.28}$$

which generates the second term of the desired formula (23.2), $B_1 = \Delta\, n_2$.

In the range $(0.15 < x < 0.25)$ there are pairons whose density n_2 decreases with increasing x. We may express χ from Eqs. (23.1) and (23.2) as

$$\chi = A(x) + B_0(x)/T - \Delta\, n_2(x)/T. \tag{23.29}$$

At $x = 0.25$, $n_2 = 0$ and Eq. (23.29) is reduced to Eq. (23.1). At $x = 0.15$ the pairon density is highest and $B_0(x)$ is small since $|\varepsilon_0 - \varepsilon_F(x)|$ is not small. Then Eq. (23.29) generates T_m; experimentally $T_m = 390$ K. Between the two extremes, there are T_m as shown in Fig. 23.1. Numerically $T_m = 390$, 260, 150, 70, 20 and 0 K at $x = 0.15$, 0.17, 0.19, 0.21, 0.23 and 0.25 respectively.

23.3 Discussion.

The susceptibility χ in $La_{2-x}Sr_xCuO_4$ above T_c exhibits unusual concentration and temperature dependence. The doping reduces the number of electrons in the copper plane, changes the O-Fermi surface, and generates "holes" in the range $(0.04 < x < 0.25)$ and "electrons" in the range $(0.25 < x < 0.33)$. The dividing point $x = 0.25$ corresponds to the inflection point of the O-Fermi surface, where the density of states, $\mathcal{N}$, is greatest. Hence the susceptibility

χ has a maximum value here. At this inflection point the density of states, $\mathcal{N}$, has a delta-function singularity, yielding a T-dependent term B_0/T for χ. This term declines in magnitude as the concentration x is distanced from $x = 0.25$. The pairons, each having charge $q = 2e$ and moving with the linear dispersion relation: $\varepsilon = (2/\pi)p$, can contribute diamagnetically in the (superconductor) range $(0.06 < x < 0.25)$. Based on the assumption that the diamagnetic current is proportional to $|q|$ and ω_0 (cyclotron frequency), we obtain the term $-\Delta n_2/T$ (for χ), whose magnitude is greatest at $x = 0.15$. The two effects, arising from the Fermi surface curvature inversion and the pairon diamagnetic current, are significant in the range $0.13 < x < 0.25$ and generate the χ-maximum at T_m. Thus all of the unusual behaviors are explained based on the independent pairon model in which pairons coexist with the conduction electrons.

Terasaki *et al.* [3] measured the susceptibilities $(\chi_\perp, \chi_\parallel)$, with the field H perpendicular to, and parallel to, the c-axis, in $La_{2-x}Sr_xCuO_4$, examining χ more closely near the optimum doping $(x = 0.15)$ where T_c is the highest. Their data for $\chi_\perp$ are essentially similar to the data shown in Fig. 23.1. They found, see Fig. 23.5, that $\chi_\perp$ at $x = 0.14, 0.15$ is smaller than $\chi_\perp$ at $x = 0.12$ in the range $(100\,K < T < 200\,K)$, an anomaly which cannot be explained if the magnetization comes from the conduction electrons only since the "hole" density smoothly increases with x. This anomaly can be explained simply in our model. The pairon density n_2 has a peak at $x = 0.15$. Hence the pairon diamagnetic contribution can make a bump in the smooth increase due to the "hole" spin.

The susceptibility of a system of moving carriers is relatively easy to treat. Since the susceptibility χ is an equilibrium property, no complicated scattering effects go into the calculation. We only need to know charge, spin (statistics) and mass (dispersion relation) of each carrier and the Fermi surface of the conduction electrons. Thus the measurement and analysis of χ gives a clear-cut information about the types of the charge-spin carriers. The spin paramagnetism, as seen from (23.3), depends on the density of states, $\mathcal{N}(\varepsilon)$. Near the inflection point $(x = 0.25)$, the Fermi surface is highly anisotropic, see Fig. 23.2. Hence we predict that the susceptibility χ in $La_{2-x}Sr_xCuO_4$ measured with the field $\mathbf{H}_a$ applied at the angle $\pi/4$ relative to the a-axis should show an anisotropy. In our calculation we assumed an isotropic (angle-averaged) density of states in Eq. (23.7). The heat capacity C, which depends on the density of states should also show a distinct maximum at $x = 0.25$.

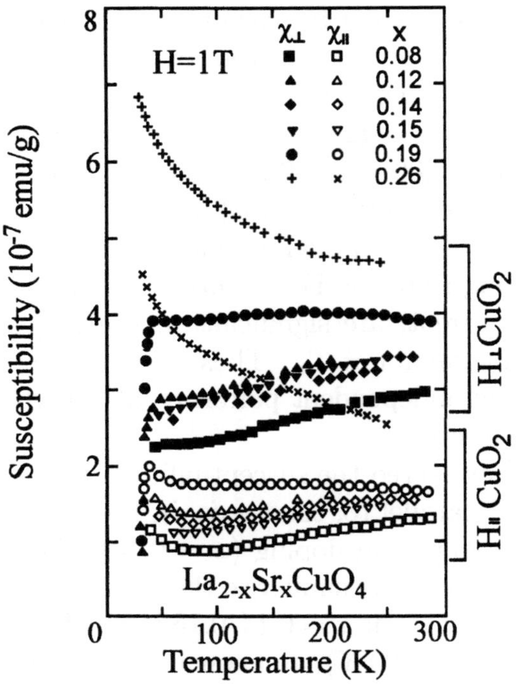

Figure 23.5: Magnetic susceptibilities of single crystal $La_{2-x}Sr_xCuO_4$ with various x under $H = 1\,T$, after Terasaki *et al.* [3]. For each χ, $\chi_\perp(T,x)$ and $\chi_\parallel(T,x)$ are measured. The upper curves (or circles) correspond to $\chi_\perp(T,x)$, while the lower ones correspond to $\chi_\parallel(T,x)$.

Consider Nd_2CuO_4 which is a cuprate of non-perovskite structure. The substitution of trivalent Nd by quadrivalent Ce changes the ionicity of the group (NdO) neighboring the copper plane. This is turn increases the number of electrons in the Cu-sublattice. Thus the doping in $Nd_{2-x}Ce_xCuO_{4-\delta}$ generates "electrons" in the Cu-Fermi sheet, which is seen by the negative R_H, see Fig. 19.1 (b). A further doping eventually changes the Cu-Fermi surface from "electron"-like to "hole"-like. This change occurs at $x = 0.18$, see Fig. 19.1 (b), which roughly corresponds to the vanishing T_c. We predict that the susceptibility χ has a maximum at $x = 0.18$, and has a temperature behavior: $\chi = A_0 + B_0/T$. The experimental check is required here.

Chapter 24

Infrared Hall Effect

A kinetic theory is developed for the Infrared (IR) Hall effect. Expressions for the dynamic conductivity $\sigma(\omega)$, $\omega =$ laser frequency, and other transport coefficients for a system of conduction electrons (or pairons) are obtained by applying the conversion rule: $\gamma_0 \to \gamma - i\omega$ to those for the static coefficients, where γ_0 (γ) are the static (dynamic) scattering rates which depend on frequency ω and temperature T. If the real (Re) and imaginary (Im) parts of $\sigma(\omega, T)$ are measured simultaneously, the ratio $Re[\sigma(\omega,T)]/Im[\sigma(\omega,T)]$ is equal to $\gamma(\omega,T)/\omega$, which *directly* gives the dynamic rate $\gamma(\omega, T)$. The ratio $Re[\cot\theta_H]/Im[\cot\theta_H] = -\gamma_H(\omega,T)/\omega$ *directly* yields the dynamic Hall rate $\gamma_H(\omega, T)$. The IR Hall effect experiments in Au and Cu give a remarkable result: $\gamma_H(\omega,T) = \gamma_{H,0}(T)$, that is, the dynamic Hall scattering rate is equal to the static rate in the mid IR $\sim 1{,}000$ cm^{-1}. The data for dynamic conductivity $\sigma(\omega)$, Hall coefficient $R_H(\omega)$ and $\cot\theta_H(\omega)$ above T_c in $YBa_2Cu_3O_7$ are analyzed based on the model in which "holes" and pairons are scattered by acoustic phonons. The low-energy pairon moving with the linear dispersion relation: $\varepsilon = (2/\pi)/v_F p$ is accelerated strongly in proportion to ε^{-1}, and hence the pairon drift velocity dominates the "hole" drift velocity. This effect, combined with the T-linear phonon scattering rate, generates a T^2-dependence of Re$[\cot\theta_H]$.

24.1 Introduction

Recently Cerne *et al.* [1] measured IR ($\sim 1{,}000$ cm^{-1}) Faraday rotation angle θ_F and circular dichroism (optical ellipticity) in Au (Cu) thin films, using

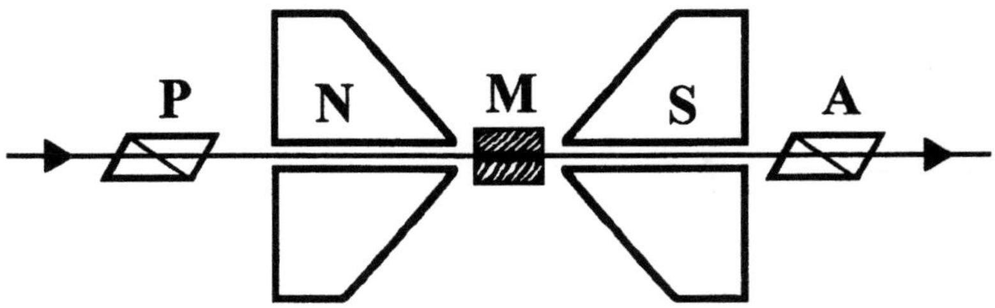

Figure 24.1: A linearly polarized laser beam is sent along the magnetic axis. After passing the material sample M, the beam at the analyzer at A shows an elliptic polarization accompanied by its major axis rotation.

sensitive polarization modulation techniques at temperatures in the range ($20\,\mathrm{K} < T < 300\,\mathrm{K}$) and magnetic fields up to 8 T. Fig. 24.1 shows a schematic sketch of the experimental set-up. A linearly polarized laser is sent along the magnetic-pole axis. The laser analyzed at A after passing the material sample M shows an elliptic polarization accompanied by its major axis rotation. In the measurements the Faraday rotation θ_F and the optical ellipticity, both of which are proportional to the sample thickness and the magnetic field, are very small, see Fig. 24.2, and θ_F is linearly related to the Hall angle θ_H.

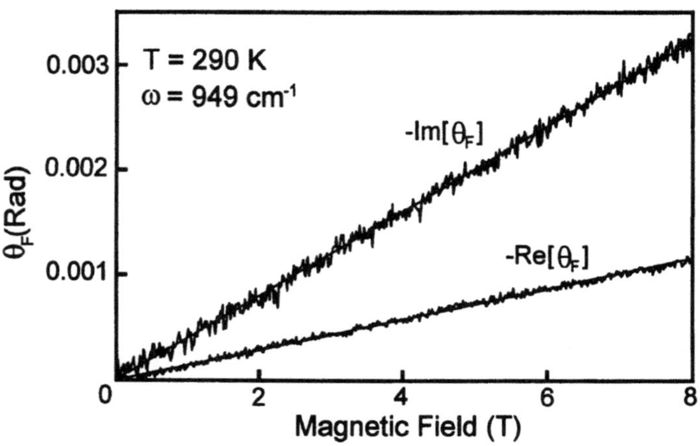

Figure 24.2:

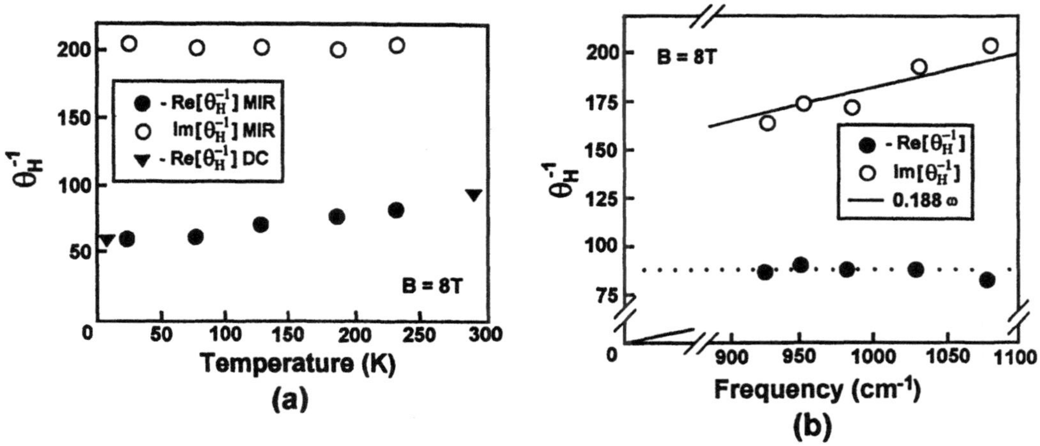

Figure 24.3: The dynamical Hall angle $\theta_H^{-1}(\omega, T)$ at magnetic field of 8 T in Au after Cerne et al. [1]. (a) The temperature dependence at 1079 cm^{-1} (b) The frequency dependence at 290 K.

Combined with zero-field conductivity data, they obtained the real (Re) and imaginary (Im) parts of the dynamic $\cot\theta_H$. In the measurements the Hall angle θ_H is very small ($\sim 10^{-3}$) so that $\tan\theta_H \simeq \theta_H$. The data for temperature T-dependent $\mathrm{Re}[\theta_H^{-1}]$ and $\mathrm{Im}[\theta_H^{-1}]$ at 1079 cm^{-1} and 8 T in Au are reproduced in Fig. 24.3 (a). $-$IR $\mathrm{Re}[\theta_H^{-1}]$ ($\bullet$) is T-linear and agrees with the measured $-$dc θ_H^{-1} ($\blacktriangledown$) while IR $\mathrm{Im}[\theta_H^{-1}]$ ($\circ$) is positive, T-independent and ω-linear, ω = laser frequency, see Fig. 24.3 (b). In summary

$$-\mathrm{Re}[\theta_H^{-1}] \propto -T, \qquad \mathrm{Im}[\theta_H^{-1}] \propto \omega. \tag{24.1}$$

Cerne *et al.* [2] extended the IR Hall effect study to $YBa_2Cu_3O_7$. The data for temperature (T)-dependent $\mathrm{Re}[\theta_H^{-1}]$ and $\mathrm{Im}[\theta_H^{-1}]$ at 8 T are reproduced in Fig. 24.4. The solid line in Fig. 24.4 (a) above $T_c = 88.2$ K represents the dc θ_H^{-1} at 8 T, which is seen to agree with the IR $\mathrm{Re}[\theta_H^{-1}]$ at 949 cm^{-1} ($\bullet$) and 1079 cm^{-1} ($\triangle$). The IR $\mathrm{Im}[\theta_H^{-1}]$ is negative, and ω-linear. In summary

$$\mathrm{Re}[\theta_H^{-1}] \propto T^2, \qquad \mathrm{Im}[\theta_H^{-1}] \propto -\omega. \tag{24.2}$$

Clearly different T-dependence of $\mathrm{Re}[\theta_H^{-1}]$ in Eqs. (24.1) and (24.2) indicate that differnt particles are involved for the two cases (Au, YBCO). We

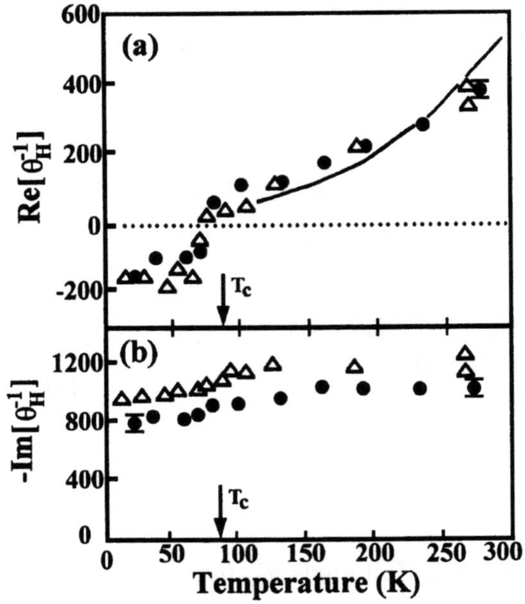

Figure 24.4: The real (a) and imaginary (b) parts of $\theta_H^{-1}(\omega, T)$ in YBa_2CuO_7 at 8 T as a function of temperature at 949 cm^{-1} (•) and 1079 cm^{-1} (△). The thin solid line in (a) shows the values for dc θ_H^{-1} obtained using the IR Hall measurements at 8 T, after Cerne at al. [6].

shall develop a kinetic theory of the IR Hall effect for a system of conduction electrons in Section 24.2 [3]. The results are used to analize the data in Au. The theory for YBCO are developed based on the independent pairon model in Section 24.3. The data in YBCO are analyzed and discussed in Section 24.4.

24.2 Theory and Experiments for Au

We first consider the static transport.

24.2.1 Conductivity

Assume a weak constant electric field **E** applied along the z-axis in a sample containing conduction electrons. Newton's equation of motion with the neglect of the scattering is $m^*(dv_z/dt) = qE$. Solving it for v_z and assuming that the acceleration persists in the mean free time τ, we obtain

$$v_d = \frac{qE}{m^*}\tau \tag{24.3}$$

for the drift velocity. The electric current density (z-component) j is

$$j = qnv_d = q^2 n (m^*)^{-1}\tau E, \tag{24.4}$$

where n is the density. Assuming Ohm's law: $j = \sigma E$, we obtain an expression for the conductivity:

$$\sigma = q^2 n (m^*)^{-1}\gamma_0^{-1}, \qquad \gamma_0 \equiv \tau^{-1}, \tag{24.5}$$

where $\gamma_0(T)$ is the static scattering rate.

24.2.2 Hall Coefficient

We take a rectangular sample having only "holes", see Fig. 20.3 (a). The current $\mathbf{j}$ runs in the z-direction. Experiments show that if a static magnetic field $\mathbf{H}$ is applied in the y-direction, the sample develops a cross (Hall) electric field $\mathbf{E}_H$ so that the Lorentz force be balanced out:

$$e(\mathbf{E}_H + \mathbf{v}_d \times \mathbf{B}) = 0 \qquad \text{or} \qquad E_H = -v_d B. \tag{24.6}$$

In the present geometry the field E_H points in the positive x-direction. We take the convention that E_H is measured relative to this direction. The Hall coefficients R_H is defined and calculated as

$$R_H \equiv \frac{E_H}{jB} = \frac{1}{qn}. \tag{24.7}$$

24.2.3 Hall Angle

The Hall angle θ_H is the angle between the current $\mathbf{j}$ and the combined field $\mathbf{E} + \mathbf{E}_H$, see Fig. 20.3 (b). Using Eq. (24.3) we obtain

$$\cot\theta_H = \frac{E}{E_H} = \frac{E}{v_d B} = \frac{m^*}{qB}\gamma_0 = \frac{\gamma_0}{\omega_c}. \tag{24.8}$$

24.2.4 Dynamic coefficients

Let us now consider the dynamic case. A monochromatic IR radiation with angular frequency ω and k-vector $\mathbf{k}$ carries an oscillating electric field represented by

$$\mathbf{E}(\mathbf{r},t) = \mathbf{E}e^{i(\mathbf{k}\cdot\mathbf{r}-\omega t)}, \qquad \mathbf{E}\cdot\mathbf{k} = 0. \tag{24.9}$$

The E-field running in the $\mathbf{k}$-direction generates a running-wave current in the same direction. The wavelength $\lambda = 2\pi/k$ is much grater than the electron wave packet. Omitting the k-dependence of the response, we assume the generalized Ohm's law:

$$\text{the running-wave part of } \mathbf{j} = \boldsymbol{\sigma}(\omega)\cdot\mathbf{E}e^{-i\omega t}, \tag{24.10}$$

where $\boldsymbol{\sigma}(\omega)$ is a dynamic conductivity tensor, which is necessarily complex.

Extending the Drude theory we set up an equation of motion for the conduction electron subject to a phonon-damping and the sinusoidally oscillating electric field:

$$\frac{dv_x}{dt} + \gamma(\omega)v_x = (m^*)^{-1}qEe^{-i\omega t}. \tag{24.11}$$

We look for an oscillatory steady-state solution ($v_x \propto e^{-i\omega t}$) and obtain the following conversion rule for the dynamic formula:

$$\gamma_0 \rightarrow \gamma(\omega) - i\omega. \tag{24.12}$$

In the large ω limit Eq. (24.11) is reduced to the original Drude equation with no γ term. The solid's response to the radiations such as light and X-ray is known to depend on frequency ω. We assume a ω-dependent rate $\gamma(\omega)$ such that

$$\gamma(\omega) \rightarrow \gamma_0 \qquad \text{as} \qquad \omega \rightarrow 0. \tag{24.13}$$

The dynamic rate γ can in general be complex. In the following development we assume a real γ. Applying the rule (24.12) to Eq. (24.5), we obtain

$$\sigma(\omega) = \frac{q^2 n}{m^*}\frac{1}{\gamma - i\omega} = \frac{q^2 n}{m^*}\frac{1}{\gamma^2+\omega^2}(\gamma + i\omega). \tag{24.14}$$

The ratio

$$\frac{\text{Re}[\sigma(\omega,T)]}{\text{Im}[\sigma(\omega,T)]} = \frac{\gamma(\omega,T)}{\omega}, \tag{24.15}$$

allows a *direct* determination of the T- and ω-dependent scattering rate $\gamma(\omega,T)$. This is significant. In contrast the static conductivity formula:

$\sigma = q^2 n(m^*)^{-1}\gamma_0^{-1}$ contains two unknowns (m^*, γ_0) and hence cannot yield γ_0 by itself.

Next, we apply the rule (24.12) to formula (24.8) and obtain

$$\cot\theta_H \simeq \theta_H^{-1} = \frac{m^*}{qB}[\gamma_H(\omega,T) - i\omega] = \frac{1}{\omega_c}[\gamma_H(\omega,T) - i\omega]. \tag{24.16}$$

Taking the ratio of the real and imaginary parts, we obtain

$$\frac{\mathrm{Re}[\theta_H^{-1}]}{\mathrm{Im}[\theta_H^{-1}]} = -\frac{\gamma_H(\omega,T)}{\omega}, \tag{24.17}$$

which gives the Hall scattering rate $\gamma_H(\omega,T)$ *directly.*

We now compare theory and experiment. In the experiment shown in Fig. 24.3 the ratio $\mathrm{Re}[\theta_H^{-1}]/\ \mathrm{Im}[\theta_H^{-1}] = -\gamma_H/\omega$, at 130 K and 8 T is 7/20. This ratio varies linearly in T from 1/4 to 2/5 as temperature is raised from 20 to 230 K. Most remarkably IR Hall effect experiments show that

$$\gamma_H(\omega,T) = \gamma_{H,0}(T). \tag{24.18}$$

That is, the dynamic rate is equal to the static rate up to the IR frequency.

The Hall coefficient R_H does not contain γ_0 and hence $R_H(\omega)$ remains to have the value $(\mathrm{qn})^{-1}$. This behavior is in agreement with the experimental data in Au and Cu [1].

The central simplifying feature of the kinetic theory of the Hall effect is the balanced force equation: $E_H = -v_d B$. Because of this relation the drift velocity v_d can be obtained directly by measuring $\cot\theta_H$ $[\equiv E/E_H = E/(v_d B)]$. In the dynamic case the applied electric field $\mathbf{E}$, and the induced quantities $(\mathbf{j}, \mathbf{v}_d, \mathbf{E}_H)$ all oscillate in time. Hence v_d may more appropriately be called the average velocity. Au (or Cu) is known to have "necks" at the Brillouin boundary [4]. Hence it is not an ideal one-carrier metal, and our formulas are not strictly applicable. As far as $\theta_H(\omega,T)$ and $R_H(\omega,T)$ are concerned our kinetic theory is in good agreement with the experiments in Au an Cu. The static scattering rate γ_0 and the Hall scattering rate γ_H are the same for the Drude model. The clearly different values for γ_0 and γ_H observed in Au an Cu are due to the non-spherical Fermi surface of these metals, and were discussed in detail by Cerne *et al.* [1]

Our main results (24.14) and (24.15) are not only applicable to fermionic electrons but also to bosonic carriers such as pairons. This is important when dealing with the IR Hall effect in cuprates, see below.

24.3 Theory for $YBa_2Cu_3O_7$

We regard the superconducting transition as a BE condensation of pairons having the linear dispersion relation:

$$\varepsilon = \omega_0 + cp, \qquad c \equiv (2/\pi) v_F. \tag{24.19}$$

In Chapter 20 we treated static conductivity σ, Hall coefficient R_H and $\cot \theta_H$, based on the independent pairon model in which pairons and conduction electrons are carriers in the normal state of cuprate superconductors. Extending the theory to the dynamic domain, we shall show that the behavior in Eqs. (24.2) can be explained based on the same model. [5]

We take a two-component system containing "holes" (1) and + pairons (2), the same system treated in Section 20.2. Let us summarize the results for the static coefficients:

- The conductivity σ

$$\sigma \equiv \sigma_1 + \sigma_2 = e^2 \frac{n_1}{m_1 \gamma_1} + \frac{4e^2 c^2 n_2}{k_B T \gamma_2} = \frac{e^2 a n_1}{T} + \frac{2e^2 b n_2}{T^2}. \tag{24.20}$$

- The Hall coefficient R_H

$$R_H = \frac{E_H}{jB} = \frac{e m_1^{-1} \gamma_1^{-1} + 2ec^2 (k_B T)^{-1} \gamma_2^{-1}}{e^2 n_1 m_1^{-1} \gamma_1^{-1} + (2e)^2 n_2 c^2 (k_B T)^{-1} \gamma_2^{-1}} = \frac{aT + b}{e(a n_1 T + 2b n_2)}. \tag{24.21}$$

- The inverse Hall angle θ_H^{-1}

$$\theta_H^{-1} = \frac{1}{[e m_1^{-1} \gamma_1^{-1} + 2ec^2 (k_B T)^{-1} \gamma_2^{-1}] B} = \frac{T^2}{eB(aT + b)}, \tag{24.22}$$

$$a \equiv \frac{\alpha_0 \hbar \omega_D}{n_a m_1 k_B v_F S_1}, \qquad b \equiv \frac{4 \alpha_0 \hbar \omega_D v_F}{\pi n_a k_B^2 S_2}. \tag{24.23}$$

Using the generalized conversion rule:

$$\gamma_{0,j} \to \gamma_j(\omega) - i\omega, \tag{24.24}$$

we obtain from Eqs. (24.20)-(24.22)

- The dynamic conductivity

$$\begin{aligned}\sigma(\omega) &= \frac{e^2 n_1}{m_1\gamma_1}\frac{1}{1-i\omega/\gamma_1}+\frac{4e^2c^2n_2}{k_BT\gamma_2}\frac{1}{1-i\omega/\gamma_2}\\ &= \frac{e^2an_1}{T}\frac{1}{1-i\omega/\gamma_1}+\frac{2e^2bn_2}{T^2}\frac{1}{1-i\omega/\gamma_2}\end{aligned} \tag{24.25}$$

- The dynamic Hall coefficient

$$R_H(\omega)=\frac{aT(1-i\omega/\gamma_1)^{-1}+b(1-i\omega/\gamma_2)^{-1}}{e[an_1T(1-i\omega/\gamma_1)^{-1}+2bn_2(1-i\omega/\gamma_2)^{-1}]}. \tag{24.26}$$

- The dynamic inverse Hall angle

$$\theta_H^{-1}(\omega)=\frac{T^2}{eB[aT(1-i\omega/\gamma_1)^{-1}+b(1-i\omega/\gamma_2)^{-1}]}. \tag{24.27}$$

24.4 Data Analysis and Discussion

The static $\cot\theta_H$ for the system of conduction electrons is, from Eq. (24.8),

$$\cot\theta_H=\frac{E}{v_dB}=\frac{m^*\gamma_0}{qB}=\frac{T}{qBa}, \tag{24.28}$$

where q is the carrier charge: $q=e$ $(-e)$ for the "electron" ("hole"). Hence the observed T^2-dependent $\cot\theta_H$ cannot be explained based on the Fermi liquid model. We can obtain the desired T^2-dependence from (24.27) if

$$\frac{aT}{b}=\frac{v_d^{(1)}}{v_d^{(2)}}=\frac{\pi S_2}{eBb}\frac{k_BT}{\varepsilon_F}\ll 1, \tag{24.29}$$

which is reasonable since $k_BT\ll\varepsilon_F$. Physically low-energy pairons are accelerated strongly in proportion to ε^{-1}, as seen from $(\varepsilon/c^2)dv_z/dt=(2e)E$, and hence $v_d^{(2)}$ dominates $v_d^{(1)}$, causing the strong inequality. Using inequality (24.29), we obtain from (24.27)

$$\theta_H^{-1}(\omega)=\frac{T^2}{eBb}(1-\frac{i\omega}{\gamma_2}). \tag{24.30}$$

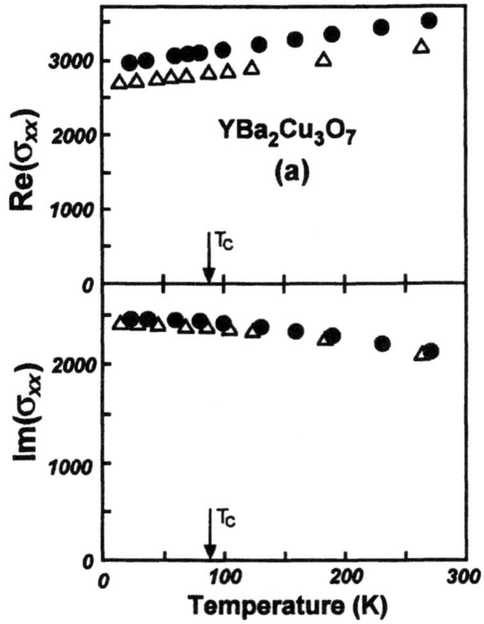

Figure 24.5: The dynamic conductivity σ in $YBa_2Cu_3O_7$ at 949 cm^{-1} (•) and 1079 cm^{-1} (△) as a function of temperature.

This result indicates that: (a) $\mathrm{Re}[\theta_H^{-1}(\omega,T)]$ is proportional to T^2, (b) it is equal to dc $\cot\theta_H$, (c) $\mathrm{Im}[\theta_H^{-1}]$ is negative and (d) it is ω-linear, all in agreement with Eqs. (24.2). At 180 K $-\mathrm{Re}[\theta_H^{-1}]/\mathrm{Im}[\theta_H^{-1}] = 1/5$, meaning that $\gamma_H/\omega = 1/5$. Combining with Eq. (24.28), this means that the pairon-phonon scattering rate $\gamma_2(\omega)$ remains equal to the static rate.

$$\gamma_2(\omega,T) = \gamma_{2,0}(T), \qquad \omega \simeq 5\gamma_2 \qquad \text{at 180 K.} \tag{24.31}$$

Using inequality (24.31), we obtain from (24.25)

$$\sigma(\omega) = \frac{e^2 n_1}{m_1}\left(\frac{\gamma_1}{\gamma_1^2+\omega^2} + \frac{i\omega}{\gamma_1^2+\omega^2}\right), \tag{24.32}$$

whose behavior is in agreement with the experimental data shown in Fig. 24.5. $\mathrm{Im}[\sigma(\omega)]$ is positive and decreases with T, while $\mathrm{Re}[\sigma(\omega)]$ increases at 949 cm^{-1} and 1079 cm^{-1}. We note that

$$\mathrm{Re}[\sigma(\omega)]/\mathrm{Im}[\sigma(\omega)] = \gamma_1/\omega = 1.4 \qquad \text{at 180 K.} \tag{24.33}$$

The conductivity σ from Eq. (24.32) is determined by the "hole"-phonon scattering rate γ_1 while the Hall angle from Eq. (24.30) is controlled by the pairon-phonon scattering rate γ_2. We found that the IR Hall scattering rates $[\gamma_1^H(\omega), \gamma_2^H(\omega)]$ are equal to the static rates $(\gamma_{1,0}^H, \gamma_{2,0}^H)$, where $\mathrm{Re}[\theta_H^{-1}]/\,\mathrm{Im}[-\theta_H^{-1}] = \gamma_2/\omega \simeq 1/5$, $\mathrm{Re}[\sigma(\omega)]/\,\mathrm{Im}[\sigma(\omega)]\gamma_1/\omega \simeq 1.4$ at 180 K.

In summary the unusual T^2-dependence of $\mathrm{Re}[\theta_H^{-1}]$ is explained microscopically. The low-energy pairon is accelerated strongly in proportion to ε^{-1} and hence the pairon drift velocity $v_d^{(2)}$ dominates the "hole" drift velocity $v_d^{(1)}$ This effect, combined with the T-dependent phonon scattering rate γ_2, generates the T^2-dependence for $\mathrm{Re}[\theta_H^{-1}]$.

Chapter 25

d-Wave Cooper Pairs

The d-wave Cooper pairs (pairons) in cuprates with strong binding along the a- and b-axes are shown to arise from the optical-phonon exchange attraction.

25.1 Introduction

Josephson tunneling experiments in underdoped cuprates [1] indicate that Cooper pairs (pairons) in cuprates are of a d-wave type with a strong binding occurring in the a- and b-crystal axes, see Fig. 25.1. Such lattice-dependent anisotropy is most likely to be explained by an intrinsically anisotropic optical-phonon-exchange attraction and Fermi surface, rather than any spin-dependent mechanism.

25.2 Phonon-Exchange Attraction

The electron-pair density matrix $\rho(\mathbf{k}_1, \mathbf{k}_2; \mathbf{k}_3, \mathbf{k}_4, t) \equiv \rho(1,2;3,4,t)$, in the presence of a Coulomb interaction v_c, changes in time following a quantum Liouville equation

$$\begin{aligned} i\hbar \frac{\partial \rho(1,2;3,4,t)}{\partial t} &= \sum_{\mathbf{k}_5} \sum_{\mathbf{k}_6} (\langle 1,2| \, v_c \, |5,6\rangle \, \rho(5,6;3,4,t) \\ &\quad - \langle 5,6| \, v_c \, |3,4\rangle \, \rho(1,2;5,6,t)). \end{aligned} \tag{25.1}$$

$$\langle 1,2| \, v_c \, |3,4\rangle \equiv 4\pi e^2 k_0 \frac{1}{Vq^2} \delta_{\mathbf{k}_1+\mathbf{k}_2,\, \mathbf{k}_3+\mathbf{k}_4} \delta_{\mathbf{k}_3-\mathbf{k}_1,\, \mathbf{q}}, \qquad k_0 \equiv \frac{1}{4\pi\epsilon_0}.$$

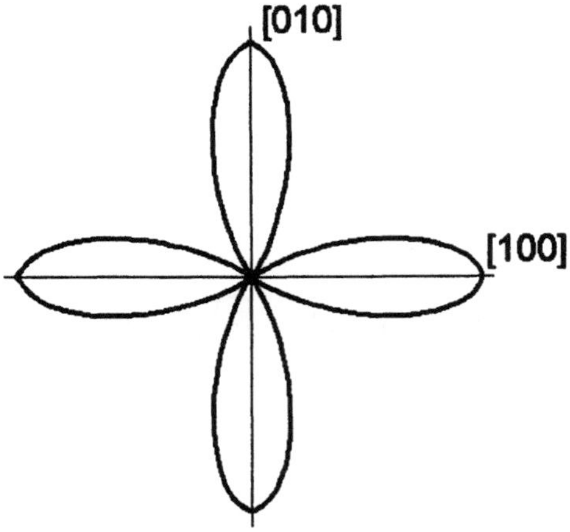

Figure 25.1: A d-wave Cooper pair with strong binding along the a- and b-axes.

Earlier in Section 4.3 we showed that if a phonon having momentum $\mathbf{q}$ and energy $\hbar\omega_q$ is exchanged, the density matrix ρ changes following the same equation with an effective interaction v_e,

$$\langle 1,2|\, v_e\, |3,4\rangle = |V_q|^2 \frac{\hbar\omega_q}{(\varepsilon_3-\varepsilon_1)^2-(\hbar\omega_q)^2}\delta_{\mathbf{k}_1+\mathbf{k}_2,\,\mathbf{k}_3+\mathbf{k}_4}\delta_{\mathbf{k}_3-\mathbf{k}_1,\,\mathbf{q}}, \qquad (25.2)$$

where $\varepsilon_j \equiv \varepsilon_{k_j}$ is the electron energy and V_q the electron-phonon interaction strength. We note that this result (25.2) is partially in agreement with the form of a pairing Hamiltonian appearing in Eq. (2.4) of the original BCS work [2]:

$$H_{int} = \sum_{\mathbf{k},\sigma}\sum_{\mathbf{k}',\sigma'}\sum_{\mathbf{q}} \frac{\hbar\omega_q\,|V_q|^2}{(\varepsilon_{\mathbf{k}}-\varepsilon_{\mathbf{k}+\mathbf{q}})^2-(\hbar\omega_q)^2}\, c^\dagger_{\mathbf{k}'+\mathbf{q}\sigma'}c^\dagger_{\mathbf{k}+\mathbf{q}\sigma}c_{\mathbf{k}\sigma}c_{\mathbf{k}'\sigma'}. \qquad (25.3)$$

We examined the effect of the exchange of the phonon with momentum $\mathbf{q}$ and energy $\hbar\omega_q$ on ρ, whence no summation with respect to $\mathbf{q}$ appears. The most important attraction between the two electrons occurs in the specific k-space region near the Fermi surface where the phonon-mediated momentum transfer $\mathbf{q}$ is nearly parallel to the surface so that $\varepsilon_3 - \varepsilon_1 = 0$. As we see in this example, the phonon-exchange attraction is direction-dependent. This is important for the d-wave pairon formation discussed below.

The nature of phonon-exchange interaction v_e depends on the type of phonons. Consider first longitudinal acoustic phonons, whose dispersion relation is linear:

$$\hbar\omega_q = c_s q. \qquad (c_s = \text{sound speed}) \tag{25.4}$$

If a deformation potential model [3] is assumed, the interaction strength V_q is

$$V_q = A(\hbar/2\omega_q)^{1/2} iq. \qquad (A = \text{constant}) \tag{25.5}$$

Using Eqs. (25.4) and (25.5), we obtain

$$|V_q|^2 \frac{\hbar\omega_q}{(\varepsilon_3 - \varepsilon_1)^2 - (\hbar\omega_q)^2} \sim -\frac{A^2\hbar^2}{2c_s^2} \equiv -v_0 \tag{25.6}$$

for the dominant attraction at 0 K ($\varepsilon_3 - \varepsilon_1 = 0$), showing that phonons of any wave lengths are equally effective. For optical phonons whose dispersion relation is given by

$$\hbar\omega_q = \varepsilon_0 \text{ (constant)}, \tag{25.7}$$

we obtain

$$|V_q|^2 \frac{\hbar\omega_q}{(\varepsilon_3 - \varepsilon_1)^2 - (\hbar\omega_q)^2} \sim -\frac{A^2\hbar^2}{2c_s^2} q^2, \tag{25.8}$$

indicating that optical phonons of shorter wavelengths (greater q) are more effective. The wavelength $\lambda \equiv 2\pi/q$ of a phonon has a lower bound $2a_0$, $a_0 =$ the lattice constant, yielding a shorter interaction range compared with the case of an acoustic phonon exchange. The ratio of the rhs of Eqs. (25.8) and (25.6), $c_s^2 q^2/\varepsilon_0^2$, is of the order unity for the maximum $q_{\text{max}} = \pi/a_0$. Hence the short-wavelength optical phonon exchange is as effective as the acoustic phonon exchange. The optical-phonon exchange pairing becomes weaker for longer wavelengths (small q).

25.3 d-Wave Pairon Formation

Let us consider the copper plane. Linear arrays of O-O and Cu-O-Cu alternate in the [100] and [010] directions, see Fig. 17.1. Thus we recognize longitudinal optical modes of oscillations along the a- an b-axes. Now let us look at the motion of an "electron" wave packet extending over unit cells. If the "electron" density is small, the Fermi surface should be a small circle as shown in the central part in Fig. 17.2. Next consider a "hole" wave packet.

If the "hole" density is small, the Fermi surface should consist of four small pockets near the Brillouin zone corners as shown in Fig. 17.2. Under the assumption of such a Fermi surface, pair creation of $\pm$ pairons by means of an optical phonon exchange can occur as shown in the figure. Here a single-phonon exchange generates the electron transition from A in the O-Fermi sheet to B in the Cu-Fermi sheet *and* the electron transition from A' to B', creating the $-$ pairon at (B, B') and the $+$ pairon at (A, A'). The optical phonon having momentum $\mathbf{q}$ nearly parallel to the a-axis is exchanged here. Likewise, the optical phonon with a momentum nearly along the b-axis help create $\pm$ pairons. But because of the location of the Fermi surface there is no pairon formation in the direction [110] and $[1\bar{1}0]$. Consequently the pairon is of a d-wave type with the dominant attraction along the a- and b-axes, see Fig. 25.1.

If the doping is increased, the O-Fermi surface grows as shown in Fig. 19.2, (b)-(d). Then, the anisotropy decreases and the pairon becomes less anisotropic. At the end of the overdoping, the O-Fermi surface undergoes a curvature inversion as in (c) and (d). Near the inflection point, the pairon is isotropic and S-wave type.

A direct way of mapping a 2D Fermi surface is to perform an angle-resolved photo-emission spectroscopy (ARPES) [4]. It would be highly desirable to see the curvature inversion by this technique.

25.4 Discussion

The attraction generated by the phonon exchange is intrinsically anisotropic. Because of the location of the Fermi surface in the optimum cuprate the exchange of an optical longitudinal phonon generates a d-wave and underdoped Cooper pair with the dominant attraction in the directions [110] and $[1\bar{1}0]$. Our model predicts that the d-wave character will be lost in the extremely overdoped sample.

Chapter 26

Connection with Other Theories

Since the discovery of a superconducting mercury in 1911 by Kamerlingh Onnes [1] a number of important theories have been developed. Our microscopic theory is guided by these theories. We shall briefly describe connections between these and ours.

26.1 Gorter-Cassimir's Two Fluid Model

Based on the analysis of the heat capacity data Gorter and Casimir [2] in 1933 proposed a two-fluid model: the *superfluid* bears the resistanceless motion, and the *normal* fluid behaves as a normal (electron) liquid. The model has been recognized to fully apply to the superconductor below T_c. It is a phenomenological theory without specifying what particles are responsible for the superfluid motion. It is widely said (and thought) that the super part originates in the superconducting (super) electrons, which is confusing. In our theory the super part is identified as the supercondensate composed of bosonically condensed pairons while the normal part arises from all other particles including non-condensed pairons, quasi-electrons and vortex lines.

The two-fluid model also fully applies to the superfluid helium, liquid He II. [3] The superfluid (frictionless fluid) arises from bosonically condensed He^4.

26.2 London-London's Theory

In 1933 Meissner and Ochsenfeld [4] discovered that the superconductor expels the applied weak magnetic field from its interior. This behavior cannot be derived from the resistanceless current, and hence it is a major property of the superconductor below T_c.

In 1935 the London brothers [5], based on the London's equation,

$$\mathbf{j}_s = -\Lambda_{el}\mathbf{A}, \qquad \Lambda_{el} \equiv e^2 m^{-1} n_s, \tag{26.1}$$

and Maxwell's equations, predicted that the magnetic field $\mathbf{B}$ does not abruptly vanish at the boundary, but it penetrates the sample a short distance λ, called a penetration depth, represented by

$$\lambda_{London} = \frac{c_0}{e}\left(\frac{m^*}{4\pi n_s}\right)^{1/2}, \qquad c_0 = \text{light speed}. \tag{26.2}$$

This prediction was later experimentally confirmed, which established the tradition that electromagnetism can, and must, be applied to the superconductor. The magnitude of the experimental penetration depth λ at the lowest temperatures is about 500 Å.

In the present theory the supercurrent arises from the motion of the condensed pairons. From this viewpoint we derived the revised London equation, see Eq. (13.35): [6]

$$\mathbf{j}_s = -\Lambda_{pairon}\mathbf{A}, \qquad \Lambda_{pairon} = 2e^2 n_0 (c_1 + c_2)^{-1} p^{-1}, \tag{26.3}$$

where p is the momentum (magnitude) of the pairon and $c_j = v_F^{(j)}/2$ (3D) $(2/\pi)v_F^{(j)}$ (2D). The revised penetration depth λ is

$$\lambda = (c_0/e)\{p/[8\pi k_0 n_0 (c_1 + c_2)]\}^{1/2}. \tag{26.4}$$

The condensed pairon density n_0 vanishes at T_c, and hence λ tends to ∞ like $n_0^{-1/2}$ as the temperature approaches T_c. Formula (26.4) contains no fictitious parameters, and hence can be used to determine (p, n_0).

Londons introduced a macrowavefunction of a running wave type to represent the supercurrent. The phase of this function is considered to be perturbed neither by small defects nor by small electric fields. This property is often called the *London rigidity*. The macrowavefunction is identified as a quasiwavefunction $\Psi(\mathbf{r})$ in the present theory, representing the state of the supercondensate. The rigidity arises from the fact that the change in the many-pairon quantum state requires a redistribution of a large number of pairons, and that the supercondensate composed of equal numbers of $\pm$ pairons is neutral and hence it is not subject to a Lorentz electric force.

26.3 Ginzburg-Landau Theory

In 1950 Ginzburg and Landau [7] introduced a revolutionary idea that the superconductor below T_c possesses a *complex order parameter*, called a GL wavefunction Ψ', just as a ferromagnet has a real order parameter (spontaneous magnetization). Based on general thermodynamic arguments GL obtained the two equations:

$$\frac{1}{2m}\left|-i\hbar\nabla - q\mathbf{A}\right|^2 \Psi'(\mathbf{r}) + \alpha\Psi'(\mathbf{r}) + \beta\left|\Psi'(\mathbf{r})\right|^2 \Psi'(\mathbf{r}) = 0, \tag{26.5}$$

$$\mathbf{j} = -\frac{iq\hbar}{2m^*}(\Psi'^*\nabla\Psi' - \Psi'\nabla\Psi'^*) - \frac{q^2}{2m^*}\Psi'^*\Psi'\mathbf{A}, \tag{26.6}$$

where m^* and q are the mass and charge of a superelectron. With the density condition:

$$\Psi'^*(\mathbf{r})\Psi'(\mathbf{r}) = n_s(\mathbf{r}) = \text{superelectron density}, \tag{26.7}$$

Eq. (26.6) for the current density $\mathbf{j}$ in the homogeneous limit ($\nabla\Psi' = 0$) is reduced to London's equation (26.1). The GL equations are quantum mechanical and nonlinear. The most remarkable results of the GL theory are the introduction of the concept of a coherence length [7] and Abrikosov's prediction of a vortex structure in a type II superconductor [8], later confirmed by experiments [9].

We derived the GL equation (26.5) from first principles based on the idea that the supercurrent arises from the motion of the condensed pairons. The GL wavefunction Ψ' can be identified as

$$\Psi'_\sigma(\mathbf{r}) = \left\langle \mathbf{r}\left|n^{1/2}\right|\sigma\right\rangle, \tag{26.8}$$

where σ denotes the condensed (momentum) state, and n the pairon density operator. The density condition (26.7) is replaced by

$$\left|\Psi'(\mathbf{r})\right|^2 = n_\sigma(\mathbf{r}) = \text{condensed pairon density}. \tag{26.9}$$

The parameter α (< 0) in Eq. (26.5) can be interpreted as the pairon condensation energy, and the parameter $\beta\,(> 0)$ represents the repulsive interpairon interaction strength. [6] The homogeneous solution of Eq. (26.5) yields a remarkable result that the T-dependent condensed pairon density $n_0(T)$ is proportional to the pairon energy gap $\varepsilon_g(T)$, see Section 26.6.

26.4 Electron-Phonon Interaction.

In 1950 Fröhlich [10] developed a theory of superconductivity based on the idea that the electron-phonon interaction is the cause of the (type I) superconductivity. (The type II superconductor was not known then). At about the same time the critical temperature T_c was found to depend on the isotopic ion mass, [11] which supported Fröhlich's idea.

There are no real phonons at 0 K. The exchange of a virtual phonon between two electrons can generate an attraction just as the exchange of a virtual pion between two nucleons generates an attractive nuclear force in Yukawa's model [12]. The treatment of such exchange force requires a quantum field theory, also called a second quantization formulation. This theory, distinct from the Schrödinger wavefunction formalism, allows one to describe processes in which the number of particles is not conserved, e.g. the creation of an electron-positron pair and the pair creation of $\pm$ pairons.

The Hamiltonian H_F representing the electron-(longitudinal) phonon interaction takes the form, see Eq. (4.37):

$$H_F = \frac{1}{2}\sum_{\mathbf{k}}\sum_{\mathbf{q}}(V_q\, c^\dagger_{\mathbf{k}+\mathbf{q}} c_{\mathbf{k}}\, a_{\mathbf{q}} + h.c.), \qquad V_q \equiv A_q(\hbar/2\omega_q)^{1/2} iq, \tag{26.10}$$

called the Fröhlich Hamiltonian. Using this and quantum perturbation method, we obtain the effective phonon exchange interaction: [13]

$$|V_q|^2 \frac{\hbar\omega_q}{(\varepsilon_{\mathbf{k}_1+\mathbf{q}} - \varepsilon_{\mathbf{k}_1})^2 - \hbar^2\omega_q^2}, \tag{26.11}$$

which is negative (attractive) if the electron energy difference before and after the transition $|\varepsilon_{\mathbf{k}_1+\mathbf{q}} - \varepsilon_{\mathbf{k}_1}|$ is less than the phonon energy $\hbar\omega_q$. The attraction is greatest when the phonon momentum $\mathbf{q}$ is parallel to the constant-energy (Fermi) surface.

26.5 The Cooper Pair

In 1956 Cooper [14] showed that the attraction, however weak, may bound a pair of electrons. He started with Cooper's equation:

$$\begin{aligned} w_q\, a(\mathbf{k},\mathbf{q}) &= [\varepsilon(|\mathbf{k}+\mathbf{q}/2|)+\varepsilon(|-\mathbf{k}+\mathbf{q}/2|)]\, a(\mathbf{k},\mathbf{q}) \\ &\quad -\frac{1}{(2\pi\hbar)^2}\, v_0 \int' d^2k' a(\mathbf{k}',\mathbf{q}), \end{aligned} \tag{26.12}$$

where w_q is the energy of a pairon, $a(\mathbf{k},\mathbf{q})$ the wavefunction and v_0 the interaction strength. The solution of Eq. (26.12) for small momenta q yields:

$$w_q = w_0 + cq < 0, \qquad w_0 = \frac{-2\hbar\omega_D}{\exp[2/v_0\mathcal{N}(0)]-1}, \tag{26.13}$$

where $c/v_F = 1/2\ (2/\pi)$ for 3 (2) D. The constant strength v_0 can be justified for the acoustic phonon exchange:

$$v_0 = |V_q|^2\,\frac{1}{\hbar\omega_q} = A_q^2\frac{\hbar q^2}{2\omega_q}\frac{1}{\hbar\omega_q} = A_q^2\frac{1}{2c_s^2}, \tag{26.14}$$

where $\omega_q = c_s q$, and c_s = the sound speed.

26.6 BCS Theory

In 1957 Bardeen, Cooper and Schrieffer [15] published an epoch-making theory of superconductivity, which is regarded as one of the most important theoretical works in the 20-th century. Starting with the BCS Hamiltonian H_0 in Eq. (7.12) containing the kinetic energies of "electrons" and "holes" and a pairing interaction Hamiltonian, and using the minimum energy principle calculation with the guessed trial ground-state ket in Eq. (7.14), they showed that the ground state energy W of the BCS system is lower than that of the Bloch system without the pairing interaction:

$$W = N_0 w_0 < 0, \qquad N_0 = \hbar\omega_D\mathcal{N}(0). \tag{26.15}$$

The minimum energy condition can be expressed in terms of the energy gap equation:

$$\Delta = v_0 {\sum_{\mathbf{k}}}' \sum_j \frac{\Delta}{2E_k^{(i)}}, \tag{26.16}$$

$$E_k^{(j)} \equiv (\varepsilon_k^{(j)2} + \Delta^{(j)2})^{1/2}, \tag{26.17}$$

where $\Delta^{(j)}$ is the quasi-electron energy gap and $E_k^{(j)}$ the energy of the quasi-electron, $j = 1(2)$ for the "electron" ("hole").

BCS extended their theory to a finite temperature, and obtained the temperature dependent energy gap equations (10.31). In the bulk limit the BCS gap equation is

$$1 = v_0 \mathcal{N}(0) \int_0^{\hbar\omega_D} d\varepsilon \frac{1}{(\varepsilon^2 + \Delta^2)^{1/2}} \tanh\left[\frac{(\varepsilon^2 + \Delta^2)^{1/2}}{2k_B T}\right]. \tag{26.18}$$

This gap Δ is temperature-dependent. The limit temperature T_c at which Δ vanishes, is given by

$$1 = v_0 \mathcal{N}(0) \int_0^{\hbar\omega_D} d\varepsilon \frac{1}{\varepsilon} \tanh\left[\frac{\varepsilon}{2k_B T_c}\right]. \tag{26.19}$$

In the weak coupling limit the critical temperature T_c is given by

$$k_B T_c \simeq 1.13\, \hbar\omega_D \exp\left[\frac{1}{v_0 \mathcal{N}(0)}\right]. \tag{26.20}$$

Formulas (26.15) and (26.20) represent two of the most important results of the BCS theory. The formula (26.15) for the ground state energy can be interpreted as follows: The greatest total number of pairons generated consistent with the BCS Hamiltonian is equal to $\hbar\omega_D \mathcal{N}(0) = N_0$. Each pairon contributes a binding energy $|w_0|$. The existence of the pairons below T_c was directly confirmed in the flux quantization experiments [16, 17] in 1961, which showed that the carrier in the supercurrent has the charge (magnitude) $2e$.

The nature of the BCS results is quite remarkable. The starting Hamiltonian H and the trial ground-state $|\Psi\rangle$ are both expressed in terms of pairon operators $(b, b^\dagger)$. But only quasi-electron variables appear in the energy gap equation, which is the minimum-energy condition. Hence it is impossible to guess even the existence of the gap Δ in the excitation energy spectrum $E_k \equiv (\varepsilon_k^2 + \Delta^2)^{1/2}$.

In the present text we recalculated the ground-state energy W_0, using the equation-of-motion method, and confirmed that formula (26.15) is the exact expression for the ground-state energy of the BCS system.

BCS assumed a Hamiltonian containing "electrons" and "holes" with a free electron model. The reason why only some, and not all, metals are superconductors was unexplained. To answer this question we must incorporate the Fermi surface of electrons in the theory. "Electrons" ("holes") are excited near the Fermi surface where the local principal curvatures are negative (positive). The reduced generalized BCS Hamiltonian written in terms of "electron" (1) and "hole" (2) variables is

$$\begin{aligned} H_0 &= \sum_{\mathbf{k}}\sum_{s} \varepsilon_k^{(1)} n_{\mathbf{k}s}^{(1)} + \sum_{\mathbf{k}}\sum_{s} \varepsilon_k^{(2)} n_{\mathbf{k}s}^{(2)} - {\sum_{\mathbf{k}}}'{\sum_{\mathbf{k}'}}' [v_{11} b_{\mathbf{k}}^{(1)\dagger} b_{\mathbf{k}'}^{(1)} \\ &+ v_{12} b_{\mathbf{k}}^{(1)\dagger} b_{\mathbf{k}'}^{(2)\dagger} + v_{21} b_{\mathbf{k}}^{(2)} b_{\mathbf{k}'}^{(1)} + v_{22} b_{\mathbf{k}}^{(2)} b_{\mathbf{k}'}^{(2)\dagger}], \end{aligned} \tag{26.21}$$

$$b_{\mathbf{k}}^{(1)} \equiv c_{-\mathbf{k}\downarrow}^{(1)} c_{\mathbf{k}\uparrow}^{(1)}, \qquad b_{\mathbf{k}}^{(2)} \equiv c_{\mathbf{k}\uparrow}^{(2)} c_{-\mathbf{k}\downarrow}^{(2)}, \tag{26.22}$$

where the primes on the last summation symbols indicate the restriction that

$$0 < \varepsilon_k^{(1)} \equiv \varepsilon_k, \varepsilon_k^{(2)} \equiv |\varepsilon_k| < \hbar\omega_D. \tag{26.23}$$

The first interaction term $-v_{11}\, b_{\mathbf{k}'}^{(1)\dagger} b_{\mathbf{k}}^{(1)}$ generates an attractive (negative sign) transition of the electron pair from $(\mathbf{k}\uparrow, -\mathbf{k}\downarrow)$ to $(\mathbf{k}'\uparrow, -\mathbf{k}'\downarrow)$. The Coulomb interaction generates a repulsive transition. The effect of this interaction is included in the strength v_{11}. Similarly, the exchange of a phonon induces an attractive transition between the "hole"-pair states, and it is represented by the term in $-v_{22}\, b_{\mathbf{k}'}^{(2)} b_{\mathbf{k}}^{(2)\dagger}$. The exchange of a phonon can also pair-create [pair-annihilate] $\pm$pairons, and the effects of these processes are represented by $-v_{12}\, b_{\mathbf{k}'}^{(1)\dagger} b_{\mathbf{k}}^{(2)\dagger}$, $[-v_{21}\, b_{\mathbf{k}}^{(1)} b_{\mathbf{k}'}^{(2)}]$. If the Fermi surface is such that both "electrons" and "holes" be generated, the pairing interaction $-v_{12}\, b_{\mathbf{k}'}^{(1)\dagger} b_{\mathbf{k}}^{(2)\dagger}$ can generate $\pm$pairons in equal numbers. Thus we can discuss how the $\pm$pairons are created in the system having a favorable Fermi surface.

The quasi-electron energy gap $\Delta(T)$ obtained as the solution of Eq. (26.18) depends on the temperature T. It is greatest at 0 K and monotonically decreases and vanishes at T_c. Photo-absorption [18] and quantum tunneling experiments [19] indicate the existence of an energy gap $\varepsilon_g(T)$ which depends on temperature. It was generally thought that the experimental gap $\varepsilon_g(T)$ represents the theoretical gap $\Delta(T)$. But this interpretation has a difficulty. The gap Δ appears in the quasi-electron energy: $E_k \equiv (\varepsilon_k^2 + \Delta^2)^{1/2}$, indicating that the quasi-electrons have the minimum excitation energy Δ relative to the Fermi energy. But the Fermi surface is blurred in the superconducting state, and hence the photo-absorption experiment is unlikely to detect the gap Δ. Our calculations show that the moving pairons have an energy gap ε_g relative to the condensed stationary pairon level. This gap $\varepsilon_g(T)$, which is T-dependent, can be detected directly in the photo-absorption and quantum tunneling experiments.

26.7 Bose-Einstein Condensation

Liquid He II and superconductors show remarkable similarities. London emphasized this fact in his well-known books [3], and proposed to explain both superfluidity and superconductivity from the BEC point of view. If a system of free bosons with mass M is considered, the critical temperature T_c for 3D is [21]

$$T_c = 3.31\, \hbar^2 \mathrm{k}_B^{-1} M^{-1} n^{2/3}, \tag{26.24}$$

where n is the boson density. If we use the mass density of the liquid helium $n = 0.145$ g cm^{-3}, we find from Eq. (26.24) that $T_c = 3.14$ K, which is very close to the observed superfluid transition temperature 2.18 K. This and other calculations led to a general consensus that the helium superfluidity is due to the BEC of weakly interacting He^4. Now consider the case of superconductivity. If we assume $2m_e$ for the bosonic mass M and use the experimental $T_c = 7.19$ K for lead, the interboson distance $r_0 = n^{-1/3}$ calculated from Eq. (26.24) is on the order of 10^{-4} cm far greater than the lattice constant, meaning an overlapping of the electron-pairs. Attempts to overcome this difficulty [22] have been largely unsuccessful. Because of this the BCS model without consideration of the bosonic nature of the pairons has been the dominant theory of the low T_c superconductors.

In the present theory we solved the Cooper equation (26.12) and obtain the linear dispersion relation (26.13). Using $\varepsilon = (1/2)v_F p$ and assuming free bosons moving in 3D we obtain [6]

$$T_c = 1.01\, \hbar \mathrm{k}_B^{-1} v_F n_0^{1/3}. \tag{26.25}$$

Note that no mass appear in this expression. The pairon size can be estimated by the BCS zero-temperature coherence length

$$\xi_0 = \hbar v_F / \pi \Delta = 0.18\, \hbar v_F / \mathrm{k}_B T_c. \tag{26.26}$$

After solving Eq. (26.25) for n_0 and introducing the interpairon distance $r_0 \equiv n_0^{-1/3}$, we obtain

$$r_0 \equiv n_0^{-1/3} = 1.01\, \hbar v_F / \mathrm{k}_B T_c = 5.61\, \xi_0, \tag{26.27}$$

indicating that the interpairon distance r_0 is several times greater than the pairon size ξ_0. Thus the B-E condensation occurs before the picture of free pairons breaks down.

The system of free pairons moving in 2D with the linear relation $\varepsilon = (2/\pi)v_F p$ (the model system for a high T_c cuprate) undergoes a B-E condensation at

$$T_c = 1.24\, \hbar k_B^{-1} v_F n_0^{1/2}. \tag{26.28}$$

The condensation of massless bosons in 2D is noteworthy. Hohenberg showed [24] that there can be no long-range order in 2D, which is derived with the assumption of an f-sum rule representing the mass conservation law. Hohenberg's theorem does not apply for massless bosons.

26.8 Josephson Theory

In 1962 Josephson [25] predicted a supercurrent tunneling through a small barrier (Josephson junction) with no energy loss. Shortly thereafter, Anderson and Rowell [26] experimentally demonstrated this Josephson tunneling. Josephson argued [25] that if a static voltage V is applied to the superconductor containing a Josephson junction, the current I is controlled by two equations:

$$I = I_0 \sin\delta, \tag{26.29}$$

$$\hbar\frac{d\delta}{dt} = 2eV, \tag{26.30}$$

where δ is the phase difference across the junction. Note that these equations predict a quite different behavior from the classical ohmic behavior: $I \propto V$. The correctness of these equations is confirmed by numerous later experiments. Mercereau and his collaborators [27] found the Josephson interference in a SQUID. Two supercurrents separated up to 1 mm can interfere just as two laser beams from the same source. This effect comes from the self-focussing power shared by these two fluxes (supercurrent and laser), meaning that the pairons move as bosons with the linear dispersion relation:

$$\varepsilon = cp, \qquad c = \text{pairon speed}. \tag{26.31}$$

The interference pattern can quantitatively be calculated, starting with Eqs. (26.29) and (26.30). Feynman [29] derived these equations, using the quantum mechanical equation of motion and assuming that the pairons move as bosons. The Josephson effect established the tradition that both fermionic and bosonic properties of the pairons must be used for the total description of the superconductivity. Eqs. (26.29 - 30) indicate that the static voltage V generates an oscillatory supercurrent with the Josephson frequency $\omega_J = 2eV\hbar^{-1}$. If a microwave with the matching frequency: $\omega \sim \omega_J$ is applied, a series of steps in the $V - I$ curve were observed by Shapiro [28]. The oscillatory supercurrent is in accord with our pictures that the supercurrent is caused by the motion of the neutral supercondensate composed of equal numbers of $\pm$ pairons.

26.9 High Temperature Superconductors

In 1986 Bednorz and Müller [30] reported the first discovery of the high-T_c cuprate superconductor (LaBaCuO, $T_c > 30\,\mathrm{K}$). In 1987 Chu, Wu and others [31] discovered YBCO with $T_c \sim 94$ K. This is a major development. The compound Nb_3Ge has $T_c = 23$ K. The BCS theory [15] and subsequent theoretical developments [32] predicted the highest T_c of about 25 K. A great number of theoretical investigations followed, re-examining the foundations of the BCS theory and developing new theories [33 - 35].

Mott and his collaborators [36] proposed a bi-polaron (boson) model to treat high-T_c cuprates. Lee and his group [37] developed a virtual bosonic model, and their calculations of T_c based on massive bosons are quite complicated. In contrast we obtained a BEC temperature $T_c = 1.24\,\hbar k_B^{-1} v_F n_0^{1/2}$ for the system of free *masless* bosons in 2 D.

The simplest cuprate superconductors are $La_{2-x}Sr_xCuO_4$ and $Nd_{2-x}Ce_xCuO_{4-\delta}$. The parents La_2CuO_4 and Nd_2CuO_4 are both antiferromagnetic insulators. Under doping they become superconductors in some concentration ranges. To understand both antiferromagnetic and superconducting phases a spin-dependent theory is required. Anderson stressed the importance of understanding the normal-state magnetotransport in cuprates and proposed a spin-dependent resonating valence bond (RVB) theory [38]. A variety of other theories are proposed, starting with spin-dependent model Hamiltonians including the d-p model [39], the Hubbard model [40], the t-J model [41].

As noted in chapter 2 the superconducting transition is a condensation in the **k**-space distinct from the magnetic transition which is a condensation in the spin angular momentum space. The **k** represents the state of the conduction electron, while the spin refers to the state of the lattice-ion core electron. Hence the electron dynamics and spin motion are weakly coupled. This is experimentally supported by the existence of the two critical temperatures, the antiferromagnetic and superconducting temperatures (T_N, T_c). We take the view that the superconductivity in the cuprates can be developed based on the spin independent generalized BCS model. Our mathematical treatment is simple. We obtained exact expressions for the ground-state energy and the superconducting temperature. In contrast the spin-dependent problems are hard to solve. No exact solutions for the ground-state and the critical temperature have yet been found. Onsager's solution of the 2D Ising model [42] is a notable exception.

26.10 Quantum Hall Effect

Experimental data by Willet *et al.* [43], reproduced in Fig. 26.1, show that the Hall resistivity ρ_H at the extreme low temperatures has plateaus at various Landau-level occupation ratios (electron density n_e/elementary flux density) $\nu = p/q$, odd q, where the resistivity ρ (nearly) vanishes. In particular at $\nu = 1/3$, the plateau in ρ_H and the drop in ρ are as distinctive as the integer Quantum Hall Effect (QHE) plateau and zero resistivity at $\nu = 1$ [44], indicating a superconducting state with an energy gap. The plateau heights are quantized in units of h/e^2. Each plateau is material- and shape-independent, indicating the basic quantum nature and stability of the QHE state. The stability arising from the energy gap is sometimes referred to as the incompressibility of a quantum liquid in literature. [45] The ground state of the hetero-junction GaAs/AlGaAs at the fractional ratios $\nu = 1/q$, odd q, can be described in terms of the Laughlin wavefunctions. [46] Zhang and others [47-48] discussed the QHE state in terms of the composite bosons, each made up of an electron and an odd number of elementary fluxes (fluxons). The same data shown in Fig. 26.1 indicate that ρ_H is linear in B at $\nu = 1/2$, indicating a Fermi-liquid state. This state can simply be described in terms of the composite fermions, each made up of an electron and two fluxons. [49]

A microscopic theory of the QHE can be developed in analogy with our theory of high T_c superconductivity. [50] We regard the fluxon as a quantum particle with half-spin and zero mass, which is in line with Dirac's theory that every quantum particle having the position as an observable is a Dirac particle with half-spin. [51] The 2D Landau energy levels: $E = (N_L + 1/2)\hbar\omega_0$,

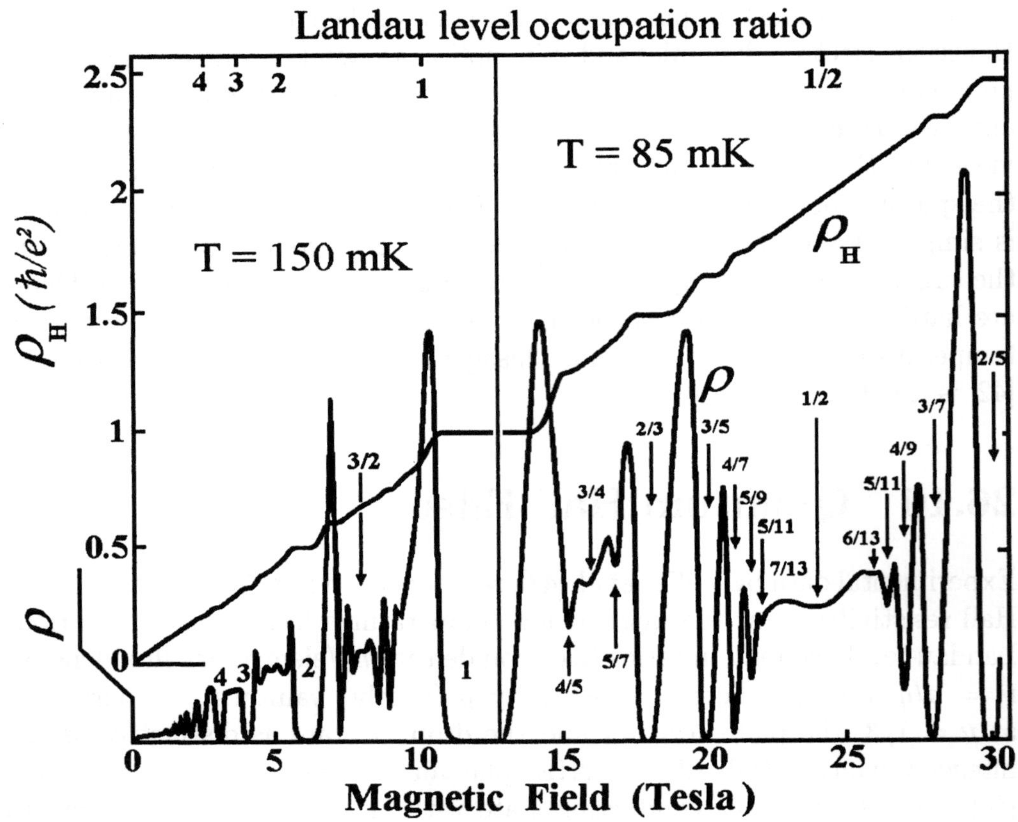

Figure 26.1: Fractional Quantum Hall Effect Experiments, after Willett *et al.* [43].

$\omega_0 \equiv eB/m^*$, with the states (N_L, k_y), $N_L = 0, 1, 2, ...$, have a great degeneracy. Exchange of a phonon between the electron and the fluxon generates an attractive transition in the degenerate states. The same exchange can also create (annihilate) electron-fluxon composites. The CM of any composite moves as a fermion (boson) if it contains an odd (even) numbers of elementary fermions. Hence the composite containing an electron and j fluxons moves as a boson (fermion) if j is odd (even). The system of c-bosons condenses below some critical temperature T_c and exhibits a superconducting state. The system of c-fermions shows a Fermi liquid behavior.

The system of free c-bosons moving with the linear dispersion relation:

$\varepsilon = (2/\pi)v_F p$ undergoes a BEC at the critical temperature

$$k_B T_c = 1.24\, \hbar v_F n_0^{1/2}, \qquad n_0 = \text{boson density}, \tag{26.32}$$

the same as Eq. (26.28) for the 2D pairons. In the principal QHE state at $\nu = 1$ fundamental (f) c-bosons, each with one fluxon, are formed. If we assume $n_0 = 10^{10}\,\text{cm}^{-2}$, $v_F = 1.36 \times 10^6$ cm s^{-1} for GaAs/AlGaAs, we obtain $T_c = 1.29$ K. Below T_c, the non-condensed c-boson has an energy gap ε_g relative to the energy of the condensed c-bosons.

In the QHE state at $\nu = p/q$, odd q, c-bosons, each with q fluxons, are formed. The c-boson carries the fractional charge

$$e^* = e/q. \tag{26.33}$$

The c-bosons condensed at the momentum directed along the sample length generate a supercurrent. The boson density n_0 is reduced compared with the case at $\nu = 1$ by the driving factor pq, as represented by

$$n_0 = n_e/(pq), \tag{26.34}$$

where n_e is the density of electrons involved in the principal QHE state. Eq. (26.34) means that both $T_c\,(\propto n_0^{1/2})$ and ε_g decreases as $(pq)^{-1/2}$ with increasing pq. This trend is in agreement with all of the twenty-plus plateaus and dips identified in Fig. 26.1. It also means that no QHE is realized for very large Nj (> 120 at 85 mK) since T_c becomes less than the observation temperature.

The spin phase transition observed at $\nu = 4/3$ and $8/5$, with tilt-field technique [52] can simply be explained, using the electron-fluxon complex model.

Chapter 27

Summary and Remarks

27.1 Summary

The five major properties of a superconductor are: zero resistance, Meissner effect, flux quantization, Josephson effects, and excitation-energy gaps. These properties all arise from the motion of a supercondensate in the conductor below T_c. In the present text we have examined these and other properties from a quantum statistical mechanical point of view.

The microscopic cause of the superconductivity is the phonon exchange attraction. Under certain conditions (see below), electrons near the Fermi surface form Cooper pairs (pairons) by exchanging phonons. Let us take a typical elemental superconductor such as lead (fcc). The virtual phonon exchange can generate an attraction if kinetic energies of the electrons involved are all close to each other. This exchange generates an attractive transition (correlation) between "electron" (or "hole") pair states whose energies are separated by twice the limit phonon energy $\hbar\omega_D$. Exchanging a phonon can also pair-create $\pm$ pairons from the physical vacuum. Phonons are electrically neutral, and hence the states of two electrons between which a phonon is exchanged, must have the same net charge before and after the exchange. Because of this, if the Fermi surface is favorable, equal numbers of $\pm$ pairons are formed in the conductor. The phonon-exchange attraction is a quantum field theoretical effect, and hence it cannot be explained by considering the potential energy alone. In fact the attraction depends on the kinetic energies of electrons. Pairons move independently with a linear dispersion (Cooper-Schrieffer) relation,

$$w_q = w_0 + cq < 0, \tag{27.1}$$

where $c = (2/\pi,\ 1/2)v_F$ for (2, 3)D, and

$$w_0 \equiv \frac{-2\hbar\omega_D}{\exp(2/v_0\mathcal{N}(0)) - 1} \tag{27.2}$$

is the ground-state energy of a pairon. Pairon's motion is very similar to photon´s. Unlike photons however pairons have charges $\pm 2e$, and the total number of $\pm$ pairons in a superconductor is limited. At 0 K the conductor may contain great and equal numbers of stationary $\pm$ pairons all condensed at zero momentum.

The most striking superconducting phenomenon is a never-decaying supercurrent ring. In the flux quantization experiment, a weak supercurrent goes around the ring, enclosing the magnetic flux. Here macroscopic numbers of pairons are condensed at a momentum

$$q_n \equiv \frac{2\pi\hbar\nu}{L}, \tag{27.3}$$

where L is the ring circumference and ν a quantum number $(0, \pm 1, \pm 2, ...)$ such that the flux Φ enclosed by the ring is $|\nu|$ times the flux quantum $\Phi_0 \equiv (h/2e)$:

$$\Phi = n\Phi_0 \equiv n\frac{h}{2e}. \tag{27.4}$$

The factor $2e$ means that the charge (magnitude) of the current-carrying particle is twice the electron charge e, supporting the BCS picture of a supercondensate composed of pairons of charge (magnitude) $2e$.

The macroscopic supercurrent generated by the supercondensate in motion is not destroyed by microscopic impurities. This condition is somewhat similar to the situation in which a flowing river (big object) is perturbed but cannot be stopped by a stick (small object). The fact that small perturbations cause no energy loss arises from the quantum nature of the superconducting state. The change in the condensed state requires redistribution of a great number of pairons. The supercurrent state can refocus by itself if the perturbation is not too strong. This is a Bose statistical effect peculiar to the condensed bosons moving with a linear dispersion relation. This self-focusing power is most apparent in the Josephson interference, where two supercurrents macroscopically separated up to 1 mm apart can interfere with each other just as two laser beams from the same source. Thus there is a close similarity between supercurrent and laser.

In the steady state, as realized in a circuit containing superconductor, resistor, and battery, all flowing currents in the superconductor are supercurrents. These supercurrents run in the thin surface layer characterized by the penetration depth λ ($\sim 500\,\text{Å}$), and they keep the magnetic field off the interior (Meissner effect). The current density j in the superconductor can be represented by

$$j = \begin{cases} \frac{1}{2}\, en_0 \,(v_F^{(2)} - v_F^{(1)}) & (3D) \\ \\ \frac{2}{\pi}\, en_0 \,(v_F^{(2)} - v_F^{(1)}), & (2D) \end{cases} \tag{27.5}$$

where n_0 is the total density of condensed pairons. The $\pm$ pairons move in the same direction, but their speeds, $v_F^{(j)}/2$, are different, and hence the net current does not vanish. If a magnetic field $\mathbf{B}$ is applied, the Lorentz-magnetic force tends to separate $\pm$ pairons. Thus there is a critical magnetic field $\mathbf{B}_c$. The supercurrent by itself generates a magnetic field, so there is a critical current. The picture of a neutral moving supercondensate explains why the superconducting state is not destroyed by the applied voltage. Since there is no net charge: $Q = 0$, no Lorentz-electric force $\mathbf{F}_E$ can act on the supercondensate: $\mathbf{F}_E \equiv QE = 0$. Thus the supercurrent is not accelerated, so it can gain no energy from the voltage.

The system of free pairons (bosons) moving with the linear dispersion relation undergoes a B-E condensation at the critical temperature T_c given by

$$k_B T_c = \begin{cases} 1.01\, \hbar v_F\, n_0^{1/3} & (3D) \\ \\ 1.24\, \hbar v_F\, n_0^{1/2}. & (2D) \end{cases} \tag{27.6}$$

The condensation transition in 3 (2)D is a phase change of second (third) order. The heat capacity in 3 D has a jump at T_c. Below T_c there is a supercondensate made up of $\pm$ pairons condensed at zero momentum. The density of condensed pairons increases to the maximum n_0 as temperature is lowered toward 0 K. The quasi-electron in the presence of the supercondensate has the energy $(\varepsilon_k^2 + \Delta^2)^{1/2}$. The gap Δ is temperature-dependent, and it reaches its maximum Δ_0 at 0 K. The maximum gap Δ_0 can be connected to the critical temperature T_c by

$$2\Delta_0 = 3.53\, k_B T_c \tag{27.7}$$

in the weak coupling limit. Phonon exchange is in action at all time and at all temperatures. Thus two quasi-electrons may be bound to form moving pairons. Since quasi-electrons have an energy gap Δ, excited pairons also have an energy gap ε_g, which is temperature-dependent. The pairon energy gap $\varepsilon_g(T)$ grows to its maximum equal to the binding energy of a Cooper pair, $|w_0|$, as the temperature approaches 0 K.

Moving pairons have negative energies, while quasi-electrons have positive energies. By the Boltzmann principle, moving pairons are therefore more numerous than quasi-electrons, and they are the predominant elementary excitations below T_c. Pairon energy gaps strongly influence the heat capacity $C(T)$ below T_c. The $C(T)$ far from T_c shows an exponential-decay-type T-dependence due to the energy gap ε_g; the maximum heat capacity $C_{\max}$ at T_c is modified by a small but non-negligible amount. The energy gap ε_g between excited and condensed pairons can be probed by quantum-tunneling experiments. The threshold voltage V_a in the I-V curve for an S-I-S system can be connected simply with ε_g:

$$V_a = \varepsilon_g(T)/e. \tag{27.8}$$

This allows a direct observation of the energy gap $\varepsilon_g(T)$ as a function of T.

Compound superconductors have optical and acoustic phonons. The Cooper pairs formed, mediated by optical phonons bridging between "electron"-like and "hole"-like Fermi surfaces, have smaller linear sizes ξ_0 ($\sim$ 50 Å). Thus these superconductors show type II behavior. The critical temperature T_c tends to be higher, since pairons can be packed more densely.

Cuprate superconductors have layered lattice structures. Conduction electrons move only in the CuO_2 planes. Since they are compounds, $\pm$ pairons can be generated with the aid of optical phonons bridging between "electron"-like and "hole"-like 2D Fermi surfaces. The pairon size is small ($\sim$ 14 Å for YBCO), and pairons may therefore be packed even more densely. The critical temperature T_c, based on the model of free massless bosons moving in 2D, is given by Eq. (27.6). The interpairon distance $r_0 \equiv n_0^{-1/2}$ is much smaller in cuprates than in elements, and the Fermi velocity v_F is smaller, making the critical temperature T_c higher. Since the pairon size is small, the

Coulomb repulsion between two electrons is not negligible. This generates two energy gaps (Δ_1, Δ_2) for quasi-electrons and two energy gaps (ε_1, ε_2) for moving pairons. Thus, the I-V curves for high T_c are asymmetric and generally more complicated.

We have treated all superconductors in a unified manner, starting with a generalized BCS Hamiltonian H and taking account of electron and phonon energy bands. The underlying assumption is that $\pm$pairons are generated from the physical vacuum by emission and absorption of virtual phonons under favorable conditions. Alkali metals like Na have nearly spherical Fermi surfaces and have "electrons" only, and hence they are not superconductors. Multivalent non-magnetic metallic elements can generate $\pm$ pairons near hyperboloidal Fermi surfaces, so they are most often superconductors.

27.2 Remarks

In the text we have discussed primarily chemically-pure, lattice-perfect, bulk-size superconductors. Many important superconducting properties arise in imperfect superconductors. We briefly discuss some of these properties in the following subsection.

27.2.1 Thin Films

If the dimension of a sample in some direction is less than the penetration depth λ, as in a thin film, the superconductor's critical temperature is a little higher than in a bulk sample. This may be explained as follows. Consider a very thin supercurrent ring. The superconducting sample tends to expel any magnetic field at the expense of the stored magnetic field energy. This expulsion is not complete because of the sample dimension; therefore stored magnetic energy density is less than in bulk, making the superconducting state more stable and rendering T_c a little higher.

27.2.2 Nonmagnetic Impurities

A small amount of non-magnetic impurities neither hinder the supercurrent nor alter the supercondensate density. This means that adding non-magnetic impurities do not change T_c drastically. The impurities however significantly affect the coherence length ξ and the penetration depth λ. These effects are often described by $\xi^{-1} = \xi_0^{-1} + l^{-1}$, $\lambda = \lambda_0(\xi_0/\xi)^{1/2}$, where l is the electron mean free path and (ξ_0, λ_0) represent the values of (ξ, λ) for pure superconductor. According to this the addition of impurities makes ξ smaller and λ greater. This is experimentally supported by the fact that alloys like Pb-In$_x$ show a type II behavior if the fraction x is made high enough.

27.2.3 Magnetic Impurities

Ferromagnetic elements such as iron (Fe) and nickel (Ni), are not superconductors. These metals of course have electrons and phonons. Thus spontaneous magnetization and the associated magnetic field are thought to be detrimental to the formation of Cooper pairs. Injection of magnetic ions in a superconductor lowers T_c significantly. This may be understood as follows. A Cooper pair is made up of electrons of up and down spins. The internal magnetic field destroys the symmetry between up and down spins; this makes the pairon formation less favorable, which lowers pairon density and the critical temperature.

27.2.4 Intermediate State

When a magnetic field $\mathbf{H}_a$ is applied along the axis of a cylindrical superconductor, surface supercurrents are generated to shield the B-field from the body with no energy loss. If the superconductor has a poor geometrical shape and/or the H_a-field is in the wrong direction, then it will be in the *intermediate state*, where normal and superconducting domains are formed side by side. The actual domain structures can be very complicated. We have avoided this complication completely in the text by assuming the ideal condition where the H_a-field is directed along the axis of a long cylindrical superconductor.

27.2.5 Critical Currents. Silsbee's Rule

How much current can be passed through a superconductor without generating resistance? This is an important question in devices and applications. As we mentioned the actual supercurrent configuration may be very complicated. It is recognized that a superconductor loses its zero resistance when at any point on the surface, the total magnetic field due to the transport current and the applied field exceeds the critical field B_c. This is often called generalized *Silsbee's rule* [1]. (The original Silsbee's rule, proposed by this author in 1916, refers to the case of zero external field.) In our condensed pairon picture, a supercurrent is generated by pairons moving in the same direction with different speeds (c_1, c_2). Any B-field tends to separate these pairons by a magnetic force. Silsbee's rule is in accord with this picture.

27.2.6 Mixed State. Pinning Vortex Lines

Type II superconductor are more useful in devices and applications than type I because the upper (or superconducting) critical field H_{c2} can be much higher than the thermodynamic critical field. Between the lower and upper critical fields (H_{c1}, H_{c2}), a type II superconductor allows a partial penetration of elementary magnetic flux lines (vortices), which forms a two-dimensional lattice. Such a state is called a *mixed state*, distinct from the intermediate state discussed earlier. Elementary vortices repel each other. If the vortex lattice is perfect, there is no net current. In the actual current-carrying state, vortices are *pinned* by various lattice imperfections, and the resulting inhomogeneous vortex configuration generates a net supercurrent. In practice lattice imperfections are purposely introduced in the fabrication processes, and the details of the flux (vortex)-pinning are very complicated.

27.2.7 Critical Currents in Type II Superconductors

If the applied magnetic field H_a is less than the lower critical field H_{c1}, the critical current tends to decrease linearly with increasing field H_a. (The same behavior is observed in type I superconductors.) The associated B-field in the surface layer tends to disrupt the motion of $\pm$ pairons by the Lorentz force, causing a linear H_a-decrease in the critical current. The practically important case however is the one where the applied field H_a is higher than H_{c1} so that vortex lines penetrate the body. Experiments indicate that the more imperfect the sample, the higher is the critical current. This behavior arises from flux-pining by imperfections; besides, the transport current appears to flow throughout the whole body. The phenomena are therefore quite complicated; but since this is very important in devices and application, extensive researches is currently being carried out.

27.2.8 Concluding Remarks

The traditional statistical mechanical theory deals with equilibrium thermodynamic properties and steady-state transport and optical properties of a macroscopic system. When a system contains super and normal domains or inhomogeneous pinning of vortex lines, theories must be developed case by case. In fact there is no unified theory dealing with such cases. Future challenging research includes seeking higher T_c-materials and raising critical currents for larger-scale applications. Such research requires a great deal of efforts and creative minds. The authors hope the elementary quantum statistical theory of superconductivity presented here will be a useful guide for the exciting future developments.

Appendix A

Second Quantization

The most remarkable fact about a system of fermions is that no more than one fermion can occupy a quantum particle state (Pauli's exclusion principle). For bosons no such restriction applies. That is, any number of bosons can occupy the same state. We shall discuss the second quantization formalism in which creation and annihilation operators associated with each quantum state are used. This formalism is extremely useful in treating of many-boson and/or many-fermion system.

A.1 Boson Creation and Annihilation Operators

The quantum state for a system of bosons (or fermions) can most conveniently be represented by a set of occupation numbers $\{n'_a\}$, where n'_a are the numbers of bosons (or fermions) occupying the quantum particle-states a. This representation is called the *occupation number representation* or simply the *number representation*. For bosons, the possible values for n'_a are zero, one, two, or any positive integers:

$$n'_a = 0, 1, 2, \ldots \qquad \text{for bosons.} \tag{A.1}$$

The many-boson state can best be represented by the distribution of particles (balls) in the states (boxes) as shown in Fig. A.1.

Let us introduce operators n_a (without prime) whose eigenvalues are given by 0, 1, 2, ... Since Eq. (A.1) is meant for each and every state a indepen-

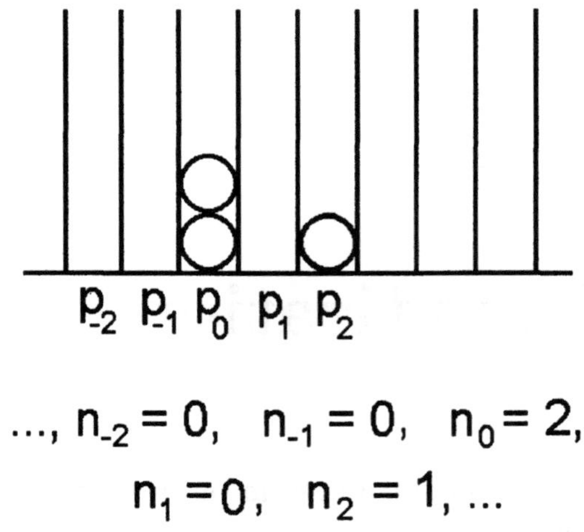

Figure A.1: A many-boson state is represented by a set of boson numbers $\{n_j\}$ occupying the states $\{p_j\}$.

dently, we assume that

$$[n_a, n_b] \equiv n_a n_b - n_b n_a = 0. \tag{A.2}$$

It is convenient to introduce complex dynamic variables η and $\eta^\dagger$ instead of directly dealing with the number operators n. We attach labels a, b, ... to the dynamic variables η and $\eta^\dagger$ associated with the states a, b, ... and assume that η and $\eta^\dagger$ satisfy the following *Bose commutation rules*:

$$[\eta_a, \eta_b^\dagger] = \delta_{ab}, \qquad [\eta_a, \eta_b] = [\eta_a^\dagger, \eta_b^\dagger] = 0. \tag{A.3}$$

Let us set

$$\eta_a^\dagger \eta_a \equiv n_a \; (= n_a^\dagger), \tag{A.4}$$

which is Hermitean. Clearly, $n_a \equiv \eta_a^\dagger \eta_a$ satisfy Eqs. (A.2).

We shall show that n_a has as eigenvalues all non-negative integers. Let n' be an eigenvalue of n (dropping the suffix a) and $|n'\rangle$ an eigenket belonging to it. By definition

$$\langle n'| \eta^\dagger \eta |n'\rangle = n' \langle n' \mid n'\rangle. \tag{A.5}$$

Now $\langle n'| \eta^\dagger \eta |n'\rangle$ is the squared length of the ket $\eta |n'\rangle$ and hence

$$\langle n'| \eta^\dagger \eta |n'\rangle \geq 0. \tag{A.6}$$

Also by definition $\langle n' \mid n' \rangle > 0$; hence from Eqs. (A.5) and (A.6), we obtain

$$n' \geq 0, \tag{A.7}$$

the case of equality occurring only if

$$\eta \, |n'\rangle = 0. \tag{A.8}$$

Consider now $[\eta, n] \equiv [\eta, \eta^\dagger \eta]$. We may use the following identities:

$$[A, BC] = B[A, C] + [A, B]C, \qquad [AB, C] = A[B, C] + [A, C]B, \tag{A.9}$$

and obtain

$$[\eta, \eta^\dagger \eta] = \eta^\dagger [\eta, \eta] + [\eta, \eta^\dagger]\eta = \eta, \qquad \eta n - n\eta = \eta. \tag{A.10}$$

Hence

$$n\eta \, |n'\rangle = (\eta n - \eta) \, |n'\rangle = (n' - 1)\eta \, |n'\rangle. \tag{A.11}$$

Now if $\eta \, |n'\rangle \neq 0$, then $\eta \, |n'\rangle$ is, according to Eq. (A.11), an eigenket of n belonging to the eigenvalue $n' - 1$. Hence for non-zero n', $n' - 1$ is another eigenvalue. We can repeat the argument and deduce that, if $n' - 1 \neq 0$, $n' - 2$ is another eigenvalue of n. Continuing in this way, we obtain a series of eigenvalues $n', n' - 1, n' - 2, \ldots.$ which can terminate *only* with the value 0 because of inequality (A.7). By a similar process, we can show from the Hermitean conjugate of Eq. (A.10): $n\eta^\dagger - \eta^\dagger n = \eta^\dagger$ that the eigenvalue of n has no upper limit [Problem A.1.1]. Hence, the eigenvalues of n are non-negative integers: $0, 1, 2, \ldots$. (q.e.d)

Let $|\phi_a\rangle$ be a normalized eigenket of n_a belonging to the eigenvalue 0 so that

$$n_a \, |\phi_a\rangle = \eta_a^\dagger \eta_a \, |\phi_a\rangle = 0. \tag{A.12}$$

By multiplying all these kets $|\phi_a\rangle$ together, we construct a normalized eigenket:

$$|\Phi_0\rangle \equiv |\phi_a\rangle \, |\phi_b\rangle \ldots \tag{A.13}$$

which is a simultaneous eigenket of all n belonging to the eigenvalues zero. This ket is called the *vacuum ket*. It has the following property:

$$\eta_a \, |\Phi_0\rangle = 0 \qquad \text{for any } a. \tag{A.14}$$

Using the commutation rules (A.3) we obtain a relation (dropping suffix a)

$$\eta(\eta^\dagger)^{n'} - (\eta^\dagger)^{n'}\eta = n'(\eta^\dagger)^{n'-1}, \tag{A.15}$$

which may be proved by induction (Problem A.1.2). Multiply Eq. (A.15) by $\eta^\dagger$ from the left and operate the result to $|\Phi_0\rangle$. Using Eq. (A.14) we obtain

$$n(\eta^\dagger)^{n'}|\phi\rangle = n'(\eta^\dagger)^{n'}|\phi\rangle, \tag{A.16}$$

indicating that $(\eta^\dagger)^{n'}|\phi\rangle$ is an eigenket belonging to the eigenvalue n'. The square length of $(\eta^\dagger)^{n'}|\phi\rangle$ is

$$\langle\phi|\,\eta^{n'}(\eta^\dagger)^{n'}|\phi\rangle = n'\langle\phi|\,\eta^{n'-1}(\eta^\dagger)^{n'-1}|\phi\rangle = \cdots = n'!. \tag{A.17}$$

We see from Eq. (A.11) that $\eta\,|n'\rangle$ is an eigenket of n belonging to the eigenvalue $n'-1$. Similarly, we can show from $[n, \eta^\dagger] = \eta^\dagger$ that $\eta^\dagger|n'\rangle$ is an eigenket of n belonging to the eigenvalue $n'+1$. Thus operator η, acting on the number eigenket, annihilates a particle while operator $\eta^\dagger$ creates a particle. Therefore, η and $\eta^\dagger$ are called *annihilation and creation operators*, respectively. From Eqs. (A.16) and (A.17) we infer that if $n_1', n_2', ...$ are any non-negative integers,

$$(n_1'!\,n_2'!\,...)^{-1/2}(\eta_1^\dagger)^{n_1'}(\eta_2^\dagger)^{n_2'}\,...\,|\Phi_0\rangle \equiv |n_1', n_2', ...\rangle \tag{A.18}$$

is a normalized simultaneous eigenket of all the n belonging to the eigenvalues $n_1', n_2', ...$ Various kets obtained by taking different n' form a complete set of kets all orthogonal to each other.

Following Dirac [1], we postulate that the quantum states for N bosons can be represented by a *symmetric ket*

$$S[|\,\alpha_a^{(1)}>|\,\alpha_b^{(2)}>\,...\,|\,\alpha_g^{(N)}>] \equiv |\alpha_a\alpha_b...\alpha_g\rangle_S\,, \tag{A.19}$$

where S is the *symmetrizing operator*:

$$S \equiv \frac{1}{\sqrt{N!}}\sum_P P \tag{A.20}$$

and P are permutation operators for the particle-indices $(1, 2, ..., N)$. The ket in Eq. (A.19) is not normalized but

$$(n_1!\,n_2!\,...)^{-1/2}\,|\alpha_a\alpha_b...\alpha_g\rangle_S \equiv |\{n\}\rangle \tag{A.21}$$

is a normalized ket representing the same state. Comparing Eqs. (A.21) and (A.18), we obtain

$$|\alpha_a \alpha_b ... \alpha_g\rangle_S = \eta_a^\dagger \eta_b^\dagger ... \eta_g^\dagger \, |\Phi_0\rangle. \tag{A.22}$$

That is, unnormalized symmetric kets $|\alpha_a \alpha_b ... \alpha_g\rangle_S$ for the system can be constructed by applying N creation operators $\eta_a^\dagger \eta_b^\dagger ... \eta_g^\dagger$ to the vacuum ket $|\Phi_0\rangle$. So far we have tacitly assumed that the total number of bosons is fixed at N'. If this number is not fixed but is variable, we can easily extend the theory to this case. Let us introduce a Hermitean operator N defined by

$$N \equiv \sum_a \eta_a^\dagger \eta_a = \sum_a n_a, \tag{A.23}$$

the summation extending over the whole set of boson states. Clearly, the operator N has eigenvalues $0, 1, 2, ...$, and the ket $|\alpha_a \alpha_b ... \alpha_g\rangle_S$ is an eigenket of N belonging to the eigenvalue N'. We may arrange kets in the order of N', i.e., zero-particle state, one-particle states, two-particle states, ... :

$$|\Phi_0\rangle, \quad \eta_a^\dagger \, |\Phi_0\rangle, \quad \eta_a^\dagger \eta_b^\dagger \, |\Phi_0\rangle, \quad \ldots . \tag{A.24}$$

These kets are all orthogonal to each other, two kets referring to the same number of bosons are orthogonal as before, and two referring to different numbers of bosons are orthogonal because they have different eigenvalues N'. By normalizing the kets, we get a set of kets like (A.21) with no restriction on $\{n'\}$. These kets form the basic kets in a representation where $\{n_a\}$ are diagonal.

Problem A.1.1. (a) Show (twice) that $n\eta^\dagger - \eta^\dagger n = \eta^\dagger$, by taking the Hermitian-conjugation of Eq. (A.10) *and* also by using Eqs. (A.9). (b) Use this relation and obtain a series of eigenvalues $n', n'+1, n'+2, ...$, where n' is an eigenvalue of n.

Problem A.1.2. Prove Eq. (A.15) by mathematical induction. Hint: use Eqs. (A.9).

A.2 Observables

We wish to express observable physical quantities (observables) for the system of identical bosons in terms of η and $\eta^\dagger$ These observables are by postulate symmetric functions of the boson variables.

An observable may be written in the form:

$$\sum_j y^{(j)} + \sum_i \sum_j z^{(ij)} + ... \equiv Y + Z + ..., \tag{A.25}$$

where $y^{(j)}$ is a function of the dynamic variables of the jth boson, $z^{(ij)}$ that of the dynamic variables of the ith and jth bosons, and so on.

We take $Y \equiv \sum_j y^{(j)}$. Since $y^{(j)}$ acts only on the ket $\left|\alpha^{(j)}\right\rangle$ of the jth boson, we have

$$y^{(j)}(\left|\alpha^{(1)}_{x_1}\right\rangle \left|\alpha^{(2)}_{x_2}\right\rangle ... \left|\alpha^{(j)}_{x_j}\right\rangle ...) = \sum_a (\left|\alpha^{(1)}_{x_1}\right\rangle \left|\alpha^{(2)}_{x_2}\right\rangle ... \left|\alpha^{(j)}_{a}\right\rangle ...) \left\langle\alpha^{(j)}_{a}\right| y^{(j)} \left|\alpha^{(j)}_{x_j}\right\rangle. \tag{A.26}$$

The matrix element $< \alpha^{(j)}_a \mid y^{(j)} \mid \alpha^{(j)}_{x_j} > \equiv < \alpha_a \mid y \mid \alpha_{x_j} >$ does not depend on the particle index j. Summing Eq. (A.26) over all j and applying operator S to the result, we obtain

$$SY\left(\left|\alpha^{(1)}_{x_1}\right\rangle \left|\alpha^{(2)}_{x_2}\right\rangle ...\right) = \sum_j \sum_a S(\left|\alpha^{(1)}_{x_1}\right\rangle \left|\alpha^{(2)}_{x_2}\right\rangle ... \left|\alpha^{(1)}_{x_1}\right\rangle \left|\alpha^{(j)}_{a}\right\rangle ...) \left\langle\alpha_a\right| y \left|\alpha_{x_j}\right\rangle. \tag{A.27}$$

Since Y is symmetric, we can replace SY by YS for the lhs. After straightforward calculations, we obtain, from Eq. (A.27),

$$\begin{aligned} Y\eta^\dagger_{x_1}\eta^\dagger_{x_2} ... \left|\Phi_0\right\rangle &= \sum_j \sum_a \eta^\dagger_{x_1}\eta^\dagger_{x_2} ... \eta^\dagger_{x_{j-1}}\eta^\dagger_a\eta^\dagger_{x_{j+1}} ... \left|\Phi_0\right\rangle \left\langle\alpha_a\right| y \left|\alpha_{x_j}\right\rangle \\ &= \sum_a \sum_b \eta^\dagger_a \sum_j \eta^\dagger_{x_1}\eta^\dagger_{x_2} ... \eta^\dagger_{x_{j-1}}\eta^\dagger_{x_{j+1}} ... \left|\Phi_0\right\rangle \delta_{bx_j} \left\langle\alpha_a\right| y \left|\alpha_b\right\rangle. \end{aligned} \tag{A.28}$$

Using the commutation rules and the property (A.14) we can show that

$$\eta_b\eta^\dagger_{x_1}\eta^\dagger_{x_2} ... \left|\Phi_0\right\rangle = \sum_j \eta^\dagger_{x_1}\eta^\dagger_{x_2} ... \eta^\dagger_{x_{j-1}}\eta^\dagger_{x_{j+1}} ... \left|\Phi_0\right\rangle \delta_{bx_j}. \tag{A.29}$$

(Problem A.2.1). Using this relation, we obtain from Eq. (A.28)

$$Y\eta_{x_1}^\dagger\eta_{x_2}^\dagger\cdots|\Phi_0\rangle = \sum_a\sum_b \eta_a^\dagger\eta_b\,\langle\alpha_a|\,y\,|\alpha_b\rangle\,(\eta_{x_1}^\dagger\eta_{x_2}^\dagger\cdots|\Phi_0\rangle). \tag{A.30}$$

Since the kets $\eta_{x_1}^\dagger\eta_{x_2}^\dagger\cdots|\Phi_0\rangle$ form a complete set, we obtain

$$Y = \sum_a\sum_b \eta_a^\dagger\eta_b\,\langle\alpha_a|\,y\,|\alpha_b\rangle\,. \tag{A.31}$$

In a similar manner Z in Eq. (A.25) can be expressed by [Problem A.2.2]

$$Z = \sum_a\sum_b\sum_c\sum_d \eta_a^\dagger\eta_b^\dagger\eta_d\eta_c\,\langle\alpha_a\alpha_b|\,y\,|\alpha_c\alpha_d\rangle, \tag{A.32}$$

$$\langle\alpha_a\alpha_b|\,y\,|\alpha_c\alpha_d\rangle \equiv \langle\alpha_a^{(1)}|\left\langle\alpha_b^{(2)}\right|z^{(12)}\left|\alpha_d^{(2)}\right\rangle|\alpha_c^{(1)}\rangle. \tag{A.33}$$

Problem A.2.1. Prove Eq. (A.29). Hint: Start with cases of one- and two-particle-state kets.

Problem A.2.2. Prove Eq. (A.32) by following those steps similar to (A.27)-(A.31).

A.3 Fermions Creation and Annihilation Operators

In this section we treat a system of identical fermions in a parallel manner.

The quantum states for fermions, by postulate, are represented by *antisymmetric kets*:

$$|\alpha_a\alpha_b\ldots\alpha_g\rangle_A \equiv A(|\alpha_a^{(1)}\rangle\left|\alpha_b^{(2)}\right\rangle\cdots|\alpha_g^{(N)}\rangle), \tag{A.34}$$

where

$$A \equiv \frac{1}{\sqrt{N!}}\sum_P \delta_P P, \tag{A.35}$$

is the *antisymmetrizing operator*, with δ_P being $+1$ or -1 according to whether P is even or odd. Each antisymmetric ket in Eq. (A.34) is characterized such that it changes its sign if an odd permutation of particle indices

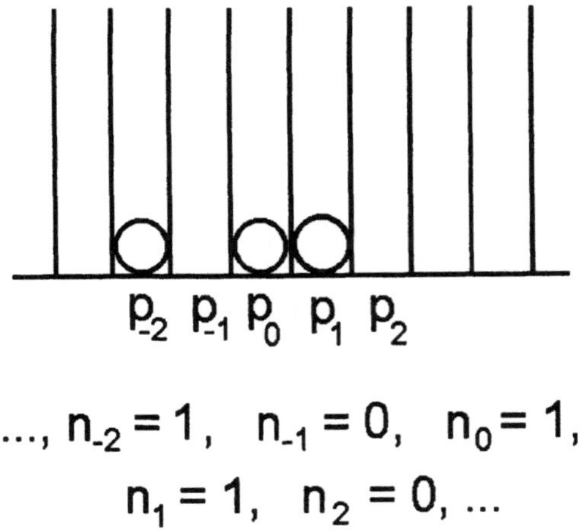

Figure A.2: A many-fermion state can be represented by the set of the numbers $\{n_j\}$ occupying the states $\{p_j \equiv 2\pi\hbar L^{-1}j\}$. Each n_j is restricted to 0 or 1.

is applied to it, and the fermion states $a, b, ..., g$ are all different. Just as for a boson system, we can introduce observables $n_1, n_2, ...$ each with eigenvalues 0 or 1, representing the number of fermions in the states $\alpha_1, \alpha_2, ...$ respectively. The many-fermion occupation-number can be represented as in Fig. A.2.

We can also introduce a set of linear operators $(\eta, \eta^\dagger)$, one pair $(\eta_a, \eta_a^\dagger)$ for each state α_a, satisfying the *Fermi anticommutation rules*:

$$\{\eta_a, \eta_b^\dagger\} \equiv \eta_a\eta_b^\dagger + \eta_b^\dagger\eta_a = \delta_{ab}, \qquad \{\eta_a, \eta_b\} = \{\eta_a^\dagger, \eta_b^\dagger\} = 0. \tag{A.36}$$

The number of fermions in the state α_a is again represented by

$$n_a = \eta_a^\dagger\eta_a = n_a^\dagger. \tag{A.37}$$

Using Eqs. (A.36), we obtain

$$n_a^2 = \eta_a^\dagger\eta_a\eta_a^\dagger\eta_a = \eta_a^\dagger(1 - \eta_a^\dagger\eta_a)\eta_a = \eta_a^\dagger\eta_a = n_a \qquad \text{or} \qquad n_a^2 - n_a = 0. \tag{A.38}$$

If an eigenket of n_a belonging to the eigenvalue n_a' is denoted by $|n_a'\rangle$, Eq. (A.38) yields

$$(n_a^2 - n_a)\,|n_a'\rangle = (n_a'^2 - n_a')\,|n_a'\rangle = 0. \tag{A.39}$$

Since $|n_a'\rangle \neq 0$, we obtain $n_a'(n_a' - 1) = 0$, meaning that the eigenvalues n_a' are either 0 or 1 as required:

$$n_a' = 0 \text{ or } 1. \tag{A.40}$$

Similarly to the case of bosons, we can show that

$$|\alpha_a\alpha_b...\alpha_g\rangle_A = \eta_a^\dagger\eta_b^\dagger \dots \eta_g^\dagger \,|\Phi_0\rangle, \tag{A.41}$$

which is normalized to unity.

Observables describing the system of fermions can be expressed in terms of operators η and $\eta^\dagger$, and the results have the same form Eqs. (A.31) and (A.32) as for the case of bosons.

In summary both states and observables for a system of identical particles can be expressed in terms of creation and annihilation operators. This formalism, called the *second quantization formalism*, has some notable advantages over the usual Schrödinger formalism. First, the permutation-symmetry property of the quantum particles is represented simply in the form of Bose commutation (or Fermi anticommutation) rules. Second, observables in second quantization are defined for an arbitrary number of particles so that the formalism may apply to systems in which the number of particles is not fixed, but variable. Third, and most importantly, all relevant quantities (states and observables) can be defined referring only to the single-particle states. This property allows one to describe the motion of the many-body system in the 3D space. This is a natural description since all particles in nature move in 3D. In fact, relativistic quantum field theory can be developed only in second quantization.

A.4 Heisenberg Equation of Motion

In the Schrödinger Picture (SP), the energy eigenvalue equation is

$$H\,|E\rangle = E\,|E\rangle, \tag{A.42}$$

where H is the Hamiltonian and E the eigenvalue. In the position representation this equation is written as

$$H(x_1, -i\hbar\partial/\partial x_1, x_2, -i\hbar\partial/\partial x_2, \ldots)\Psi(x_1, x_2, \ldots) = E\Psi, \tag{A.43}$$

where Ψ is the wave function for the system. We consider a one-dimensional motion for conceptional and notational simplicity. [For a three-dimensional motion, (x, p) should be replaced by $(x, y, z, p_x, p_y, p_z) = (\mathbf{r}, \mathbf{p})$.] If the number of electrons N is large, the wave unction Ψ contains many electron variables $(x_1, x_2, \ldots)$. This complexity needed in dealing with many electron variables can be avoided if we use the second quantization formulation and the Heisenberg Picture (HP), which will be shown in this section.

If the Hamiltonian H is the sum of single-particle Hamiltonians:

$$H = \sum_j h^{(j)}, \tag{A.44}$$

this Hamiltonian H can be represented by

$$H = \sum_a \sum_b \langle \alpha_a | \, h \, | \alpha_b \rangle \, \eta_a^\dagger \eta_b \equiv \sum_a \sum_b h_{ab} \, \eta_a^\dagger \eta_b, \tag{A.45}$$

where η_a ($\eta_a^\dagger$) are annihilation (creation) operators associated with particle-state a and satisfying the Fermi commutation rules.

In the HP a variable $\xi(t)$ changes in time, following the *Heisenberg equation of motion*:

$$-i\hbar \frac{d\xi(t)}{dt} = [H, \xi] \equiv H\xi - \xi H. \tag{A.46}$$

Setting $\xi = \eta_a^\dagger$, we obtain

$$-i\hbar \frac{d\eta_a^\dagger}{dt} = [H, \eta_a^\dagger], \tag{A.47}$$

whose Hermitian conjugate is given by

$$i\hbar \frac{d\eta_a}{dt} = ([H, \eta_a^\dagger])^\dagger = -[H, \eta_a]. \tag{A.48}$$

By quantum postulate the physical observable ξ is Hermitian: $\xi^\dagger = \xi$. Variables η_a and $\eta_a^\dagger$ are *not* Hermitian, but both obey the *same* Heisenberg equation of motion.

We introduce Eq. (A.45) into Eq. (A.47), and calculate the commutator $[H, \eta_a^\dagger]$. In such a commutator calculation, identities (A.9) and the following identities:

$$[A, BC] = \{A, B\}C - B\{A, C\}, \qquad [AB, C] = A\{B, C\} - \{A, C\}B, \tag{A.49}$$

are very useful. Note: The negative signs on the right-hand terms in Eqs. (A.49) occur when the cyclic order is destroyed. We obtain from Eqs. (A.47) and (A.48)

$$-i\hbar \frac{d\eta_a^\dagger}{dt} = \sum_c \sum_b h_{cb} \, [\eta_c^\dagger \eta_b, \eta_a^\dagger] = \sum_c \sum_b h_{cb} \, \eta_c^\dagger \{\eta_b, \eta_a^\dagger\} = \sum_c h_{ca} \, \eta_c^\dagger \tag{A.50}$$

$$i\hbar\frac{d\eta_a}{dt} = \sum_c h_{ac}\,\eta_c. \tag{A.51}$$

Equation (A.50) means that the change of the one-body operator $\eta_a^\dagger$ is determined by the one-body Hamiltonian h. This is the main advantage of working in the HP. Equations (A.50)-(A.51) are valid for *any* single-particle states $\{a\}$.

In the field operator language Eq. (A.51) reads

$$i\hbar\frac{\partial\psi(\mathbf{r},t)}{\partial t} = h\left(\mathbf{r}, -i\hbar\partial/\partial\mathbf{r}\right)\psi(\mathbf{r},t), \tag{A.52}$$

which is formally identical to the Schrödinger equation of motion for a particle.

If the system Hamiltonian H contains an interparticle interaction

$$V = \frac{1}{2}\int d^3r\int d^3r'\, v(\mathbf{r}-\mathbf{r}')\,\psi^\dagger(\mathbf{r},t)\psi^\dagger(\mathbf{r}',t)\psi(\mathbf{r}',t)\psi(\mathbf{r},t), \tag{A.53}$$

the evolution equation for $\psi(\mathbf{r},t)$ is nonlinear (Problem A.4.2):

$$\begin{aligned} i\hbar\frac{\partial\psi(\mathbf{r},t)}{\partial t} &= h\left(\mathbf{r}, -i\hbar\partial/\partial\mathbf{r}\right)\psi(\mathbf{r},t) \\ &\quad + \int d^3r'\, v(\mathbf{r}-\mathbf{r}')\,\psi^\dagger(\mathbf{r}',t)\psi(\mathbf{r}',t)\psi(\mathbf{r},t). \end{aligned} \tag{A.54}$$

In quantum field theory the basic dynamical variables are particle-field operators. The quantum statistics of the particles are given by the Bose commutation or the Fermi commutation rules satisfied by the field operators. The evolution equations of the field operators are intrinsically nonlinear when the interparticle interaction is present.

Problem A.4.1. Verify that the equation of motion Eq. (A.50) holds for bosons.

Problem A.4.2. Use Eq. (A.46) to verify Eq. (A.54).

REFERENCES

Chapter 1

1. H. Kamerlingh Onnes, *Akad. V. Wetenschappen* (Amsterdam) **14,** 113 (1911).
2. W. Meissner and R. Ochsenfeld, *Naturwiss*, **21,** 787 (1933).
3. J. File and R. G. Mills, *Phys. Rev. Lett.* **10,** 93 (1963).
4. B. S. Deaver and W. M. Fairbank, *Phys. Rev. Lett.* **7,** 43 (1961).
5. R. Doll and M. Näbauer, *Phys. Rev. Lett.* **7,** 51 (1961).
6. B. D. Josepson, *Phys. Lett.* **1,** 251 (1962).
7. P. W. Anderson and J. M. Rowell, *Phys. Rev. Lett.* **10,** 486 (1963).
8. R. C. Jaklevic, J. Lambe, J. E. Marcereau and A. H. Silver, *Phys. Rev.* **A 140,** 1628 (1965).
9. R. E. Glover III and M. Tinkham, *Phys. Rev.* **108,** 243 (1957); M. A. Biondi and M. Garfunkelm, *Phys. Rev.* **116,** 853 (1959).
10. I. Giaever, *Phys. Rev. Lett.* **5,** 147, 464 (1960).
11. N. E. Phillips, *Phys. Rev.* **114,** 676 (1959).
12. I. Giaever and K. Megerle, *Phys. Rev.* **122,** 1101 (1961).
13. C. A. Reynolds *et al., Phys. Rev.* **78,** 487 (1950); E. Maxwell, *Phys. Rev.* **78,** 477 (1950).
14. J. G. Bednorz and K. A. Müller, *Z. Phys.* B **64,** 189 (1986).
15. M. K. Wu *et al.*, *Phys. Rev. Lett.* **58,** 908 (1987).
16. P. A.. M. Dirac, *Principles of Quantum Mechanics*. 4th Ed., (Oxford University Press, London, 1958).
17. P. Ehrenfest and J. R. Oppenheimer, *Phys. Rev.* **37,** 333 (1931); H. A. Bethe and R. F. Bacher, *Rev. Mod. Phys.* **8,** 193 (1936).
18. F. London, *Nature,* **141,** 643 (1938); *Superfluids*, I and II, (Dover, New York 1964).
19. A. Einstein, *Sits. Ber. Berl. Akad.* **3** (1925).
20. L. D. Landau, *J. Exptl. Theoret. Physics.* (USSR) **30,** 1058 (1956); ibid, **32,** 59 (1957).
21. A. Haug, *Theoretical Solid State Physics*, Vol. I, (Pergamon,Oxford 1972,); C. Kittel, *Introduction to Solid State Physics.* 6-th ed., (Wiley, New York 1986); N. W. Ashcroft and N. D. Mermin, *Solid State Physics*, (Saunders, Philadelphia, 1976); W. Harrison, *Solid State Theory*, (Dover, New York, 1980).

Chapter 2

1. K. Huang, *Statistical Mechanics*, 2nd ed. (Wiley, New York, 1987).
2. C. J. Gorter and H. B. G. Casimir, *Physica*, **1,** 306 (1934).
3. V. L. Ginzburg and L. D. Landau, *J. Exp. Theor. Phys.* (USSR) **20,** 1064 (1950).
4. B. S. Deaver and W. M. Fairbank, *Phys. Rev. Lett.* **7,** 43 (1961); R. Doll and M. Näbauer, *Phys. Rev. Lett.* **7,** 51 (1961).
5. R. C. Jaklevic, J. Lambe, J. E. Marcereau and A. H. Silver, *Phys. Rev.* **A 140,** 1628 (1965).
6. P. Ehrenfest and J. R. Oppenheimer, *Phys. Rev.* **37,** 333 (1931); H. A. Bethe and R. F. Bacher, *Rev. Mod. Phys.* **8,** 193 (1936).
7. S. Fujita and D. L. Morabito, *Mod. Phys. Lett.* **12,** 753 (1998).
8. A. Einstein, *Sits. Ber. Berl. Akad.* **3** (1925).
9. M. H. Anderson *et al., Science* **269,** 198 (1995).
10. K. B. Davis *et al., Phys. Rev. Lett.* **75,** 3969 (1995).
11. L. N. Cooper, *Phys. Rev.* **104,** 1189 (1956); J. R.Schrieffer, *Theory of Superconductivity* (Addison-Wesley, Redwood City, CA, 1964), pp. 38-39.
12. S. Fujita, *J. Supercond.* **4,** 297 (1991).

13. J. Bardeen, L. N. Cooper and J. R. Schrieffer, *Phys. Rev.* **108**, 1175 (1957).
14. I. Giaever, *Phys. Rev. Lett.* **5**, 147, 464 (1960).
15. J. G. Bednorz and K .A. Müller, *Z. Phys*. B, **64**,189 (1986).
16. T. Ishiguro and K. Yamaji, *Organic Superconductors* (Springer, New York, 1990).
17. S. Fujita, T. Kimura and Y. Zheng, *Found. Phys.* **21**, 1117 (1991).
18. P. C. Hohenberg, *Phys. Rev.* **158**, 383 (1967).

Chapter 3

1. F. Bloch, *Z. Phys.* **52**, 555 (1928).
2. L. D. Landau, *J. Exptl. Theoret. Phys.* (U.S.S.R.) **30**, 1058 (1956).
3. W. A. Harrison, *Phys. Rev.* **118**, 1190 (1960).
4. W. A. Harrison, *Solid State Theory* (Dover, New York, 1979).
5. T. L. Loucks and P. H. Cutler, *Phys. Rev.* **133**, A 819 (1964).
6. D. Schönberg and A.V. Gold, D. "*Physics of Metals* - 1" in Electrons, ed. J. M. Ziman, (Cambridge University Press, Cambridge, UK, 1969), p. 112.
7. C. Kittel, *Introduction to Solid State Physics*, 6h ed. (Wiley, New York 1986).
8. N. W. Ashcroft and N. D. Mermin, *Solid State Physics* (Saunders, Philadelphia 1976).
9. S. Fujita, S. Godoy and D. Nguyen, *Found. Phys.* **25**, 1209 (1995).
10. P. A.. M. Dirac, *Principles of Quantum Mechanics.* 4h ed. (Oxford University Press, London, 1958).
11. e.g. see L. P. Eisenhart, *Introduction to Differential Geometry,* (Princeton University Press, Princeton, 1940).
12. T. W. B. Kibble, *Classical Mechanics,* (McGraw-Hill, Maidenhead, England, 1966), pp. 166-171.

Chapter 4

1. T. W. B. Kibble, *Classical Mechanics,* (McGraw-Hill, Maidenhead, England, 1966), pp. 166-171.
2. C. B. Walker, *Phys. Rev.* **103**, 547 (1956).
3. L. Van Hove, *Phys. Rev.* **89**, 1189 (1953).
4. H. Fröhlich, *Phys, Rev.* **79**, 845 (1950); *Proc. R. Soc.* London **A 215**, 291 (1950).
5. R. P. Feynman, *Statistical Mechanics* (Addison-Wesley, Reading, MA. 1972); *Quantum Electrodynamics* (Addison-Wesley, Reading, MA.1961).
6. S. Fujita and S. Godoy, *Quantum Statistical Theory of Superconductivity*, (Plenum, New York, 1996), pp. 150-153.
7. L. N. Cooper, *Phys. Rev.* **104**, 1189 (1956).

Chapter 5

1. L. N. Cooper, *Phys. Rev.* **104**, 1189 (1956).
2. J. Bardeen, L. N. Cooper and J. R. Schrieffer, *Phys. Rev.* **108**, 1175 (1957).
3. D. Douglas Jr. and R. W. Schmitt, *Rev. Mod. Phys.* **36**, 1-331 (1964); R. D. Park, ed., *Superconductivity* (Marcel Dekker, NewYork, 1969).

Chapter 6

1. L. N. Cooper, *Phys. Rev.* **104**, 1189 (1956).

Chapter 7

1. J. Bardeen, L. N. Cooper and J. R. Schrieffer, *Phys. Rev.* **108**, 1175 (1957).
2. S. Fujita, *J. Supercond.* **4**, 297 (1991); ibid **5**, 83 (1992); S. Fujita and S. Watanabe, *J. Supercond.* **5**, 219 (1992); ibid **6**, 75 (1993); S. Fujita and S. Godoy, *J. Supercond.* **6**, 373 (1993).
3. L. D. Landau, *Sov. Phys.* JETP **3**, 920 (1957); ibid **5**, 101 (1957); ibid **8**, 70 (1959).
4. L. N. Cooper, *Phys. Rev.* **104**, 1189 (1956).
5. J. R. Schrieffer, *Theory of Superconductivity*, (Benjamin, NewYork, 1964).
6. C. J. Gorter and H. G. B. Casimir, *Phys. Z.* **35**, 963 (1934); *Z. Tech. Phys.* **15**, 539 (1934).
7. Y. Onuki, H. Suematsu and S. Tamura, *J. Phys. Chem. Solids* **38**, 419 and 431 (1977) (experiment); S. Watanabe and S. Fujita, *J. Phys. Chem. Solids* **52**, 985 (1991) (theory).

Chapter 8

1. P. A. M. Dirac, *Principles of Quantum Mechanics*, 4th ed. (Oxford University Press, London, 1958), p. 211, pp. 136-138, p. 37, pp. 253-257.
2. W. Pauli, *Phys. Rev.* **58**, 716 (1940).
3. P. Ehrenfest and J. R. Oppenheimer, *Phys. Rev.* **37**, 331 (1931).
4. H. A. Bethe and R. Jackiw, *Intermediate Quantum Mechanics*, 2d. ed. (Benjamin, New York, 1968), p. 23.
5. L. N. Cooper, *Phys, Rev.* **104**, 1189 (1956).
6. R. P. Feynman, R. B. Leighton and M. Sands, *Feynman Lectures on Physics,* vol. III (Addison-Wesley, Reading, MA. 1965), p. 21-7, p. 21-8.
7. S. Fujita and D. L. Morabito, *Int. J. Mod. Phys.* **12**, 2139 (1998).
8. J. Bardeen, L. N. Cooper and J. R. Schrieffer, *Phys. Rev.* **108**, 1175 (1957).
9. S. Fujita, *Introduction to Non-Equilibrium Quantum Statistical Mechanics*, (Krieger, Malabar, FL. 1983), pp. 9-11.
10. R. P. Feynman, *Statistical Mechanics* (Addison-Wesley, Redwood City, CA, 1972), p. 304.
11. B. D. Josephson, *Phys. Lett.* **1**, 251 (1992); *Rev. Mod. Phys.* **36**, 216 (1964).
12. S. Fujita and S. Godoy, *Quantum Statistical Theory of Superconductivity*, (Plenum, New York, 1996), pp. 122-124; S. Fujita, J. Supercond. **4**, 297 (1991); **5**, 83 (1992).
13. D. M. Dennison, *Proc. Roy. Soc.* London **115**, 483 (1927); T. Hori, *Zeits. f. Physik* **44**, 834 (1927).
14. E. Fermi and C. N. Yang, *Phys. Rev.* **76**, 1739 (1949).
15. M. Gell-Mann and Y. Ne'eman, *The Eightfold Way* (Benjamin, Reading, MA, 1964).

Chapter 9

1. S. Fujita, T. Kimura and Y. Zheng, *Found. Phys.* **21**, 1117 (1991).
2. P. C. Hohenberg, *Phys. Rev.* **158**, 383 (1967).
3. J. Bardeen, L. N. Cooper and J .R. Schrieffer, *Phys. Rev.* **108**, 1175 (1957).
4. A. B. Pippard, *Proc. Roy. Soc.* (London) **A 216**, 547 (1953).

Chapter 10

1. J. Bardeen, L. N. Cooper and J. R. Schrieffer, *Phys. Rev.* **108**, 1175 (1957).

Chapter 11

1. S. Fujita and S. Watanabe, *J. Supercond.* **5**, 219 (1992).

2. J. Bardeen, L. N. Cooper and J. R. Schriefer, *Phys. Rev.* **108**, 1175 (1957).

Chapter 12

1. I. Giaever and K. Megerle, *Phys. Rev.* **122**, 1101 (1961).
2. I. Giaever, *Phys. Rev. Lett.* **5**, 147 (1960); **5**, 464 (1960).
3. I. Giaever, H. R. Hart and K. Megerle, *Phys. Rev.* **126**, 941 (1961).
4. J. Bardeen, *Phys. Rev. Lett.* **6**, 57 (1961).
5. R. E. Glover III and M. Tinkham, *Phys. Rev. Lett.* **108** , 243 (1957); M. A. Biondi and M. Garfunkel, *Phys. Rev.* **116**, 853 (1959).
6. B. N. Taylor and E. Burstein, *Phys. Rev. Lett.* **10**, 14 (1963); C. J. Adkins, *Phil, Mag.* **8**, 1051 (1963); P. Townsend and J. Sutton, *Proc. Phys. Soc.* (London) **78**, 309 (1961).

Chapter 13

1. J. File and R. G. Mills, *Phys. Rev. Lett.* **10**, 93 (1963).
2. J. Bardeen, L. N. Cooper and J. R. Schrieffer, *Phys. Rev.* **108**, 1175 (1957).
3. B. S. Deaver and W. M. Fairbank, *Phys. Rev. Lett.* **7**, 43 (1961); R. Doll and M. Näbauer, *Phys. Rev. Lett.* **7**, 51 (1961).
4. L. Onsager, *Phyl. Mag.* **43**, 1006 (1952).
5. L. Onsager, *Phys. Rev. Lett.* **7**, 50 (1961); N. Byers and C. N. Yang, *Phys. Rev. Lett.* **7**, 46 (1961).
6. F. London and H. London, *Proc. Roy. Soc.* (London) **A 149**, 71 (1935); *Physica* **2**, 341 (1935); F. London, *Superfluids*, Vol I, (Wiley, New York 1950).

Chapter 14

1. V. L. Ginzburg and L. D. Landau, *J. Exp. Theor. Phys.* (USSR) **20**, 1064 (1950).
2. L. D. Landau and E. M. Lifshitz, *Statistical Physics*, part I, 3d ed., (Pergamon Press, Oxford, England, 1980) pp. 171-174.
3. F. and H. London, *Proc. Roy. Soc.* (London) A, **149**, 71 (1935); *Physica* **2**, 341 (1935).
4. P. G. de Gennes, *Superconductivity of Metals and Alloys*, (Addison-Wesley, Redwood City, CA, 1989), pp. 176-209.
5. A.. A. Abrikosov, *Sov. Phys.* JETP **5**, 1174 (1957).
6. H. Träuble and U. Essmann, *J. Appl. Phys.* **39**, 4052 (1968).
7. B. S. Deaver and W. M. Fairbank, *Phys. Rev. Lett.* **7**, 43 (1961); R. Doll and M. Näbauer, *Phys. Rev. Lett.* **7**, 51 (1961) (experiment).
8. L. Onsager, *Phys. Rev. Lett.* **7**, 50 (1961); N. Byers and C. N. Yang. *Phys. Rev. Lett.* **7**, 46 (1961) (theory).
9. L. N. Cooper, *Phys. Rev.* **104**, 1189 (1956).
10. J. Bardeen, L. Cooper and J. R. Schrieffer, *Phys. Rev.* **108**, 1175 (1957).
11. S. Fujita and S. Godoy, *Int. J. Mod. Phys.* **B 12**, 99 (1998); *J. Supercond.* **6**, 373 (1993).
12. I. Giaever, *Phys. Rev. Lett.* **5**, 147 (1960); **5**, 464 (1960); I. Giaever and K. Megerle, *Phys, Rev.* **122**, 1101 (1961).
13. P. A. M. Dirac, *Proc. Roy. Soc.* (London) A, **114**, 243 (1927).
14. P. Jordan and E. Wigner, *Zeits. f. Phys.* **47**, 631 (1928); P. Jordan and O. Klein, *Zeitz. f. Phys.* **45**, 751 (1927).
15. L. P. Gorkov, *Sov. Phys.* JETP **7**, 505 (1958); ibid **9**, 1364 (1959); **10**, 998 (1960).
16. N. R. Werthamer, *Phys. Rev.* **132**, 663 (1963); *Rev. Mod. Phys.* **36**, 292 (1964); T. Tewardt, *Phys. Rev.* **132**, 595 (1963); B. B. Goodman, *Rev. Mod. Phys.* **36**, 12 (1964).
17. S. Fujita and S. Watanabe, *J. Supercond.* **5**, 219 (1992).

18. R. E. Glover, III and Tinkham, *Phys. Rev.* **108**, 243 (1957); M. A. Biondi and M. Garfunkel, *Phys, Rev.* **116**, 853 (1959).
19. J. R. Schrieffer, *Theory of Superconductivity*, (Addison-Wesley, Redwood City, CA, 1964), p. 33, p. 44.

Chapter 15

1. D. Josephson, *Phys. Lett.* **1**, 251 (1962); *Rev. Mod. Phys.* **36**, 216 (1964).
2. P. W. Anderson and J. M. Rowell: *Phys. Rev. Lett.* **10**, **486** (1963).
3. R. C. Jaklevic, J. Lambe, J. E. Marcereau and A. H. Silver, *Phys. Rev.* **A 140**, 1628 (1965).
4. E. A. Ueling and G. E. Uhlenbeck, *Phys. Rev.* **43**, 552 (1933).
5. S. Tomonaga, *Zeits. f. Physik*, **110**, 573 (1938).
6. R. P. Feynman, R. B. Leighton and M. Sands, *Feynman Lectures on Physics*, vol. III, (Addison-Wesley, Reading, MA, 1965), p. 21-8.
7. R. P. Feynman, *Statistical Mechanics*, (Addison-Wesley, Redwood City, CA., 1972).
8. S. Shapiro, *Phys. Rev. Lett.* **11**, 80 (1963).
9. A. Hartland, *Precis. Meas. Fundam. Constants*, **2**, 543 (1981).
10. D. H. Menzel; *Fundamental Formulas of Physics*, (Dover, New York., 1960), p. 59.
11. A. A. Abrikosov, *J. Exptl. Theor. Phys.* (USSR) **32**, 1442 (1957).

Chapter 16

1. B. T. Matthias, *Progress in Low Temperature Physics*, ed. C. J. Gorter, Vol. **2** (North-Holland, Amsterdam, 1957), p. 138; B. T. Matthias *et al.*, *Rev. Mod. Phys.* **36**, 155 (1964).
2. D. Saint-James, E. D. Thomas and G. Sarma, *Type II Superconductivity*, (Pergamon, Oxford, 1969).
3. A. A. Abrikosov, *J. Exp. Theor. Phys.* (USSR), **5**, 1174 (1957).
4. V. L. Ginzburg and L. D. Landau, *J. Exp. Theor. Phys.* (USSR), **20**, 1064 (1950).
5. U. Essmann and H. Träuble, *Phys. Lett.* **A 24**, 526 (1967).
6. T. McConville and B. Serin, *Rev. Mod. Phys.* **36**, 112 (1964).
7. A. D. B. Woods *et al.*, *Phys. Rev.* **131**, 1025 (1963).
8. H. Yukawa, *Proc. Math. Soc.* Japan, **17**, 48 (1935).

Chapter 17

1. J. G. Bednorz and K. A. Müller, *Z. Phys. B. Cond. Matt.* **64**, 189 (1986).
2. J. W. Halley, ed., *Theory of High-Temperature Superconductivity* (Addison-Wesley, Redwood City, CA. 1988).
3. S. Lundquist *et al.*, eds., *Towards the Theoretical Understanding of High-T Superconductivity*, (World Scientific, Singapore 1988).
4. D. M. Ginsberg, ed., *Physical Properties of High-Temperature Superconductors* (World Scientific, Singapore 1989).
5. S. A. Wolf and W. Z. Kresin, *Novel Superconductivity* (Plenum, New York, 1989); K. Kitazawa and T. Ishiguro, eds. *Advances in Superconductivity*, (Springer, Tokyo, 1989).
6. M. K. Wu *et al.*, *Phys. Rev. Lett.* **58**, 908 (1987).
7. See, e.g. Ginsberg's overview, Ref. 4, pp. 1-38. 8.
8. D. E. Farrell *et al.*, *Phys. Rev.* **B 42**, 6758 (1990).
9. S. Godoy and S. Fujita, *J. Eng. Sci.* **29**, 1201 (1991).
10. J. H. Kang, R. T. Kampwirth and K. E. Gray, *Appl. Phys. Lett*, **52**, 2080 (1988); M. J. Naughton *et al.*, *Phys. Rev.* **B 38**, 9280 (1988).
11. L. Onsager, *Phil. Mag.* 43, 1006 (1952).

12. J. Wosnitza *et al.*, *Phys. Rev. Lett.* **67**, 263, (1991).
13. Y. Maeno, H, Hashimoto, K. Yoshida, S. Nishizaki, T. Fujita, J. G. Bednorz and F. Lichtenberg, *Nature* **372**, 532 (1994).

Chapter 18

1. J. G. Bednorz and K. A. Müller, *Z. Phys. B. Cond. Matt.* **64**, 189 (1986).
2. J. W. Halley, ed., *Theory of High-Temperature Superconductivity* (Addison-Wesley, Redwood City, CA. 1988); S. Lundquist *et al.*, eds., *Towards the Theoretical Understanding of High-T_c Superconductivity*, (World Scientific, Singapore 1988);. S. A. Wolf and D. M. Ginsberg, ed., *Physical Properties of High-Temperature Superconductors* (World Scientific, Singapore, 1989)-(series); W. Z. Kresin, *Novel Superconductivity* (Plenum, New York, 1989); K. Kitazawa and T. Ishiguro, eds. *Advances in Superconductivity*, (Springer, Tokyo, 1989).
3. P. W. Anderson, *Theory of Superconductivity in High-T_c Cuprates*, (Princeton University Press, Princeton, NJ, 1997); J. R. Waldram *Superconductivity of Metals and Cuprates*, (Intitute of Physics Publishing, Bristol, UK, 1996).
4. S. Fujita and D. L. Morabito, *Mod. Phys. Lett.* **B 12,** 1061 (1998).
5. S. Fujita and S. Watanabe, *J. Supercond.* **5,** 219 (1992).
6. S. Fujita and D. L. Morabito, *Int. J. Mod. Phys.* **B 21,** 2139 (1998).
7. J. Bardeen, L. N. Cooper and J. R. Schrieffer, *Phys. Rev.* **108,** 1175 (1957).
8. R. A. Fisher, J. E. Gordon, and N. E. Phillips, *J. Supercond.* **1**, 231 (1988).
9. J. W. Loram, K. A. Mirza, J. R. Cooper and W. Y. Liang, *J. Supercond.* **7**, 347 (1994).
10. T. Ekino *et al.*, *Physica* **C 218**, 387 (1993); P. J. M. van Bentum *et al.*, *Phys. Rev*, **B 36**, 843 (1987); F. Frangi *et al.*, *Sol. State Commun.* **81**, 599 (1992).

Chapter 19

1. J. G. Bednorz and K. A. Müller, *Z. Phys.* **B 64** 189 (1986).
2. J. B. Torrance *et al.*, *Phys. Rev. Lett.* **61**, 1127 (1988); M. W. Shafer, T. Penney and B. L. Olsen, *Phys. Rev.* **B 36**, 4047 (1987); R. B. van Dover *et al.*, *Phys. Rev.* **B 35**, 5737 (1987).
3. H. Takagi, S. Uchida and Y. Tokura, *Phys. Rev. Lett.* **62**, 1197 (1989); H. Takagi, *Kotai Butsuri* **25**, 736 (1990).
4. J. Bardeen, L. N. Cooper and J. R. Schrieffer, *Phys. Rev.* **108,** 1175 (1957).
5. R. P. Feynman, R. B. Leighton and M. Sands, *Feynman Lectures on Physics*, Vol. III (Addison-Wesley, Reading. MA. 1965) p. 21-7, p. 21-8.
6. B. D. Josephson, *Phys. Lett.* **1,** 251 (1962); *Rev. Mod. Phys.* **36,** 216 (1964).
7. (experiment); B. S. Deaver and W. M. Fairbank, *Phys. Rev. Lett.* **7**, 43 (1961); R. Doll and M. Näbauer, *Phys. Rev. Lett.* **7**, 51 (1961); (theory) L. Onsager, *Phys. Rev. Lett.* **7**, 50 (1961); N. Byers and C. N. Yang, *Phys. Rev. Lett.* **7**, 46 (1961).
8. R. C. Jaklevic, J. Lambe, J. E. Marcereau and A. H. Silver, *Phys. Rev.* **A 140**, 1628 (1965).
9. J. R. Schrieffer, *Theory of Superconductivity* (Addison-Wesley, Redwood City, CA, 1964), p. 33, pp. 49-50.
10. P. C. Hohenberg, *Phys. Rev.* **158,** 383 (1967).
11. R. A. Ogg Jr., *Phys. Rev.* **69,** 243 (1946).
12. F. London, *Superfluids*, vols. **1** and **2** (Dover, New York, 1964).
13. M. R. Schafroth, S. T. Butler and J. M. Blatt. *Helv. Phys. Acta*, **30,** 93 (1957).
14. A. J. Leggett, in *Modern Trends in Theory of Condensed Matter*, eds. by A. Pekalski and Przystawa (Springer, 1980), pp. 14-17.
15. P. Nozieres and S. Schmit-Rink, *J. Low Temp. Phys.* **59**, 195 (1985).
16. H. Takagi *et al.*, *Phys. Rev.* **B 40,** 2254 (1989).

Chapter 20

1. Y. Iye, in *Physical Properties of High Temperature Superconductors* III, D. M. Ginzberg, ed. (Word Sc., 1992).
2. J. G. Bednorz and K. A. Müller, *Z. Phys.* **B 64**, 189 (1996).
3. H. Takagi, *Kotai Butsuri*, **25**, 736 (1990).
4. J. B. Torrance *et al.*, *Phys. Rev. Lett.* **61**, 1127 (1998); H. Takagi, S. Uchida and Y. Tokura, *Phys. Rev. Lett.* **62**, 1197 (1989); T. Tokura, H. Takagi and S. Uchida, *Nature* **337**, 345 (1989).
5. H. Nakano, *Prog. Theoret. Phys.* **15**, 77 (1956); R. Kubo, *J. Phys. Soc.* Japan **12**, 570 (1957).
6. S. Fujita, Y-G. Kim and Y. Okamura, *Mod. Phys. Lett.* **B 14**, 495 (2000).
7. S. Fujita, Y-G. Kim, *Mod. Phys. Lett.* **B 14**, 505 (2000).
8. e.g. N. W. Ashcroft and N. D. Mermin, *Solid State Physics* (Saunders, Philadelphia, 1976), pp. 239-240, 133-141.
9. T. R. Chien *et al.*, *Phys. Rev. Lett.* **67**, 2088 (1991).

Chapter 21

1. I. Terasaki *et al.*, *Phys. Rev.* **B 52**, 16246 (1995); I. Terasaki, Y. Sato and S. Tajima, *Phys. Rev.* **B 55**, 15300 (1997); I. Terasaki, Y. Sato and S. Tajima, *J. Kor. Phys. Soc.* **31**, 23 (1997).
2. T. Kimura *et al.*, *Physica* **C 192**, 247 (1992).
3. S. Fujita, Y. Tamura and Suzuki, *Mod. Phys. Lett.* **B**, (submitted)
4. S. Godoy and S. Fujita, *J. Eng. Sci.* **29**, 1201 (1991).
5. S. Fujita, *Physica* **51**, 601 (1971).
6. J. G. Bednorz and K. A. Müller, *Z. Phys.* **B 64**, 189 (1996).
7. Y. Maeno, H, Hashimoto, K. Yoshida, S. Nishizaki, T. Fujita, J. G. Bednorz and F. Lichtenberg, *Nature* **372**, 532 (1994).

Chapter 22

1. e.g. P. L. Rossiter and J. Bass, *Metals and Alloys*, in *Encyclopedia of Applied Physics* **10**, VCH Publ., 163-197 (1994).
2. I. Terasaki, Y. Sato, S. Tajima, S. Miyamoto and S. Tanaka, *Physica* **C 235**, 1413 (1994); I. Terasaki, Y. Sato, S. Tajima, S. Miyamoto and S. Tanaka, *Phys. Rev.* **B 52**, 16246 (1995).
3. S. Fujita, T. Obata, T. Shane and D. Morabito, *Phys. Rev.* **B 63**, 54402 (2001).
4. N. W. Ashcroft and N. D. Mermin, *Solid State* Physics, (Saunders, Philadelphia, 1976), pp. 256-258, pp. 290-293.
5. S. Fujita, H-C. Ho and D. L. Morabito, *Int. J. Mod. Phys.* **B 14**, 2223 (2000).
6. I. Tersasaki, Y, Sato and S. Tajima, *J. Kor. Phys. Soc.* **31**, 23 (1997).
7. S. Fujita, H-C. Ho and S. Godoy, *Mod. Phys.* Lett. **B 13, 689**, (1999).

Chapter 23

1. H. Takagi, T. Ido, S. Ishibashi, M. Uota, S. Uchida and Y. Tokura, *Phys. Rev.* **B 40**, 2254 (1989).
2. J. B. Torrance, A. Bezinge, A. I. Nazzal, T. C. Huang, S. S. P. Parkin, D. T. Keane, S. J. La Placa, P. M. Horn, and G. A. Held, *Phys. Rev.* **B 40, 8872** (1989).
3. I. Terasaki, M. Hase, A. Maeda, K. Uchinokura, T. Kimura, K. Kishio, I. Tanaka and H. Kojima, *Physics* C **193**, 365 (1992).
4. W. Pauli, *Zeits. f. Phys.* **41**, 81 (1927).

5. e.g. L. D. Landau and E. M. Lifshitz, *Statistical Physics*, part I, 3d ed., (Pergamon Press, Oxford, England, 1980), pp. 171-174.
6. S. Fujita, T. Obata, T. F. Shane and D. L. Morabito, *Phys. Rev.* **B 63**, 54402 (2001).
7. L. Onsager, *Phil. Mag.* **43**, 1006 (1952).

Chapter 24

1. J. Cerne, D. C. Schmadel, M. Grayson, G. S. Jenkins, J. R. Simpson and H. D. Drew, *Phys. Rev.* **B 61**, 8133 (2000).
2. J. Cerne, M. Grayson, D. C. Schmadel, G. S. Jenkins, H. D. Drew, H. Hughes, J. S. Preston and P. J. Kung. *Phys. Rev. Lett.* **84**, 3418 (2000).
3. S. Fujita, Y-G. Kim and Y. Okamura, *Mod. Phys. Lett.* **B 14**, 495 (2000).
4. N. W. Ashcroft and N. D. Mermin, *Solid State Physics* (Saunders, Philadelpia, 1976), pp. 225 and 240.
5. S. Fujita and Y-G. Kim, *Mod. Phys. Lett.* **B 14**, 505 (2000).
6. J. Cerne, private communication.

Chapter 25

1. D. A. Wollman *et al.*, *Phys. Rev. Lett.* **71**, 2134 (1993); C. C. Tsuei *et al.*, *Nature* **386**, 481 (1997).
2. J. Bardeen, L. N. Cooper and J. R. Schrieffer, *Phys. Rev.* **108**, 1175 (1957).
3. W. A. Harrison, *Solid State Theory* (Dover, New York, 1980), pp. 390-397.
4. D. S. Dessau *et al.*, *Phys. Rev. Lett.* **71**, 2781 (1993); Z-X. Shen *et al.*, *Phys. Rev. Lett.* **70**, 1553 (1993).

Chapter 26

1. H. Kamerlingh Onnes, *Akad. V. Wetenschappen* (Amsterdam) **14**, 113 (1911).
2. B. J. Gorter and H. B. G. Casimir, *Physica* **1**, 306 (1934).
3. F. London, *Nature* (London) **141**, 643 (1938); *Superfluids*, I and II (Dover, New York, 1964).
4. W. Meissner and R. Ochsenfeld, *Naturwiss.* **21**, 787 (1933).
5. F. London and H. London, *Proc. Roy. Soc.* (London) **A 149**, 71 (1935); *Physica* **2**, 341 (1935).
6. S. Fujita and S. Godoy, *I. J. Mod. Phys.* **B 12**, 49 (1998).
7. V. L. Ginzburg and L. D. Landau, *J. Exp. Theor. Phys.* (USSR) **20**, 1064 (1950).
8. A. A. Abrikosov, *Sov. Phys.* JETP **5**, 1174 (1957).
9. H. Träuble and U. Essmann, *J. Appl. Phys.* **39**, 4052 (1968).
10. H. Fröhlich, *Phys, Rev.* **79**, 845 (1950); *Proc. Roy. Soc.* London **A 215**, 291 (1950).
11. A. A. Reynolds, B. Serin, W. H. Wright and N. B. Nesbitt, *Phys. Rev.* **78**, 487 (1950); E. Maxwell, *Phys. Rev.* **78**, 477 (1950).
12. H. Yukawa, *Proc. Phys. Math. Soc.* Japan, **17**, 48 (1935).
13. S. Fujita and D. L. Morabito, *Mod. Phys. Lett.* **B 12**, 1061, (1998).
14. L. N. Cooper, *Phys. Rev.* **104**, 1189 (1956).
15. J. Bardeen, L. N. Cooper and J. R. Schrieffer, *Phys. Rev.* **108**, 1175 (1957).
16. A. S. Deaver and W. M. Fairbank, *Phys. Rev. Lett.* **7**, 43 (1961); R. Doll and M. Näbauer, *Phys. Rev. Lett.* **7**, 51 (1961) (experiment).
17. L. Onsager, *Phys. Rev. Lett.* **7**, 50 (1961); N. Byers and C. N. Yang. *Phys. Rev. Lett.* *7*, 46 (1961) (theory).
18. R. E. Glover III and M. Tinkham: *Phys. Rev. Lett.* **108**, 243 (1957); M. A. Biondi and M. Garfunkel, *Phys. Rev.* **116**, 853 (1959).
19. I. Giaever, *Phys. Rev. Lett.* **5**, 147 (1960); **5**, 464 (1960); I. Giaever, H. R. Hart and K. Megerle, *Phys. Rev.* **126**, 941 (1961).
20. R. Doll and M. Näbauer, *Phys. Rev. Lett.* **7**, 51 (1961).

21. A. Einstein, *Sits. Ber. Berl. Akad.* **3** (1925).
22. M. R. Schafroth, S. T. Butler and J. M. Blatt. *Helv. Phys. Acta,* **30**, 93 (1957); A. J. Leggett, in *Modern Trends in Theory of Condensed Matter,* eds. by A. Pekalski and Przystawa (Springer, 1980), pp.14-17; P. Nozieres and S. Schmit-Rink, *J. Low Temp. Phys.* **59**, 195 (1985).
23. S. Fujita, T. Kimura and Y. Zheng, *Found. Phys.* **21**, 1117 (1991).
24. P. C. Hohenberg, *Phys. Rev.* **158**, 383 (1967).
25. B. D. Josephson, *Phys. Lett.* **1**, 251 (1962); *Rev. Mod. Phys.* **36**, 216 (1964).
26. P. W. Anderson and J. M. Rowell, *Phys. Rev. Lett.* **10**, 486 (1963).
27. R. C. Jaklevic, J. Lambe, J. E. Marcereau and A. H. Silver, *Phys. Rev.* **A 140**, 1628 (1965).
28. S. Shapiro, *Phys. Rev. Lett.* **11**, 80 (1963).
29. R. P. Feynman, R. B. Leighton and M. Sands, 1965, *Feynman Lectures on Physics*, vol. III, p. 21-8; R. P. Feynman,1972, *Statistical Mechanics,* Addison-Wesley, Redwood City, CA.
30. J. G. Bednorz and K. A. Müller, *Z. Phys. B. Cond. Matt.* **64**, 189 (1986).
31. M. K. Wu *et al.*, *Phys. Rev. Lett.* **58**, 908 (1987).
32. W. L. McMillan, *Phys. Rev.* **167**, 331 (1968); P. B. Allen and R. C. Dynes, *Phys. Rev.* **12**, 905 (1975).
33. J. W. Halley, ed., *Theory of High-Temperature Superconductivity* (Addison-Wesley, Redwood City, CA. 1988); S. Lundquist *et al.*, eds., *Towards the Theoretical Understanding of High-T Superconductivity*, Vol. 14, (World Scientific, Singapore 1988).
34. C. M. Ginsberg, ed., *Physical Properties of High-Temperature Superconductors* (World Scientific, Singapore 1989)-(series).
35. S. A. Wolf and D. M. Ginsberg, ed., *Physical Properties of High-Temperature Superconductors* (World Scientific, Singapore 1989)-(series); W. Z. Kresin, *Novel Superconductivity* (Plenum, New York, 1989); K. Kitazawa and T. Ishiguro, eds. *Advances in Superconductivity*, (Springer, Tokyo, 1989).
36. A. S. Alexandrov and N. F. Mott, *Int. J. Mod. Phys.* **B 8**, 2075 (1994); A. S. Alexandrov, *Phys. Rev.* **B 38**, 925 (1988); N. F. Mott, *Phys. Rev. Lett.* **71**, 1075 (1993).
37. R. Friedberg and T. D. Lee, *Phys. Rev.* **B 40**, 1745 (1986).
38. P. W. Anderson, *Science,* **235**, 1196 (1987).
39. V. J. Emery, *Phys. Rev. Lett,* **58**, 2794 (1987); H. Kamimura, S. Matsuno, Y. Suva and H. Ushio, *Phys. Rev. Lett.* **77**, 723 (1996); A. Sano, M. Eto and H. Kamimura, *Int. J. Mod. Phys.* **B 11**, 3733 (1997).
40. J. E. Hirsch, *Phys. Rev.* **B 31**, 4403 (1985); *Phys. Rev. Lett.* **25**, 1317 (1985).
41. F. C. Zhang and T. M. Rice, *Phys. Rev.* **B, 37**, 3759 (1989).
42. L. Onsager, *Phys. Rev.* **65**, 117 (1944).
43. R. Willett, J. P. Eisenstein, H. L. Störmer, D. C. Tsui, A. C. Gossard and J. H. English, *Phys. Rev. Lett.* **59**, 1776 (1987); H. L. Störmer, D. C. Tsui and A. C. Gossard, *ibid*, **48**, 1559 (1982).
44. K. von Klitzing, G. Dorda and M. Pepper, *Phys. Rev. Lett.* **45**, 494 (1980).
45. R. E. Prange and S. M. Girvin, eds., *Quantum Hall Effect*, (Springer-Verlag, Berlin, 1988).
46. R. B. Laughlin, *Phys. Rev. Lett.* **50**, 1395 (1983).
47. S. C. Zhang, T. H. Hansson and S. Kivelson, *Phys. Rev. Lett.* **62**, 82 (1989); R. B. Laughlin, *ibid*, **60**, 2677 (1989); S. Kivelson, D. H. Lee and S-C Zhang, *Sci. Amer.* **274**, 86 (1996).
48. S. M. Girvin and A. H. MacDonald, *Phys. Rev. Lett.* **58**, 1252 (1987); N. Read, *ibid*, **62**, 86 (1989); R. Shankar and G. Murthy, *ibid*, **79**, 4437 (1997).
49. J. K. Jain, *Phys. Rev. Lett.* **63**, 199 (1989); *Phys. Rev.* B **40**, 8079, (1989); *ibid*, B **41**, 7653 (1990).
50. S. Fujita, Y. Tamura and A. Suzuki, *Mod. Phys. Lett.* **B** 15, (2001).
51. P. A. M. Dirac, *Principles of Quantum Mechanics*, 4th ed. (Oxford University Press, 1958), p. 267.
52. R. G. Clark *et al.*, Phys. Rev. Lett. **62**, 1536 (1989); J. P. Eisenstein, H. L. Stormer, L. Pfeiffer and K. W. West, *ibid*, **62**, 1540 (1989).

Chapter 27

1. F. B. Silsbee, J. Wash. *Acad. Sci.* **6**, 597 (1916).

Appendix

1. P. A.. M. Dirac, *Proc. Roy. Soc.* (London) **117**, 610 (1928).
2. P. Ehrenfest and J. R. Oppenheimer *Phys. Rev.* **37**, 333 (1931); H. A. Bethe and R. Jackiw, *Intermediate Quantum Mechanics*, 2nd. ed., (Benjamin, New York, 1968), pp. 23.
3. W. Pauli, *Phys. Rev.* **58**, 716 (1940).

Bibliography

SUPERCONDUCTIVITY

Introductory and Elementary Books

Lynton, E. A.: *Superconductivity*, (Methuen, London, 1962).
Vidali, G.: *Superconductivity*, (Cambridge University Press, Cambridge, England, 1993).

General Textbooks at about the same level as the present text.

Feynman, R. P., Leighton R. B. and Sands, M.: *Feynman Lectures on Physics*. Vol 3, (Addison-Wesley, Reading, MA, 1965) pp. 1-19.
Feynman, R. P.: *Statistical Mechanics*, (Addison-Wesley, Reading, MA, 1972), pp. 265-311.
Rose-Innes, A. C. and Rhoderick, E. H.: *Introduction to Superconductivity*, 2d ed., (Pergamon, Oxford, England, 1978).

More Advanced Texts and Monographs

Abrikosov, A. A.: *Fundamentals of the Theory of Metals*, A. Beknazarov, trans. (North Holland-Elsevier, Amsterdam, 1988).
Gennes, P.: *Superconductivity of Metals and Alloys*, (Benjamin, Menlo Park, CA. 1966).
Rickayzen, G.: *Theory of Superconductivity*, (Interscience, New York, 1965).
Saint-James, D., Thomas, E. J. and Sarma, G.: *Type II Superconductivity*, (Pergamon, Oxford, England, 1969).
Schafroth, M. R.: *Solid State Physics*, Vol. 10, eds. F. Seitz and D. Turnbull, (Academic, New York, 1960) p. 488.
Schrieffer, J. R.: *Theory of Superconductivity*, (Benjamin, New York, 1964).
Tilley, D. R. and Tilley, J.: *Superfluidity and Superconductivity*, 3d ed., (Adam Hilger, Bristol, England, 1990).
Tinkham, M.: *Introduction to Superconductivity*, (McGraw-Hill, New York, 1975).

High-Temperature Superconductivity

Anderson, P. W.: *Theory of Superconductivity in High-T_c Cuprates*, (Princeton University Press, Princeton, NJ, 1997)
Burns, G.: *High-Temperature Superconductivity*, an Introduction, (Academic, New York, 1992).
Ginsberg D. M., ed., *Physical Properties of High-Temperature Superconductors* (World Scientific, Singapore 1989)-(series).
Halley, J. W.: ed., *Theory of High-Temperature Superconductivity* (Addison-Wesley, Redwood City, CA. 1988).
Kresin, V. Z. and Wolf S. A.: *Fundamentals of Superconductivity*, (Plenum, New York, 1990).
Kresin, W. Z.: *Novel Superconductivity* (Plenum, New York, 1989).
Lindquist, S.: *et al.*, eds., *Towards the Theoretical Understanding of High-T_c Superconductivity*, Vol. 14, (World Scientific, Singapore 1988).

Lynn, J. W., ed., *High Temperature Superconductivity* (Springer-Verlag, New York, 1990).
Owens, F. J. and Poole, C. P.: *New Superconductors*, (Plenum, New York, 1996).
Phillips, J. C.: *Physics of High-T_c Superconductors*, (academic Press, san Diego, CA, 1989).
Poole, C. P., Farach, H. A. and Creswick, R. J.: *Superconductivity*, (Academic, New York, 1995).
Sheahen, T. P.: *Introduction to High-Temperature Superconductivity*, (Plenum, New York, 1994).
Waldram J. R.: *Superconductivity of Metals and Cuprates*, (Intitute of Physics Publishing, Bristol, UK, 1996).

BACKGROUND

Solid State Physics

Ashcroft, N. W. and Mermin, N. D.: *Solid State Physics*, (Saunders, Philadelphia, 1976).
Harrison, W. A. : *Solid State Theory*, (Dover, New York, 1979).
Haug, A.: *Theoretical Solid State Physics*, I, (Pergamon, Oxford, England, 1972).
Kittel, C.: *Introduction to Solid State Physics*, 6th ed. (Wiley, New York, 1986).

Mechanics

Goldstein, H.: *Classical Mechanics*, (Addison Wesley, Reading, MA, 1950).
Kibble, T. W. B.: *Classical Mechanics*, (McGraw-Hill, London, 1966).
Marion, J. B.: *Classical Dynamics*, (Academic, New York, 1965).
Symon, K. R.: *Mechanics*, 3d ed. (Addison-Wesley, Reading, MA, 1971).

Quantum Mechanics

Alonso, M. and Finn, E. J.: *Fundamental University Physics, III Quantum and Statistical Physics*, (Addison-Wesley, Reading, MA, 1989).
Dirac, P. A. M.: *Principles of Quantum Mechanics*, 4th ed. (Oxford University Press, London, 1958).
Gasiorowitz, S.: 1974, *Quantum Physics*, (Wiley, New York, 1974).
Liboff, R. L.: *Introduction to Quantum Mechanics*, (Addison-Wesley, Reading, MA, 1992).
McGervey, J. D.: *Modern Physics*, (Academic Press, New York, 1971).
Pauling, L. and Wilson, E. B.: *Introduction to Quantum Mechanics*, (McGraw-Hill, New York, 1935).
Powell, J. L. and Crasemann, B.: *Quantum Mechanics*, (Addison-Wesley, Reading, MA, 1961).

Electricity and Magnetism

Griffiths, D. J.: *Introduction to Electrodynamics*, 2d ed. (Prentice-Hall, Englewood Cliffs, NJ, 1989).
Lorrain, P. and Corson, D. R.: *Electromagnetism*, (Freeman, San Francisco, 1978).
Wangsness, R. K.: *Electromagnetic Fields*, (Wiley, New York, 1979).

Thermodynamics

Andrews, F. C.: *Thermodynamics: Principles and Applications*, (Wiley, New York, 1971).
Bauman, R. P.: *Modern Thermodynamics with Statistical Mecanics*, (Macmillan, New York, 1992).
Callen, H. B.: *Thermodynamics*, (Wiley, New York, 1960).
Fermi, E.: *Thermodynamics*, (Dover, New York, 1957).
Pippard, A. B.: *Thermodynamics:Applications*, (Cambridge University Press, Cambridge, England, 1957).

Statistical Physics (undergraduate)

Baierlein, R.: *Thermal Physics,* (Cambridge U. P., Cambridge, UK, 1999).
Carter, A. H.: *Classical and Statistical Thermodynamics*, (Prentice-Hall, Upper Saddle River, NJ, 2001).
Fujita, S.: *Statistical and Thermal Physics,* I and II, (Krieger, Malabar, FL, 1986).
Kittel, C. and Kroemer, H.: *Thermal Physics*, (Freeman, San Francisco, CA, 1980).
Mandl, F.: *Statistical Physics*, (Wiley, London, 1971).
Morse, P. M.: *Thermal Physics*, 2d ed. (Benjamin, New York, 1969).
Reif, F.: *Fundamentals of Statistical and Thermal Physics*, (McGraw-Hill, New York, 1965).
Rosser, W. G. V.: *Introduction to Statistical Physics*, (Horwood, Chichester, England, 1982).
Terletskii, Ya. P.: *Statistical Physics*, N. Froman, trans. (North-Holland, Amsterdam, 1971).
Zemansky, M. W.: *Heat and Thermodynamics*, 5th ed. (McGraw-Hill, New York, 1957).

Statistical Physics (graduate)

Davidson, N.: *Statistical Mechanics*, (McGraw-Hill, New York, 1969).
Feynman, R. P.: *Statistical Mechanics*, (Benjamin, New York, 1972).
Finkelstein, R. J.: *Thermodynamics and Statistical Physics*, (Freeman, San Francisco, CA, 1969).
Goodstein, D. L.: *States of Matter*, (Prentice-Hall, Englewood Cliffs, NJ).
Heer, C. V.: *Statistical Mechanics, Kinetic Theory, and Stochastic Processes*, (Academic Press, New York, 1972).
Huang, K.: *Statistical Mechanics*, 2d ed. (Wiley, New York, 1972).
Isihara, A.: *Statistical Physics*, (Academic, New York, 1971).
Kestin, J. and Dorfman, J. R.: *Course in Statistical Thermodynamics*, (Academic, New York, 1971).
Landau, L. D. and Lifshitz, E. M.: *Statistical Physics*, 3d ed. Part 1, (Pergamon, Oxford, England, 1980).
Lifshitz, E. M. and Pitaevskii, L. P.: *Statistical Physics*, Part 2, (Pergamon, Oxford, England, 1980).
McQuarrie, D. A.: *Statistical Mechanics*, (Harper and Row, New York, 1976).
Pathria, R. K.: *Statistical Mechanics*, (Pergamon, Oxford, England, 1972).
Robertson, H. S.: *Statistical Thermodynamics*, (Prentice Hall, Englewood Cliffs, NJ.).
Wannier, G. H.: *Statistical Physics*, (Wiley, New York, 1966).

INDEX

Fundamental Theories of Physics

Series Editor: Alwyn van der Merwe, University of Denver, USA

1. M. Sachs: *General Relativity and Matter.* A Spinor Field Theory from Fermis to Light-Years. With a Foreword by C. Kilmister. 1982 ISBN 90-277-1381-2
2. G.H. Duffey: *A Development of Quantum Mechanics.* Based on Symmetry Considerations. 1985 ISBN 90-277-1587-4
3. S. Diner, D. Fargue, G. Lochak and F. Selleri (eds.): *The Wave-Particle Dualism.* A Tribute to Louis de Broglie on his 90th Birthday. 1984 ISBN 90-277-1664-1
4. E. Prugovečki: *Stochastic Quantum Mechanics and Quantum Spacetime.* A Consistent Unification of Relativity and Quantum Theory based on Stochastic Spaces. 1984; 2nd printing 1986 ISBN 90-277-1617-X
5. D. Hestenes and G. Sobczyk: *Clifford Algebra to Geometric Calculus.* A Unified Language for Mathematics and Physics. 1984 ISBN 90-277-1673-0; Pb (1987) 90-277-2561-6
6. P. Exner: *Open Quantum Systems and Feynman Integrals.* 1985 ISBN 90-277-1678-1
7. L. Mayants: *The Enigma of Probability and Physics.* 1984 ISBN 90-277-1674-9
8. E. Tocaci: *Relativistic Mechanics, Time and Inertia.* Translated from Romanian. Edited and with a Foreword by C.W. Kilmister. 1985 ISBN 90-277-1769-9
9. B. Bertotti, F. de Felice and A. Pascolini (eds.): *General Relativity and Gravitation.* Proceedings of the 10th International Conference (Padova, Italy, 1983). 1984 ISBN 90-277-1819-9
10. G. Tarozzi and A. van der Merwe (eds.): *Open Questions in Quantum Physics.* 1985 ISBN 90-277-1853-9
11. J.V. Narlikar and T. Padmanabhan: *Gravity, Gauge Theories and Quantum Cosmology.* 1986 ISBN 90-277-1948-9
12. G.S. Asanov: *Finsler Geometry, Relativity and Gauge Theories.* 1985 ISBN 90-277-1960-8
13. K. Namsrai: *Nonlocal Quantum Field Theory and Stochastic Quantum Mechanics.* 1986 ISBN 90-277-2001-0
14. C. Ray Smith and W.T. Grandy, Jr. (eds.): *Maximum-Entropy and Bayesian Methods in Inverse Problems.* Proceedings of the 1st and 2nd International Workshop (Laramie, Wyoming, USA). 1985 ISBN 90-277-2074-6
15. D. Hestenes: *New Foundations for Classical Mechanics.* 1986 ISBN 90-277-2090-8; Pb (1987) 90-277-2526-8
16. S.J. Prokhovnik: *Light in Einstein's Universe.* The Role of Energy in Cosmology and Relativity. 1985 ISBN 90-277-2093-2
17. Y.S. Kim and M.E. Noz: *Theory and Applications of the Poincaré Group.* 1986 ISBN 90-277-2141-6
18. M. Sachs: *Quantum Mechanics from General Relativity.* An Approximation for a Theory of Inertia. 1986 ISBN 90-277-2247-1
19. W.T. Grandy, Jr.: *Foundations of Statistical Mechanics.* Vol. I: *Equilibrium Theory.* 1987 ISBN 90-277-2489-X
20. H.-H von Borzeszkowski and H.-J. Treder: *The Meaning of Quantum Gravity.* 1988 ISBN 90-277-2518-7
21. C. Ray Smith and G.J. Erickson (eds.): *Maximum-Entropy and Bayesian Spectral Analysis and Estimation Problems.* Proceedings of the 3rd International Workshop (Laramie, Wyoming, USA, 1983). 1987 ISBN 90-277-2579-9
22. A.O. Barut and A. van der Merwe (eds.): *Selected Scientific Papers of Alfred Landé.* [*1888-1975*]. 1988 ISBN 90-277-2594-2

Fundamental Theories of Physics

23. W.T. Grandy, Jr.: *Foundations of Statistical Mechanics.* Vol. II: *Nonequilibrium Phenomena.* 1988 ISBN 90-277-2649-3
24. E.I. Bitsakis and C.A. Nicolaides (eds.): *The Concept of Probability.* Proceedings of the Delphi Conference (Delphi, Greece, 1987). 1989 ISBN 90-277-2679-5
25. A. van der Merwe, F. Selleri and G. Tarozzi (eds.): *Microphysical Reality and Quantum Formalism, Vol. 1.* Proceedings of the International Conference (Urbino, Italy, 1985). 1988 ISBN 90-277-2683-3
26. A. van der Merwe, F. Selleri and G. Tarozzi (eds.): *Microphysical Reality and Quantum Formalism, Vol. 2.* Proceedings of the International Conference (Urbino, Italy, 1985). 1988 ISBN 90-277-2684-1
27. I.D. Novikov and V.P. Frolov: *Physics of Black Holes.* 1989 ISBN 90-277-2685-X
28. G. Tarozzi and A. van der Merwe (eds.): *The Nature of Quantum Paradoxes.* Italian Studies in the Foundations and Philosophy of Modern Physics. 1988 ISBN 90-277-2703-1
29. B.R. Iyer, N. Mukunda and C.V. Vishveshwara (eds.): *Gravitation, Gauge Theories and the Early Universe.* 1989 ISBN 90-277-2710-4
30. H. Mark and L. Wood (eds.): *Energy in Physics, War and Peace.* A Festschrift celebrating Edward Teller's 80th Birthday. 1988 ISBN 90-277-2775-9
31. G.J. Erickson and C.R. Smith (eds.): *Maximum-Entropy and Bayesian Methods in Science and Engineering.* Vol. I: *Foundations.* 1988 ISBN 90-277-2793-7
32. G.J. Erickson and C.R. Smith (eds.): *Maximum-Entropy and Bayesian Methods in Science and Engineering.* Vol. II: *Applications.* 1988 ISBN 90-277-2794-5
33. M.E. Noz and Y.S. Kim (eds.): *Special Relativity and Quantum Theory.* A Collection of Papers on the Poincaré Group. 1988 ISBN 90-277-2799-6
34. I.Yu. Kobzarev and Yu.I. Manin: *Elementary Particles. Mathematics, Physics and Philosophy.* 1989 ISBN 0-7923-0098-X
35. F. Selleri: *Quantum Paradoxes and Physical Reality.* 1990 ISBN 0-7923-0253-2
36. J. Skilling (ed.): *Maximum-Entropy and Bayesian Methods.* Proceedings of the 8th International Workshop (Cambridge, UK, 1988). 1989 ISBN 0-7923-0224-9
37. M. Kafatos (ed.): *Bell's Theorem, Quantum Theory and Conceptions of the Universe.* 1989 ISBN 0-7923-0496-9
38. Yu.A. Izyumov and V.N. Syromyatnikov: *Phase Transitions and Crystal Symmetry.* 1990 ISBN 0-7923-0542-6
39. P.F. Fougère (ed.): *Maximum-Entropy and Bayesian Methods.* Proceedings of the 9th International Workshop (Dartmouth, Massachusetts, USA, 1989). 1990 ISBN 0-7923-0928-6
40. L. de Broglie: *Heisenberg's Uncertainties and the Probabilistic Interpretation of Wave Mechanics.* With Critical Notes of the Author. 1990 ISBN 0-7923-0929-4
41. W.T. Grandy, Jr.: *Relativistic Quantum Mechanics of Leptons and Fields.* 1991 ISBN 0-7923-1049-7
42. Yu.L. Klimontovich: *Turbulent Motion and the Structure of Chaos.* A New Approach to the Statistical Theory of Open Systems. 1991 ISBN 0-7923-1114-0
43. W.T. Grandy, Jr. and L.H. Schick (eds.): *Maximum-Entropy and Bayesian Methods.* Proceedings of the 10th International Workshop (Laramie, Wyoming, USA, 1990). 1991 ISBN 0-7923-1140-X
44. P. Pták and S. Pulmannová: *Orthomodular Structures as Quantum Logics.* Intrinsic Properties, State Space and Probabilistic Topics. 1991 ISBN 0-7923-1207-4
45. D. Hestenes and A. Weingartshofer (eds.): *The Electron.* New Theory and Experiment. 1991 ISBN 0-7923-1356-9

Fundamental Theories of Physics

46. P.P.J.M. Schram: *Kinetic Theory of Gases and Plasmas.* 1991 ISBN 0-7923-1392-5
47. A. Micali, R. Boudet and J. Helmstetter (eds.): *Clifford Algebras and their Applications in Mathematical Physics.* 1992 ISBN 0-7923-1623-1
48. E. Prugovečki: *Quantum Geometry.* A Framework for Quantum General Relativity. 1992 ISBN 0-7923-1640-1
49. M.H. Mac Gregor: *The Enigmatic Electron.* 1992 ISBN 0-7923-1982-6
50. C.R. Smith, G.J. Erickson and P.O. Neudorfer (eds.): *Maximum Entropy and Bayesian Methods.* Proceedings of the 11th International Workshop (Seattle, 1991). 1993 ISBN 0-7923-2031-X
51. D.J. Hoekzema: *The Quantum Labyrinth.* 1993 ISBN 0-7923-2066-2
52. Z. Oziewicz, B. Jancewicz and A. Borowiec (eds.): *Spinors, Twistors, Clifford Algebras and Quantum Deformations.* Proceedings of the Second Max Born Symposium (Wrocław, Poland, 1992). 1993 ISBN 0-7923-2251-7
53. A. Mohammad-Djafari and G. Demoment (eds.): *Maximum Entropy and Bayesian Methods.* Proceedings of the 12th International Workshop (Paris, France, 1992). 1993 ISBN 0-7923-2280-0
54. M. Riesz: *Clifford Numbers and Spinors* with Riesz' Private Lectures to E. Folke Bolinder and a Historical Review by Pertti Lounesto. E.F. Bolinder and P. Lounesto (eds.). 1993 ISBN 0-7923-2299-1
55. F. Brackx, R. Delanghe and H. Serras (eds.): *Clifford Algebras and their Applications in Mathematical Physics.* Proceedings of the Third Conference (Deinze, 1993) 1993 ISBN 0-7923-2347-5
56. J.R. Fanchi: *Parametrized Relativistic Quantum Theory.* 1993 ISBN 0-7923-2376-9
57. A. Peres: *Quantum Theory: Concepts and Methods.* 1993 ISBN 0-7923-2549-4
58. P.L. Antonelli, R.S. Ingarden and M. Matsumoto: *The Theory of Sprays and Finsler Spaces with Applications in Physics and Biology.* 1993 ISBN 0-7923-2577-X
59. R. Miron and M. Anastasiei: *The Geometry of Lagrange Spaces: Theory and Applications.* 1994 ISBN 0-7923-2591-5
60. G. Adomian: *Solving Frontier Problems of Physics: The Decomposition Method.* 1994 ISBN 0-7923-2644-X
61. B.S. Kerner and V.V. Osipov: *Autosolitons.* A New Approach to Problems of Self-Organization and Turbulence. 1994 ISBN 0-7923-2816-7
62. G.R. Heidbreder (ed.): *Maximum Entropy and Bayesian Methods.* Proceedings of the 13th International Workshop (Santa Barbara, USA, 1993) 1996 ISBN 0-7923-2851-5
63. J. Peřina, Z. Hradil and B. Jurčo: *Quantum Optics and Fundamentals of Physics.* 1994 ISBN 0-7923-3000-5
64. M. Evans and J.-P. Vigier: *The Enigmatic Photon.* Volume 1: The Field $\boldsymbol{B}^{(3)}$. 1994 ISBN 0-7923-3049-8
65. C.K. Raju: *Time: Towards a Constistent Theory.* 1994 ISBN 0-7923-3103-6
66. A.K.T. Assis: *Weber's Electrodynamics.* 1994 ISBN 0-7923-3137-0
67. Yu. L. Klimontovich: *Statistical Theory of Open Systems.* Volume 1: A Unified Approach to Kinetic Description of Processes in Active Systems. 1995 ISBN 0-7923-3199-0; Pb: ISBN 0-7923-3242-3
68. M. Evans and J.-P. Vigier: *The Enigmatic Photon.* Volume 2: Non-Abelian Electrodynamics. 1995 ISBN 0-7923-3288-1
69. G. Esposito: *Complex General Relativity.* 1995 ISBN 0-7923-3340-3

Fundamental Theories of Physics

70. J. Skilling and S. Sibisi (eds.): *Maximum Entropy and Bayesian Methods.* Proceedings of the Fourteenth International Workshop on Maximum Entropy and Bayesian Methods. 1996 ISBN 0-7923-3452-3
71. C. Garola and A. Rossi (eds.): *The Foundations of Quantum Mechanics Historical Analysis and Open Questions.* 1995 ISBN 0-7923-3480-9
72. A. Peres: *Quantum Theory: Concepts and Methods.* 1995 (see for hardback edition, Vol. 57) ISBN Pb 0-7923-3632-1
73. M. Ferrero and A. van der Merwe (eds.): *Fundamental Problems in Quantum Physics.* 1995 ISBN 0-7923-3670-4
74. F.E. Schroeck, Jr.: *Quantum Mechanics on Phase Space.* 1996 ISBN 0-7923-3794-8
75. L. de la Peña and A.M. Cetto: *The Quantum Dice.* An Introduction to Stochastic Electrodynamics. 1996 ISBN 0-7923-3818-9
76. P.L. Antonelli and R. Miron (eds.): *Lagrange and Finsler Geometry.* Applications to Physics and Biology. 1996 ISBN 0-7923-3873-1
77. M.W. Evans, J.-P. Vigier, S. Roy and S. Jeffers: *The Enigmatic Photon.* Volume 3: Theory and Practice of the $\boldsymbol{B}^{(3)}$ Field. 1996 ISBN 0-7923-4044-2
78. W.G.V. Rosser: *Interpretation of Classical Electromagnetism.* 1996 ISBN 0-7923-4187-2
79. K.M. Hanson and R.N. Silver (eds.): *Maximum Entropy and Bayesian Methods.* 1996 ISBN 0-7923-4311-5
80. S. Jeffers, S. Roy, J.-P. Vigier and G. Hunter (eds.): *The Present Status of the Quantum Theory of Light.* Proceedings of a Symposium in Honour of Jean-Pierre Vigier. 1997 ISBN 0-7923-4337-9
81. M. Ferrero and A. van der Merwe (eds.): *New Developments on Fundamental Problems in Quantum Physics.* 1997 ISBN 0-7923-4374-3
82. R. Miron: *The Geometry of Higher-Order Lagrange Spaces.* Applications to Mechanics and Physics. 1997 ISBN 0-7923-4393-X
83. T. Hakioğlu and A.S. Shumovsky (eds.): *Quantum Optics and the Spectroscopy of Solids.* Concepts and Advances. 1997 ISBN 0-7923-4414-6
84. A. Sitenko and V. Tartakovskii: *Theory of Nucleus.* Nuclear Structure and Nuclear Interaction. 1997 ISBN 0-7923-4423-5
85. G. Esposito, A.Yu. Kamenshchik and G. Pollifrone: *Euclidean Quantum Gravity on Manifolds with Boundary.* 1997 ISBN 0-7923-4472-3
86. R.S. Ingarden, A. Kossakowski and M. Ohya: *Information Dynamics and Open Systems.* Classical and Quantum Approach. 1997 ISBN 0-7923-4473-1
87. K. Nakamura: *Quantum versus Chaos.* Questions Emerging from Mesoscopic Cosmos. 1997 ISBN 0-7923-4557-6
88. B.R. Iyer and C.V. Vishveshwara (eds.): *Geometry, Fields and Cosmology.* Techniques and Applications. 1997 ISBN 0-7923-4725-0
89. G.A. Martynov: *Classical Statistical Mechanics.* 1997 ISBN 0-7923-4774-9
90. M.W. Evans, J.-P. Vigier, S. Roy and G. Hunter (eds.): *The Enigmatic Photon.* Volume 4: New Directions. 1998 ISBN 0-7923-4826-5
91. M. Rédei: *Quantum Logic in Algebraic Approach.* 1998 ISBN 0-7923-4903-2
92. S. Roy: *Statistical Geometry and Applications to Microphysics and Cosmology.* 1998 ISBN 0-7923-4907-5
93. B.C. Eu: *Nonequilibrium Statistical Mechanics.* Ensembled Method. 1998 ISBN 0-7923-4980-6

Fundamental Theories of Physics

94. V. Dietrich, K. Habetha and G. Jank (eds.): *Clifford Algebras and Their Application in Mathematical Physics*. Aachen 1996. 1998 ISBN 0-7923-5037-5
95. J.P. Blaizot, X. Campi and M. Ploszajczak (eds.): *Nuclear Matter in Different Phases and Transitions*. 1999 ISBN 0-7923-5660-8
96. V.P. Frolov and I.D. Novikov: *Black Hole Physics*. Basic Concepts and New Developments. 1998 ISBN 0-7923-5145-2; Pb 0-7923-5146
97. G. Hunter, S. Jeffers and J-P. Vigier (eds.): *Causality and Locality in Modern Physics*. 1998 ISBN 0-7923-5227-0
98. G.J. Erickson, J.T. Rychert and C.R. Smith (eds.): *Maximum Entropy and Bayesian Methods*. 1998 ISBN 0-7923-5047-2
99. D. Hestenes: *New Foundations for Classical Mechanics (Second Edition)*. 1999 ISBN 0-7923-5302-1; Pb ISBN 0-7923-5514-8
100. B.R. Iyer and B. Bhawal (eds.): *Black Holes, Gravitational Radiation and the Universe*. Essays in Honor of C. V. Vishveshwara. 1999 ISBN 0-7923-5308-0
101. P.L. Antonelli and T.J. Zastawniak: *Fundamentals of Finslerian Diffusion with Applications*. 1998 ISBN 0-7923-5511-3
102. H. Atmanspacher, A. Amann and U. Müller-Herold: *On Quanta, Mind and Matter Hans Primas in Context*. 1999 ISBN 0-7923-5696-9
103. M.A. Trump and W.C. Schieve: *Classical Relativistic Many-Body Dynamics*. 1999 ISBN 0-7923-5737-X
104. A.I. Maimistov and A.M. Basharov: *Nonlinear Optical Waves*. 1999 ISBN 0-7923-5752-3
105. W. von der Linden, V. Dose, R. Fischer and R. Preuss (eds.): *Maximum Entropy and Bayesian Methods Garching, Germany 1998*. 1999 ISBN 0-7923-5766-3
106. M.W. Evans: *The Enigmatic Photon Volume 5: O(3) Electrodynamics*. 1999 ISBN 0-7923-5792-2
107. G.N. Afanasiev: *Topological Effects in Quantum Mecvhanics*. 1999 ISBN 0-7923-5800-7
108. V. Devanathan: *Angular Momentum Techniques in Quantum Mechanics*. 1999 ISBN 0-7923-5866-X
109. P.L. Antonelli (ed.): *Finslerian Geometries A Meeting of Minds*. 1999 ISBN 0-7923-6115-6
110. M.B. Mensky: *Quantum Measurements and Decoherence Models and Phenomenology*. 2000 ISBN 0-7923-6227-6
111. B. Coecke, D. Moore and A. Wilce (eds.): *Current Research in Operation Quantum Logic*. Algebras, Categories, Languages. 2000 ISBN 0-7923-6258-6
112. G. Jumarie: *Maximum Entropy, Information Without Probability and Complex Fractals*. Classical and Quantum Approach. 2000 ISBN 0-7923-6330-2
113. B. Fain: *Irreversibilities in Quantum Mechanics*. 2000 ISBN 0-7923-6581-X
114. T. Borne, G. Lochak and H. Stumpf: *Nonperturbative Quantum Field Theory and the Structure of Matter*. 2001 ISBN 0-7923-6803-7
115. J. Keller: *Theory of the Electron*. A Theory of Matter from START. 2001 ISBN 0-7923-6819-3
116. M. Rivas: *Kinematical Theory of Spinning Particles*. Classical and Quantum Mechanical Formalism of Elementary Particles. 2001 ISBN 0-7923-6824-X
117. A.A. Ungar: *Beyond the Einstein Addition Law and its Gyroscopic Thomas Precession*. The Theory of Gyrogroups and Gyrovector Spaces. 2001 ISBN 0-7923-6909-2
118. R. Miron, D. Hrimiuc, H. Shimada and S.V. Sabau: *The Geometry of Hamilton and Lagrange Spaces*. 2001 ISBN 0-7923-6926-2

Fundamental Theories of Physics

119. M. Pavšič: *The Landscape of Theoretical Physics: A Global View*. From Point Particles to the Brane World and Beyond in Search of a Unifying Principle. 2001 ISBN 0-7923-7006-6
120. R.M. Santilli: *Foundations of Hadronic Chemistry*. With Applications to New Clean Energies and Fuels. 2001 ISBN 1-4020-0087-1
121. S. Fujita and S. Godoy: *Theory of High Temperature Superconductivity*. 2001 ISBN 1-4020-0149-5

KLUWER ACADEMIC PUBLISHERS – DORDRECHT / BOSTON / LONDON

MIX
Papier aus verantwortungsvollen Quellen
Paper from responsible sources
FSC® C105338

If you have any concerns about our products,
you can contact us on
ProductSafety@springernature.com

In case Publisher is established outside the EU,
the EU authorized representative is:
Springer Nature Customer Service Center GmbH
Europaplatz 3, 69115 Heidelberg, Germany

Printed by Libri Plureos GmbH
in Hamburg, Germany